AF658558

Introductory Ornamental Horticulture

Introductory Ornamental Horticulture

J.S. ARORA
Former Head
Department of Floriculture and Landscaping
Punjab Agricultural University
Ludhiana (Panjab)

KALYANI PUBLISHERS
LUDHIANA • NEW DELHI • NOIDA (UP) • HYDERABAD • CHENNAI
KOLKATA • CUTTACK • GUWAHATI • KOCHI • BENGALURU

KALYANI PUBLISHERS

Head Office

B-1/1292, Rajinder Nagar, **Ludhiana**-141 008 • Ph : 0161-2760031, 2745756
E-mail : kalyanibooksraj@yahoo.co.in • Website : www.kalyanipublishers.co.in

Administration Office

4779/23, Ansari Road, **Daryaganj,** New Delhi-110 002
Ph : 011-23271469, 23274393, 23278688
E-mail : kalyani_delhi@kalyanipublishers.in

Works

B-16, Sector-8, **NOIDA** (U.P.)

Branches

1, Mahalakshmi Street, T. Nagar, **Chennai**-600 017 • Ph : 044-24344684
Gopabandhu Lane, Behind Govt. Bus Stand, Badambadi,
Cuttack-753 012 (Odisha) • Ph : 0671-2311391
3-5-1108, Narayanaguda, **Hyderabad**-500 029 • Ph : 040-24750368
10/2B, Ramanath Mazumdar Street, **Kolkata**-700 009 • Ph : 033-22416024
Arunalaya, 1st Floor, Saraswati Road, Pan Bazar, **Guwahati**-781 001 • Ph : 0361-2731274
Koratti Parambil House, Convent Road, **Kochi**-682 035 • Ph : 0484-2367189
No. 24 & 25, 1st Floor, Hameed Shah Complex, Cubbonpet Main Road, **Bengaluru**-560 002

P-31243 1
R-33454 1
S-37680 1
U-41439 1

18+382+8 (4 Colour Plates) = 408 Pages 28x40 25.5 Forms

First Edition, 1990
Fourth Revised Edition, 2003
Fifth Revised, Edition, 2007
Sixth Revised Edition, 2008
Seventh Revised Edition, 2012
Eighth Revised Edition, 2016
Ninth Revised Edition, 2020
Reprinted, 2021, 2023

Reprinted, 2024

Typesetting at : **Kalyani Publishers**

ISBN : 978-93-89987-43-0

PRINTED IN INDIA
At Process World India, C-90, Sec-63, NOIDA

Dedicated
to
Harinder, Soneeka
and Amrinder

Preface

Ninth Edition

It is a matter of great pleasure that all eight editions of 'Introductory Ornamental Horticulture' have been very well received by the students, amateurs, nurserymen, hobbyist and city dwellers. However, with increasing urbanization, role of horticulture in environmental amelioration and indoor plants have attained greater significance than ever before. Keeping this in view, six new chapters Lilium, Dahlia, Bird of Paradise, Postharvest handling and storage of cut flowers, Vertical garden and Xeriscape garden have been added. Many new plants introduced by nurseries have been added and scientific names of many plants have been updated in accordance with latest accepted nomenclature. Information on control measures of insect pests and pathogens has been updated with newly introduced and more effective pesticides. The latest statistical figures have been also included. Thus, it is hoped that this ninth edition will also be well received by the students, nurserymen, amateurs, hobbyist and city dwellers and help in solving their day to day gardening problems.

In the last but not least, I am highly thankful to the editorial department of M/s Kalyani Publishers for their untiring efforts to release the book in its revised form within short period with a new attractive outlook.

2020 **J.S. ARORA**

Preface

First Edition

Ornamental horticulture is a branch of horticulture which deals with commercial growing of flowers, ornamental plants and beautification of surroundings. This subject assumes great importance in improving total environment and checks visual pollution by way of beautification. Keeping the importance of this subject in view, it is being taught in B.Sc. (Ag.) in Agricultural Universities. In many Agricultural Universities, M.Sc. and Ph.D. degrees are also being awarded. Punjab Agricultural University, is also one of them.

The study of ornamental horticulture comprises of two parts *i.e.,* the knowledge of growing of plants like annuals, shrubs, climbers, trees, bulbous, succulents and cactus, shade loving plants, lawns etc. and their use in beautification programme. Beautification involves the principles of art, and gardening styles.

The information on above aspects is scattered in different books and a great difficulty was being felt by undergraduate students. Therefore, an effort has been done to bring the class lectures in the form of book which have been taught more than fifteen years. It is hoped that the book will be a great help in understanding the introductory aspect of ornamental horticulture. The flowering time of most of the flowering plants has been mentioned as it has been observed under Panjab conditions. Author believes that it will serve a great help to undergraduate students in preparing for degree. Author is grateful to students, flower and garden lovers of Panjab for their constant encouragement.

1990 **J.S. ARORA**

Contents

INTRODU [illegible]

Flowers [illegible]
in India man [illegible]
Flowers are used for various purposes [illegible]
and social functions, weddings, [illegible]
flowers is ve[illegible]
Flowers [illegible]
Carnation [illegible]
gold for vulg[illegible]
ios for messe[illegible]
Stock for lux[illegible]

What Is F[illegible]

Floricult[illegible]
cultivation of [illegible]
ones but also [illegible]
export of cu[illegible]
medicines, etc[illegible]

Present St[illegible]

Besides [illegible]
value as cut [illegible] as cut blooms, [illegible]
making of [illegible] products like [illegible]
is not as [illegible]
under floriculture [illegible]

1

Present Position and Future Scope of Horticulture in India

INTRODUCTION

Flowers are associated with mankind from the dawn of civilization. It is said that in India man is born with flowers, lives with flowers and finally dies with flowers. Flowers are used for various purposes in our day to day life like worshipping, religious and social functions, wedding, interior decoration and self adornment. Saying it with flowers is very common and different flowers are used to convey the human feelings. Flowers commonly used for such purposes are : Rose for love; Pansy for thoughts; Carnation (white) for women's love; French marigold for jealousy/sorrow; African marigold for vulgar minds; Narcissus for self esteem; Daffodil for regard; Amaryllis for pride; Iris for message; Snap dragon for presumption; Jasmine for amiability; Lily for purity; Stock for luxury; Sweet pea for departure; etc.

What is Floriculture ?

Floriculture is the aesthetic branch of horticulture which deals not only with cultivation of ornamentals, annuals, biennials, and perrnnial plants including potted ones but also their marketing. This also includes marketing for local, distant markets and export of cut flowers, live plants and their economic products like scents, oils and medicines, etc.

Present Status of Floriculture in India

Besides beauty and aesthetic values of flowers, they are important for their economic value as sale of flowers (loose as well as cut blooms), extraction of essential oils and making of economic products like gulkand. The present position of floriculture in India is not as bright and lucrative as it is in European countries and the U.S.A. The area under floricultural crops in different states of India is shown in Table 1.1.

TABLE 1.1
Area and production of flowers in different States/UT's during 2017-18

State/UT	*Area ('000 ha)*	*Loose Flowers ('000 MT)*	*Cut Flowers (Lakh No.)*
Jammu & Kashmir	49.58	29.70	0.39
Kerala	38.25	0.08	28.16
Tamil Nadu	34.23	482.52	0.00
Karnataka	31.36	230.96	92.90
West Bengal	26.78	74.88	204.47
Andhra Pradesh	25.74	428.95	0.00
Uttar Pradesh	21.22	46.42	65.20
Gujarat	20.43	152.16	0.00
Madhya Pradesh	19.79	245.55	37.24
Chhatisgarh	13.18	47.45	179.58
Others	7.33	1.17	0.52
Odisha	6.61	24.91	48.79
Haryana	5.49	56.73	4.46
Maharashtra	5.49	29.08	56.99
Assam	5.06	33.93	56.08
Telangana	3.71	17.92	5.20
Rajasthan	3.49	7.07	0.00
Punjab	2.07	10.67	0.00
Uttarakhand	1.53	2.54	11.72
Jharkhand	0.80	4.47	8.61
Himachal Pradesh	0.64	12.35	10.56
Bihar	0.55	5.54	0.12
Mizoram	0.24	0.00	2.55
Sikkim	0.24	16.50	0.09
Manipur	0.16	0.49	0.00
Nagaland	0.07	0.00	8.79
Meghalaya	0.01	0.00	0.34
All India	**324.00**	**1962.03**	**822.75**

Source : NHB Detabase 2018.

Sale of Flowers, Live Plants, Bulbs, Seeds, etc. It is an organised business in metropolitan towns of India but in small towns it is handled by a fewer merchants.

Flowers, live plants, bulbs, seeds, etc. are sold in local market, produced locally or at distant places and are exported. Flowers like rose, gladiolus and tuberose as cut flowers are flown to distant markets from their production centres. For example, gladiolus and roses from Kashmir to Delhi, tuberose from Kolkata to Delhi or Mumbai. Loose flowers like marigold, jasmine, aster, gaillardia, kaner, malti, etc. are frequently taken to big cities from distant places. Jasmine flowers are regularly exported from South India to gulf countries. Similarly live plants, bulbs and seeds are regularly exported. From India, export of flowers, bulbs and live plants was as follows :

Year	*Value (' in Crores)*
2003-04	249.55
2004-05	221.11
2005-06	301.45
2006-07	652.70
2007-08	340.14
2008-09	368.81
2009-10	294.46
2010-11	296.04
2011-12	365.32
2012-13	423.45
2013-14	455.91
2014-15	460.77
2015-16	477.76
2016-17	546.71
2017-18	507.31
2018-19	571.41

Source : DGCIS Annual Export Data 2015.

INTERNATIONAL SCENARIO

There is a great increase in demand of floricultural products with increasing income and globalization of economy. Netherlands, Italy, Germany and Japan have strong tradition for growing and consumption of flowers. Netherlands is the leading country with an area of around 30,000 hectares under flower crops in open conditions and around 3,500 hectare of flower crops under protected cultivations. However, new production centres are developing in Latin America-Columbia, Costa Rica, Chile; Africa-Kenya, Rhodesia, Morocco, South Africa etc. Asia-Israel, India, Sri Lanka, etc. to meet the internal demand as well as export. These countries produce quality product at low cost, off course, with high cost of investment.

Total world export of floriculture products in 2018 was reported to be of US $ 19,822.15 Million (Table 1.2). The Netherlands remained world's largest exporter of floricultural products with a share of 44.54% of the total export, other major exporters of floricultural products are Columbia, Germany, Italy, Ecuador, Kenya, Belgium, Spain, China etc.

TABLE 1.2
Major Exporters of Floricultural Products

Ranking	*Country*	*2014 ('000 US $)*
1	Netherland	8,827.85
2	Colombia	1,380.42
3	Germany	1,075.75
4	Italy	930.09
5	Ecuador	914.09
6	Kenya	852.09
7	Belgium	631.83
8	Spain	455.3
9	China	422.85
10	Canada	422.16
27	India	79.95
	Others	3,829.77
	Total	**19,822.15**

Source : APEDA.

Consumption of floricultural products is rising both in developed and developing countries. Annual global consumption of flowers varies by source and range from US$ 40-60 billion. Germany was the largest importing country for floricultural products (Table 1.3). During 2018 it imported floriculture products of worth US $ 3027.50 Million. Second largest importing country was Netherlands which imported floricultural products worth of US $ 2508.54 Million followed by USA (245.24 Million US $) and UK(1755.01 Million US $).

TABLE 1.3
Major importers of floricultural products

Ranking	*Country*	*2014 ('000 US $)*
1	Germany	3,027.50
2	Netherland	2,508.54
3	U SA	2,435.24
4	U K	1,755.01
5	France	1,300.61
6	Japan	605.8
7	Italy	603.15
8	Russia	599.22
9	Switzerland	593.95
10	Belgium	504.86
56	India	24.15
	Others	6,117.25
	Total	20,075.28

Source : APEDA.

In internal markets, generally the price of flowers though is dependent of the demand and supply, yet varies as follows :

	Loose Flowers	***Cut bloom***
Rose	₹ 40-50/kg	₹ 6-8 dozen (field grown)
	—	₹ 36-48 dozen (greenhouse)
Gladiolus	—	₹ 24-36/dozen
Tube rose	₹ 60-80/kg	₹ 15-18/dozen
Gerbera	—	₹ 10-70/10 pieces
Carnation	—	₹ 60-70/20 pieces
Orchid	—	₹ 120-150/10 pieces
Chrysanthemum	₹ 40-50/kg	₹ 40-50/dozen (standard Incurved or Reflexed type)
Annual Chrysanthenum	₹ 5-40/kg	
Aster	₹ 10-12/kg	
Marigold	₹ 5-15/kg	
Jasminum sambac	₹ 40-60/kg	
Ervatamia divaricata	₹ 40-50/kg	
Cut Foliage		
Asparagus plumosus	₹ 50-150/kg	
Fern	₹ 10-15/20 pieces	
Cycas revoluta (single leaf)	₹ 18-24/dozen	
Areca Palm leaves	₹ 18-24/dozen	

Economic Products. The products like gulkand of rose, rose water, rose otto and perfumes of other flowers like tuberose, *Pandanus* (Keora), molsari, juhi, chameli, etc. are made and sold in the market. These essential oils are used in soap and cosmetic industry. There is a localized production in certain pockets where it is being done since long. For extraction of essential oils, the places like Kannauj and Hathras (UP), Coimbatore (Tamil Nadu), Haldighati (Rajasthan), Amritsar (Punjab) are well known in the country.

Ornamental Nurseries. The production of propagating material like seeds, bulbs, live plants is being done conventionally and these materials are being sold. Successful nurserymen are making handsome money. There are a few nurseries reputed for true-to-type and reliable plant material.

Sale of Garden Tools, Pots and other Ornaments. For growing flowering plants, raising plants in pots, certain garden tools like secateurs, hedge shears, khurpa, plant lifter, mower are required constantly. Hence, the money is earned by selling these items. Similarly pots are always in demand. Earlier earthen-pots were dominating but now concrete and plastic pots are being manufactured and sold. Moss stick is another important item which is in good demand. Other materials like stone lanterns, sun dial, bird's nests, and ducks made up of plastic are used as replicas in gardens and are also in demand.

Research on Floricultural Crops. To establish an industry, the role of research is well known. In India the research on floricultural crops is still in infancy in comparison to western countries. Research on limited scale is being conducted in Agricultural Universities and Horticultural Institutes. In many Universities teaching in floriculture and landscaping has been strengthened and many graduates are passing out. Technical persons are now becoming available.

To give a boost to floriculture industry in the country, ICAR (Indian Council of Agricultural Research New Delhi) has upgraded All India Coordinated Research Project on Floriculture to Directorate Floricultural Research at IARI (Indian Agricultural Research, Institute New Delhi). In 11th Five Year Plan, its outlay of Directorate is ₹ 2009.27 lakhs which is likely to be enhanced 2-3 times in next five year plan. Many new varieties have been released which are getting popular on commercial basis, *e.g.*, in roses—Raktgandha, Arjun (for cut bloom export) and many varieties for garden decoration and exhibition; gladiolus-Suchittra, Mayur, Poonam, Nazrana, Chrysanthemum-Kundan, Baggi, Santi, Basanti, Birbal Sahni, etc.

Future Scope. In India, floricultural industry is not so developed as it should have been. Therefore, there is immense scope for its expansion. If floriculture is developed on industry pattern scientifically, it will flourish and generate tremendous money and employment.

Increasing Demand of Cut Blooms, Live Plants. The demand of cut flowers and live plants is increasing day by day with the increasing standard of living, aesthetic sense and awareness in the people. Although it is possible to grow flowers round the year in one part or the other of the country depending upon season, yet the production of quality bloom under controlled environment is essential which will fetch high price. It has been experienced that in North India during end of November to end of February, quality cut blooms are produced under unheated plastic houses than in open.

Quality Seed Production. There is always a good demand for high quality seed. Open pollinated seed of many winter, summer and rainy season annual is being produced in about 600-800 ha for export to different countries. Now a days, F_1 hybrids due to uniformity, earliness, bigger flowers and profuse flowering, are becoming popular. In many flowers like petunia, pansy, marigold, antirrhinum, nicotiana, carnation etc. F_1 hybrids are common. The production cost of these F_1 hybrids is a little more but they fetch good price. Bulbs, corms and tubers can also be produced and exported from India to Denmark, Sweden, France, West Germany, gulf countries, etc.

Perfumeries. Most of the fragrant flowers like rose, tuberose, jasmine etc. can be grown easily, so extraction plants can be set up to meet the increasing demand of cosmetic and soap industry. Flower growers and industry will get benefit. The per cent oil content in different flowers is as follows :

Rosa bourboriana	0.04
Rosa damascena	0.05
Polyanthes tuberosa	0.1
Jasminum auriculatum	0.29
J. grandiflorum	0.24-0.30
J. sambac	0.14

Bonsai development. Bonsai are mini plants which have great importance in modern landscaping of homes, restaurants, hotels etc. Plants like bougainvillea, rubber tree, quamquats, pomegranate, *Ficus benghalensis, Ficus infectoria, Ficus religiosa, Pinus longifolia* etc. can be used for making bonsai and sold in local as well as in foreign markets.

Favourable Factors for Floricultural Industry in India

Climate. The varied climate of India makes it most suitable for growing flowers in one part or the other so that flowers can be produced the year round. For example, roses, gladiolus, carnation, chrysanthemum, etc. are successfully grown in northern plains during winter months, whereas during summer months these can be grown in the hills. The flowering period can be extended by controlling environment through economical modified structures. Growing of wide range of flowers is also possible in different parts of India.

India being native of many flowers opens up new vistas for exploiting the commercial growing of rare flowers for export.

Soil. Like climate, the soil of different climatic zones is very favourable for growing different types of flowers. Suitable soil pockets can also be selected for growing of flowers.

Labour. Comparatively cheap and easily available labour in India is another favourable factor for growing flowers at cheaper rate. The labourers are employed for plucking flowers, emasculation, pollination, seed collection and cleaning operations.

Other facilities like transport and storage are also adequately available.

***

2

History of Gardening in India

The origin of planet earth took place about 500 million years ago. Whether the life first appeared in the form of blood or chlorophyll is still not known clearly. The simpler forms of flora and fauna passed through evolutionary processes and developed into complex systems. Dominance of flowering plants started during Cenozoic age (60-70 million years) and at the same time dominance of mammals and insects started. Man appeared on this planet about 1,50,000 years ago. At an early stage, man might have felt the necessity of cool and shady places near water sources for rest after tiring job of hunting and this gave the concept of gardening.

According to Vavilov, Indo-Malaya region is one of the important centers of origin of plants. Many flowering plants are strictly-native of India. Among them important ones are: orchids, rhododendrons, muskrose *(Rosa moschata)*, balsam *(Impatiens balsamina)*, primula *(Primula denticulata, P. rosea)*, blue popply *(Maconopsis* sp.), gloriosa lily *(Gloriosa superba)*, tulip *(Tulipa stellata* and *tulipa aitchisonii)*. Among trees which are native of India are : Sita ashoka *(Saraca asoca)*, pipal *(Ficus religiosa)*, banyan *(Ficus benghalensis)*, nag kesar *(Mesua ferrea)*, amaltas *(Cassia fistula)*, kadam *(Neonauclea purpurea)*, Swaran champa *(Michelia champaka)*, neem *(Azadirchata indica)*, palas *(Butea monosperma)*, arjun tree *(Terminalia arjuna)*, kachnar *(Bauhinia variegata)*, Karanj *(Pongamia glabra)*, Bhendi *(Thespesia populnea)*, Chalta *(Dillenia indica)*, Barna *(Cratavea roxburghii)*, *Sterculia colorata*, etc. Indigenous shrubs are: Bela *(Jasminum sambac)*, Pili chameli *(Jasminum humile)*, Rukmani *(Ixora coccinea, I. parviflora, I. barbata)*, Har shrinagar *(Nyctanthus arbortristis)*, *Holmskioldia sanguinea, Hamiltonia suaveolens, Clerodendron inerme*, Chandni *(Ervatamia coronaria)*, *Crossandra infundibiliformis, Plumbago rosea*, etc. Important native climbers are : *Passiflora* sp., *Clitoria tarneata, Porana paniculata, Clematis montana*, Madhvi lata *(Hiptage benghalensis)*, etc.

The history of systematic gardening in India is as old as civilization of Indus or Harappa which existed between 2500 B.C. and 1750 B.C. People were living in well planned dwellings. Roads cut across one another almost at right angles. The trees were associated with the human life in many ways is quite clear from the seals available of that period. The picture of God is represented in the middle of branch of pipal. Besides pipal, mango and willow have also been depicted on the seals. Harappan pots were generally decorated with the designs of trees. The cult of tree and animal worship was at its peak. The tree of pipal *(Ficus religiosa)* and banyan *(Ficus benghalensis)* served mankind and fauna in many ways. These trees provided shelter for night stay, their fruits were eaten, leaves were used to cover the naked bodies. These trees were considered the symbol of fertility. That is why till today the trees are being worshipped. In every village the trees especially the *Ficus religiosa* and *F. benghalensis* are planted for worship as well as for shade. This all confirms that trees, climbers and other ornamental plants were associated with the Harappan civilization.

Aryans came in India about 1600 B.C. They came through Iran, Afghanistan and reached the plains. The country at that time was called as 'Aryavarta' which means the country of lotus and sunshine because the lakes were studded with lotus flowers and there were wide open spaces. Land was covered with the trees of palash laden with scarlet flowers. Lotus being a native of India was found every where. These Aryans were literary people and brought with them the four vedas *viz.* Rig Veda, Artharva Veda, Yuzur Veda and Sam Veda and the Puranas. They were also the ardent lovers of nature and had aesthetic sense. The use of flowers in religious and social ceremonies was started by Aryans. They appreciated the beauty of flowering plants; lakes, mountains, forests, etc. and named their kids after the flowers like kamal, Champa, Madhvi, Bela, Chameli, Rukmani, etc. The forests of sal *(Shorea robusta)* and flame of the forest *(Butea monosperma)* were admired by Aryans. River Ganga and Yamuna were given the status of mother which flow down from Himalaya through plains. Similarly, cow was given the status of *Kamu Dhenu.* The Lord Vishnu and Goddess Luxmi were associated with Lotus. According to a legend, during sagar manthan, Goddess Luxmi appeared from the sea in sitting posture on lotus. The different parts of the body of Goddess Luxmi have been compared with the different parts of lotus. *Ficus benghalensis* was worshipped as Kalpvriksha which will fulfil all the human desires. Kadamba was associated with Lord Krishna; Semal with Siva; Bauhinia with Saraswati—goddess of wisdom; Amaranthus with Goddess Kali and yellow amaltas with prosperity in trade.

The life of Buddha was associated with a number of trees from birth to death. He was born in 563 B.C. under the tree of Ashoka at Lumbini in the grove of sal. In this grove, the trees of palash were also there; hence, all the three trees were associated with the birth of Buddha. Birth place of Buddha has been described by Hiuen Tsang who visited the place in 630 A.D. According to a description, there was a bathing tank of Sakya Muni filled with clear water, lotus and lilies. In the north of this tank, there was a tree of Ashoka and under this tree Lord Buddha was born. He attained enlightenment under the tree of Pipal at Gaya and preached first sermon of his religion under mango and banyan grove at Sarnath (Varanasi). When Lord Buddha visited Vaishali, Amrapalli—

presented a park known as Amravana which was dominated by trees. The death place of Buddha has also been described by Hiuen Tsang. There is a sal grove on the bank of river Ajitavati in Kusi Nagara. Its bark is greenish blue and its leaves are white shining and very lustrous. There are four trees pairs of equal height. This is the place where Buddha died.

Buddhist monks beautified their monasteries and vihars by planting trees and other flowering plants on a large scale for making surroundings peaceful, a place of solace and ideal for meditation. The planting of trees alongside the roads was intensified by emperor Ashoka after embracing Buddhist religion during 264-227 B.C. His daughter and son also spread the art of planting trees around vihars and alongside roads alongwith Buddhist religion in neighbouring countries. The messages of Buddhist religion were written on stupas (long and wide stones) which were placed near Buddha monasteries and alongside roads. On these stupas alongwith sermons, pipal tree, wheel and lotus were also carved. Symbolically lotus represented the birth of Buddha; pipal—enlightenment, wheel—sermon and stupa—emancipation. Carving of lotus flowers, buds and leaves was also very common on medallions. The art of gardening developed during this time was introduced in neighbouring east countries like Thailand, Korea, China and Japan alongwith religion.

Gardening during Epic Era

Detailed account of Lord Rama's life and status of gardening at that time has been mentioned in Ramayana written by Valmiki and Tulsidas. Other books related on this subject also throw light on the status of gardening. At that time, Ayodhya city was having wide streets, large houses, noble palaces, richly decorated temples and gardens. These gardens were planted with fruit trees, flowers, lakes were full of lotuses and different kinds of birds. Roads were planted with shady trees. During exile, Lord Rama and Sita observed a number of trees and were fascinated by their beautiful flowers. One of such trees was Ashoka. According to Skanda Purana, Sita was charmed by the beautiful forest of Ashoka and said to Rama that it is the month of Baisakh. Let us stay in this wood and take bath in the river. Such groves are found where Kosi river leaves the mountains.

Another book of Epic era, 'Mahabharat' was written by Saint Vyasa. The detailed events of war between Kauravas and Pandavas have been narrated alongwith the vegetation and various flowering plants. The pleasure gardens were planted with flowering trees, which flowered round the year. The famous tree of this era was Kadamba *(Nauclea cadamba)* which is associated with Lord Krishna. Underneath the trees of Kadamba, Lord Krishna played *ras lila* with milk maids. Other flowering trees which have been described are mango, Ashoka, coral tree, Sal, Nag champa, Champa, Palmyra, screw pine, oleander etc. The climbers like *Jasminum grandi-florum* and *J. muliflorum* have also been described. Ponds were used to be filled with lotus and lilies.

Great poet Kalidas has described the numerous flowering plants of that era in number of his books. In Kumar Sambhav, trees like Ashoka, Kalpvriksha, Shirish flower,

Devdaru, *Butea monosperma,* Prajapati *(Nyctanthes arbortristis),* Molsari *(Mimuspos elengi),* Bhoj patri, lotus and lilies have been mentioned. In Abhigyan Shakuntlum plants like Shirish, Saptparni, Chameli, Juhi, Madhvilata, lotus and lilies have been described. In another famous book Meghdoot-Cloud Messenger—which is written in the form of poetry in Sanskrit, the route and vegetation from Ujjain to Himalaya (Alkapuri) have been described. In a description the use of flowers by women folk of Himalaya has been narrated as follows :

**Play lotus adorns the hand*
Plaited kund buds in the tresses dark
Sirisha on the delicate ears
and fresh kurubkas on the braids bedeck the women
The countenance is beautiful by
the golden Lodhra pollen
and Kadambas blossomed at the touch
decorates the line on the head
left by parting the hair.

Merriest festival of this period was 'Basant Panchami'. This festival is celebrated in the memory of Lord Kamadeva—God of Love. This is the month of spring season when temperature warms up after cold winter. Life activities are at full swing in this season. Everybody used to be in gay mood. All around happiness used to dominate. People used to greet each other with exchange of flowers. Young maids used to decorate themselves with the flowers, sing and dance in gardens. Competitions of poetry, songs, drama and sports were organised by state and prizes distributed to the best.

Vatsyayana (A.D. 300-400) described four kinds of gardens which were made for the kings, queens, courtiers and ministers. Pramododyan used to be private garden of the king and the queen; Udyan—where the kings passed their leisure time in playing chess, enjoying dance, listening songs and jokes by singers and court jesters. Vrikshavatika was the garden for ministers and courtiers where they made merry with courtesans. There was a special garden dedicated to Lord Indra. These gardens had water pool studded with lotuses and lilies, flowering and fragrant shrubs, trees and climbers. Swinging under the trees of Neem and Pipal during monsoon months used to be a common site.

Famous poet Bana Bhatta described the number of flowering plants in his famous book the Harsh Charita*. These flower plants were growing predominantly in the gardens in the vicinity of the palaces. Water pools were studded with red lotus and blue water lilies. Trees like Arjun, Banyan, Sal, Sapt parna, Champaka, flame of the forest, Vakula *(Mimuspos elengi),* Karanj, Kadamba, Ashoka, Indian coral tree, siris, shrubs like Malati,

* In Meghdoot translated by Dr. Rangay Rangav and Ramavtar Tyagi published by Atma Ram and Sons, New Delhi.

Harshringar, Gurhal (China roses), Rukmani (*Ixora bandhuca)* and climbers like Ketaki, Madhvi lata have been described. In one beautiful description, association of human feelings with the plants has been described like this : "Beloved mango, take thought of to yourself, your mother is seeking another home. I am going darling jasmine cluster, bid me farewell-without me, sister pomegranate by the house, you are now to be defenceless—for give me red Ashoka—my kicks and sins in plucking your sprays for ear ornaments. I see you seraglio vakula—wayward child as you are become through those mouthfuls of wine-clasp me tight dear Priyangu climber; I am passing beyond your reach friend mango at porch—you must render me funeral libation of water..."

Gardening during Mogul Era

Babur attacked India five times and in his fifth attempt defeated Ibrahim Lodhi at Panipat and established dynastic rule in Delhi in 1525. He did not intend to return back after conquering Delhi. Therefore, he started shaping his kingdom according to his likings. Babur had high aesthetic sense and was fond of gardens. Thus he made gardens at Panipat and Agra. Aram bagh at Agra is still being maintained nicely by Archeological Department, alongwith other gardens built by Mogul descendants. Mogul gardens are synonymous of formal style of gardening which is an extended plan of primitive Persian gardens, *i.e., Char Bagh Design.* The style of gardening was preserved by Mogul descendants and many gardens on similar style were made. Several modifications like addition of two or more terraces, with a central canal of running water with flowing fountains, water-fall in the form of Chadar (water sheet) and baradari structure, were made. The square or rectangular flower beds and trees like chinar and cypress are special features of Mogul gardens. Another important contribution was the introduction of exotic plants like chinar, cypress and flowers like rose, carnation, narcissus, daffodils, lilies, tulip etc.

Broadly, Mogul gardens can be grouped into two categories *viz. (i)* Pleasure gardens of king and queen, and *(ii)* Tomb gardens. Former gardens were made with the sole purpose of pleasure of king, queen and family members, whereas latter gardens were attached to tombs of king or queen for giving peace to dead person.

For some time Humayun was out of power and Delhi was ruled by Sher Shah Suri. He made grand trunk road from Lahore to Calcutta and planted shady trees along both sides of the roads. After Humayun, Akbar (1556-1605) was a very popular king and ruled very successfully. He made Fatehpur Sikri (Agra) and tomb garden, Sikandra (Agra). Among Mogul kings, Jahangir (1605-1627) and Shah Jahan (1627-1658) made several gardens at different locations. Famous gardens built by Jahangir are Shalimar, Achhabal, Varinag (at Kashmir), Itmad-ud-Daulah (Agra) and Dilkusha garden at Lahore. His wife Noor Jahan discovered 'Otto' of rose while taking bath in water tank filled with rose flowers. The contributions of Shah Jahan in gardening were immense and famous

* Bana Bhatta 1961. The Harsh Charita, Motilal Banarsi Dass, Delhi, p. 149.

gardens like Shalimar (Lahore), Chashma Shahi (Srinagar) and gardens in Taj Mahal, Red fort (Agra), Red fort (Delhi) were built. Aurangzeb made a garden at Aurangabad. Fadai Khan, a general of Aurangzeb, made a breathtaking garden at picturesque site of Pinjore. Mogul Emperors also made several sarais (Aram bagh) alongwith roads for taking rest during transit from one place to the other.

Till today, these gardens are example of their style and enjoy high esteem in the hearts of people. Being architecturally superb, aesthetically designed and harmoniously planted, they create breathtaking sight.

British Period

East India Company came to trade with India but finally paved the way for the establishment of British rule in India. During the British era, there was a lot of activity in gardening by Britishers and Indian kings, King Hyder Ali and his son Tipu Sultan were famous amateur horticulturists of South India. King Hyder Ali established most famous Lal Bagh garden in Bangalore. Plants were imported from Mauritius and other countries and were maintained at Lal Bagh. With the fall of Tipu Sultan it was managed by Britishers. In North India, Maharaja Ranjit Singh made an impressive garden at Amritsar. King Bhupinder Singh and his son Yadvindra Singh of Patiala, well known amateur horticulturists, contributed a lot for the improvement of horticulture. Famous Baradari garden at Patiala and fruit orchard at Chail were established. They also introduced latest varieties of fruit plants, strawberries, gladiolus and carnations etc.

Britisher's memorable gift to Indian gardens was making of herbaceous border and lawn. Concentrated efforts were initiated to improve the gardening in three ways *viz. (i)* introduction of exotic plants from England and other countries, *(ii)* establishment of royal-agri-horticulture societies and botanical gardens, and *(iii)* compilation of local flora of different regions. India was rich in colourful and fragrant trees, shrubs and climbers but was lacking in colourful annuals like phlox, verbena, larkspur, nusturtium, dahlia, corn flower, pansy, aster, antirrhinum etc. which were introduced; other plants like Bougainvillea, golden shower, gulmohr, nili gulmohr etc. were also introduced. A number of botanical gardens were established in different parts of the country where exotic plants were introduced, assessed and maintained. Important ones are Royal Agri-horticulture Society Garden, Calcutta; Botanical Garden, Sibpore, Calcutta; Lloyd Botanical Garden, Darjeeling (West Bengal); Botanical Garden, Saharanpur; National Botanic Gardens, Lucknow; Botanical Garden of the Forest Research Institute, Dehradun (Uttarakhand); Lal Bagh Botanical Garden, Bangalore; Government Botanical Garden, Ootacamund (Tamil Nadu); Botanical Garden, Coimbatore (Tamil Nadu) Biyand Park at Kodai Kanal (Tamil Nadu). In these botanical gardens valuable plant material has been planted family wise and labelled. This helped in preservation of plant material. The utilization of Indian flora in European gardens was also done in a big way. The plants like orchids, rhododendron, primula etc. were further used in breeding programme. Several tropical plants decorated European gardens.

Post-Independence

There have been radical changes in the field of ornamental gardening during post-independence period. Remarkable achievements have been made at all three important fronts *viz.* conscious planning for improving total environment, commercial floriculture, and teaching and research of ornamental horticulture at graduate and post-graduate levels.

Several gardens in different cities have been laid out to provide active and passive recreational facilities and to improve the environment. Important gardens are Budha Jayanti Garden, Delhi; Rose Garden, Chandigarh and Ludhiana. The main feature of these gardens are undulating grassy lawns, serpentine pathways and colourful plant material. These gardens are deviated from traditionally Mogul gardens in their layout. As a general rule, landscaping of public and private buildings has become an integral part of planning. State Departments like Urban Development, Archeological and Tourism are actively busy improving total environment of buildings by conscious planning and planting. Green belts are being provided in industrial towns to reduce air pollution. It is becoming customary to engage technical trained persons for such purposes.

Keeping the immense scope of commercial floriculture in view, much attention is being given to exploit it fully. The cut-flower trade in big cities like Bombay, Calcutta, Bangalore, Madras, Delhi, Pune, Ludhiana, etc. has developed tremendously during the last two decades. Now farmers earn good income by growing flowers around big cities. The growing of certain flowers and production of planting material like corms, bulbs and pot plants have been specialized in different agro-climatic zones, as detailed here. *Hilly region* (Srinagar, Darjeeling, Nainital, Himachal Pradesh)—gladiolus, narcissus and other bulbous plants; *North Indian plains*—roses, gladiolus, carnation, marigold, other seasonal flowers; *Western ghats* (Pune, Nasik)—rose, gladiolus, marigold, aster; *South India*—tuberose, Jasmine, marigold, crossandra, barleria and foliage plants; *Eastern region*—chrysanthemum, tuberose, aster and lotus and foliage plants.

In order to make growing of flowers into a successful proposition, it should be based on scientific research. To give boost to floriculture industry in the country, ICAR (Indian Council of Agricultural Research, New Delhi) has upgraded All India Coordinated research Project on Floriculture to Directorate Floricultural Research at IARI (Indian Agricultural Research Institute, New Delhi) whose first director is Dr. Ramesh Kumar who is able and dynamic person. In 11th Five Year Plan, its outlay of Directorate is ₹ 2009.27 lakhs which is likely to be enhanced 2-3 times in next five year plan. There are 23 coordinated centers situated all over India in different agro-climatic zones in which 16 centres are budgetary, 4 are institutional and 3 are voluntary. The directorate's vision is to harness the research and development activities if flower crops and landscape gardening for promotion of domestic and export markets whereas mission is to carry out research, impart education, conduct out reach programme in floriculture and landscaping with national and international partners for enhancing production, productivity and profitability besides alleviating the rural poverty. The main flower crops on which concentrated research work is being done are : Rose, Carnation, Chrysanthemum, Gladiolus, Orchids, Gerbera, Lilium, Alstroemeria, Tuberose, Daffodil, Anthurium. The crops have been allotted to different centre according to their climatic conditions.

National Horticulture Board was established by the Government of India in 1984 as an autonomous society to develop high quality horticultural farms in identified belts which in turn will act, as hubs for developing commercial horiculture by adopting high tech horticulture techniques including post harvest management infrastructure. It also helps in improving linkages between horticulture producers and marketers to increase producer's share in consumer price. It has helped in establishment of poly houses for growing flowers in different states of India by providing subsidies and arranging the facilities. This has brought a revolution in production of green house cut flowers like Rose, Gerbera, Carnation, Orchid, Lilium, Calla Lily, Anthurium, etc.

During Tenth Five Year Plan (2005-06) National Horticulture mission has been established in ministry of Agriculture in which Government of India provides 100% assistance to the states for development in the States. The share of Government of India will be 85% and states will contribute 15%. It is provide a holistic growth of the horticulture in the country. It also help in enhancing horticulture productivity, farms and creating employment opportunities for skilled, unskilled persons and specially for unemployed youth. Many floriculture projects have come up in various states with the help of National Horticulture Mission. Other than this, different institutes like Indian Agricultural Research Institute, New Delhi; Indian Institute of Horticultural Research, Bangalore; National Botanical Research Institute, Lucknow; Bhabha Atomic Research Centre, Bombay and Agricultural Universities like Punjab Agricultural University, Ludhiana; University of Agricultural Sciences, Bangalore; Dr. Y.S. Parmar University of Horticulture and Forestry, Solan; Bidhan Chandra Krishi Vishwavidalaya, Kalyani, are the places where scientific research on floriculture is being conducted.

Availability of technically trained persons is also very important factor for making a programme successful. Till 1971, no university was providing education upto M.Sc. level in floriculture and landscaping. Punjab Agricultural University, Ludhiana, was the first Indian University to start imparting education in floriculture and landscaping. Now the facility is available at other universities like Solan, Kalyani, Coimbatore, Pune, Kanpur, I.A.R.I., etc. The School of Town Planning and Architecture, New Delhi, imparts postgraduate course in landscaping. Strengthening of such facilities is essential to make floriculture industry prosperous.

* * *

3
Annuals

Annuals or seasonals are the group of plants which complete their life cycle in one season or in one year. In such a limited period, they complete the process of life like germination, growth, flowering, seed formation and after this plants wither out. They are easily grown plants. These plants very widely in form, habit of growth, flower colour and flower size. They beautify the surroundings and exhibit a good show of blooms at low cost and labour. They bring a change in the look of garden with the change in the season and keep gardeners busy in raising them throughout the year. The annuals are grown for various situations and purposes in the garden like for pots, hanging baskets, bed, edging, rock garden, as herbaceous border or in shrubberies, shady situation, etc.

Annuals are divided into three groups according to their growing season.

1. Summer Annuals. These plants grow luxuriantly and produce flowers under high temperature. They have ability to survive extremely high temperature of North Indian Plains. The seeds are sown in end of February or beginning of March and seedlings are transplanted in end of March-April. There are limited number of annuals which can be grown in this season *e.g.,* Zinnia, Kochia, Portulaca, Tithonia, Gaillardia, Gomphrena, Sunflower, *Cosmos sulphreus,* etc.

2. Rainy Season Annuals. These annuals are grown in rainy season and can withstand heavy rains, and high humidity in atmosphere than other annuals. Seeds are sown in June and seedlings are transplanted in July. For example balsam, Cock's Comb, Amaranthus, Gaillardia, etc.

3. Winter Annuals. Most of the winter annuals have been introduced by Britishers in India. These annuals are able to tolerate comparatively low temperature and, hence, are comfortably grown in winter season and bloom best during this season. There is a large number of winter annuals available in wide range of flower form, size and colour. They bring a riot of colour in gardens. There are annuals which emit sweet fragrance and make garden fragrant. These winter annuals are sown in September and transplanted in October in plains whereas in hills, these are planted in February-March and July August.

Different annuals are suited to different situations and therefore, right type of annuals should be selected for particular purpose. The selection of these flowers can be made according to following purposes.

(*i*) ***Bedding purpose.*** Dahlia, Marigold, Phlox, Verbena, Pansy, Carnation, Sweet William, Sweet Sultan, Petunia, Ice Plant, Candytuft, Zinnia, Balsam, Portulaca, Gomphrena, Gaillardia, etc.

(*ii*) ***For fragrant flowers.*** Mignonette, Carnation, Sweet Pea, Sweet Sultan, Sweet William, Sweet Alyssum, Stock etc.

(*iii*) ***For cut flowers.*** Carnation, Sweet Pea, Sweet William, Aster, Larkspur, Helichrysum, Lupin, Antirrhinum, Sweet Sultan, Corn Flower, etc.

(*iv*) ***For loose flower.*** Marigold, Annual Chrysanthemum, Aster, Sunflower, Zinnia, Gaillardia, etc.

(*v*) ***For hanging basket.*** Daisy, Nasturtium, Verbena, Phlox, Sweet Alyssum, Portulaca.

(*vi*) ***For Shady situation.*** Saliva and Cineraria.

(*vii*) ***For rock garden.*** Ice Plant, Nasturtium, Verbena, Phlox, Gamolepis, etc.

(*viii*) ***For screening purpose.*** Holly Hock and Sweet Pea.

(*ix*) ***For peculiar shape.*** Clianthus.

(*x*) ***For Pots.*** Carnation, Antirrhinum, Aster, Petunia, etc.

(*xi*) ***For dry flowers.*** Statice, Helichrysum, Acroclinum, Nigella, Lady's Lace.

HERBACEOUS BORDER

The planting of annuals in the border of a plot is called as herbaceous border. Due to wide range of flowers available in winter it becomes a site of attraction in the garden. The concept of herbaceous border has been introduced in Indian gardens by Britishers and now it has become a common feature. The herbaceous border can be of two types *i.e.* Single and Double face. Single face herbaceous border is made when border is situated on one side of plot and having some background like wall or shrubbery border. Annuals are arranged in the border according to their height *i.e.* tall plants in back; medium in center and dwarf in front line. Double face border is more ideal when border is to be made in between a big plot. In such border, there is no background and tall annuals are planted in the center; medium and dwarf on both sides in descending order so that beauty can be watched from both sides.

The selection of site for herbaceous border is very important and desired picturesque effect may not be obtained if site is not selected properly. The site should be open sunny and facing towards south side so that plants get sufficient sunlight for maximum hours of the day. If this direction is not available, it should face east and last choice is for the west.

Back and Fore Grounds. To make an effective herbaceous border, there should be suitable back ground as well as fore ground. Back ground may be of dwarf flowering trees, shrubbery border or back ground made by training of sweet peas, brick wall in comb nation of both. Fore ground is generally of turf grass which may or may not have other garden ornaments like lily pool, sun dial, statue etc. Green turf in front of herbaceous border is a living green carpet which serves a place for enjoying beauty of flowers while resting.

Size of the Border. The size of border depends upon the availability of space. The border can be of any length but width should be about 1-1.5 m or more depending upon the area of fore ground available. Therefore, the width can be adjusted accordingly.

Colour Seheme of the Border. The arrangement of these annuals according to colour scheme conveys the mood and feelings of the gardener. Dominance of a particular colour in border is suggestive quality of that colour. Hence, the arrangement of these annuals should be done very carefully. In nature green colour being the colour of foliage dominates in garden throughout the year. Warm colour like red dominants in winter whereas yellow colour in spring. There are three main colour schemes according to which these annuals can be arranged.

1. Monochromatic scheme
2. Analogous or harmonious colour scheme
3. Contrast colour scheme

Monochromatic colour scheme restricts the use of single colour or its shades which may be available in the same or different annuals. For example, if blue colour is to be used, annuals like blue corn flower, blue larkspur, ageratum, anchusa, etc. can be used. For yellow colour annuals like yellow dahlia, yellow antirrhinum, yellow calendula, yellow nasturtium, yellow marigold, yellow annual chrysanthemum, yellow coreopsis etc. can be used. The difficulty may be in getting the seeds of pure colours. According to analogous colour scheme, annuals are arranged according to the nearest wave length of colour. This arrangement harmonizes with the surrounding beds and, hence, it is called a harmonious colour scheme. The arrangement of colours will follow the order as white, creamish yellow, light yellow and finally deep yellow which will follow the same but in descending order. If red colour is to be used the order will be light pink, deep pink, light red, deep red and crimson and colours will descend in the same order. In contrast colour scheme, the opposite colour of colour wheel are used and plants of contrast colours are planted accordingly *e.g.* red dahlia against the background of green or next to blue corn flowers creates a good contrast. For yellow colour, violet, blue and red are contrast. Yellow gamolepis, red phlox or petunia and white alyssum can be planted for a brilliant contrast. The width of the border is roughly divided into three parts in proportion of 5 : 3 : 2 or 6 : 3 : 1 for tall, medium and dwarf plants. Similarly, the plots are made lengthwise. The front line of border is not a straight line, but it is made curvaceous and thus plots become irregular in shape. In these plots planting is also done irregularly at an approximate distance so that the border looks natural at the time of bloom.

GROWING AND MANAGEMENT OF ANNUALS

Raising of nursery. In general seeds of annuals are small and are sown in controlled space for better handling. Seeds are sown in pots, boxes and on raised nursery beds depending upon the quantity of seeds to be sown. These nursery beds are prepared 10-15 cm above the ground. Width is kept about 1 m which facilitates the operation of weeding, watering and thinning. The length of nursery bed should be about 2-3 meter. The soil should be well prepared by incorporating well rotten farmyard manure (10-15 kg/sq m.). To check the infection of soil-borne diseases in nursery, soil should be drenched with 0.2% Brassicol or Captan solution. Soil can be sterilized by formaline (2%) and covering with polythene. Small seeds are mixed with bulk material like sand for even distribution. Seeds are sown in lines which should be 5-6 cm apart and depth should be about 0.5 cm. The depth can be increased with the increase in size of seeds. After sowing, seeds should be covered with sieved mixture of farmyard manure and soil. Watering should be done twice or thrice a day. The bed should be covered with news paper or light cloth to ensure darkness which ensures better germination. The outer surface of nursery beds should be dusted with Chlorpyrifos 1.5% to check the entry of ants which otherwise carry away the seeds. The seeds start germinating within 2-3 days and then covering is removed. It takes about one month to grow the seedlings sufficiently and are ready for transplanting. By this time, they have 3-4 true leaves.

Preparation of Land and Transplanting of Seedling. Before laying out the plan, land should be well prepared by digging or ploughing and mixing five kg. well rotten farmyard manure, 10 g each of N, P_2O_5 and K_2O per sq. m (40 g calcium ammonium nitrate, 62 g single super phosphate and 16 g of muriate of potash, respectively). It is advisable that before transplanting, light irrigation should be given which helps in better establishment of seedlings. According to plan, layout should be done. The seedlings should be transplanted in the cool hours of evening. The tall, medium and dwarf annuals are planted at the distance 30-40 cm, 15-20 cm and 10-12 cm respectively. After transplanting the seedling, watering should be done immediately.

MANAGEMENT OF THE ANNUALS

These annuals are quite tender and require proper attention throughout their life. Regular weeding and hoeing is essential for the development of these seedlings into healthy plants. If weeds are not removed well in time they compete with the main plants, check the growth and affect the flower production. Regular water supply is also essential for the successful growing of these annuals. It should be done according to the requirement which is influenced by season and soil type. In rainy season, generally irrigation is not required except during dry spell. If rains are not well distributed, irrigation should be done accordingly. During winter season due to prevalence of low temperature frequent irrigation is not required but is done at the interval of 10-12 days. During summer season, frequent irrigations are required due to high temperature which should be done at an interval for 4-5 days. In light soil frequent irrigation is required than in heavy soil. One month after transplanting of seedlings, another dose of 10 g N (40 g calcium ammonium nitrate or 20 g urea) should be broadcast.

Seed Collection and Storage

Seed collection of annuals is an important aspect which needs scientific knowledge and special care in handling and storage. Maintenance of the purity of varieties which is highly technical, requires the knowledge whether crop is self, self to partial cross or cross pollinated. Self pollinated annuals which do not need any isolation distance are Lupins, Sweet pea, and Salvia. The annuals which are self to partial cross pollinated require the isolation distance of 50-100 metres, are: Antirrhinum, Larkspur, Linaria, Phlox, and Pansy. Majority of the annuals are cross pollinated which require an isolation ditance of more than 400 metres *e.g.* Balsam, Zinnia, Gomphrena, Cock's Comb, Sunflower, Portulaca, Gaillardia, Sweet alyssum, Calendula, Corn flower, Sweet Sultan, Chrysanthemum, Cosmos, Dahlia, Helichrysum, Candytuft, Eschscholtzia, Marigold Nasturtium, Verbena, etc. To produce seeds of different varieties large area for proper isolation is required. In some of the nurseries, sweet pea trained as an artificial curtain upto 2 or 2.5 m high is used to provide an isolation distance for self to cross pollinated crops. In limited space other devices like bagging and growing under glass houses are also used. Other than isolation, removal of off types is also an important operation which should be done before flowering or flower opening. The improvement in seed set is also achieved in certain crops like large pansy and carnation by brushing. Now-a-days, the F_1 hybrids are very popular due to uniformity in plant height, flower size, floriferousness and bigger size of bloom. Similarly F_2 hybrids are very popular in some crops like pansy, antirrhinum which bring out colour carnival in the garden. These F_1 hybrids are produced every year by crossing two selected lines. The crossing involves emasculation, collection of pollen and pollination which is done manually; that is why the cost of F_1 hybrids is high. In crops like marigold, male sterile line is used as one of the parents which saves above operation done by manual labour.

Seed collection should be done at proper maturity stage before splitting of seed pods. Seeds are generally thrashed by hand, winnowed, cleaned and screened. Dry and inert material, small, deformed and broken seeds should be removed from bulk. Seeds should be treated with insecticide to prevent the pest infestation during storage. Seeds packed in muslin bags with silica gel are stored best in air tight tins and glass jars. Cardboard boxes can also be used.

There are some difficulties in seed production business of ornamental crops in comparison to vegetable or field crops. The demand of annuals is not very large and seeds are small in size which does not permit the use of machines in handling. Thus, the cost of handling and packing becomes high.

DESCRIPTION OF ANNUALS

Summer and Rainy Season Annual

Amaranthus : *Amaranthus* sp. (Amarantaceae)

It is native of tropical Asia and tropical America. It's generic name consists of two Greek words, *A*-not and *maraino*-fading, and is due to an illusion to the lasting character of the calyx and floral bracts. Plants are tall (90-120 cm) and are very ornamental due to colourful foliage. Therefore, it is grown in groups or in rows but should not be mixed with other annuals. In mixed planting, colourful foliage dominates over colourful flowers. Planting should be done in sunny situation where foliage develops rich colour. There are many species which are grown in gardens as rainy season annuals. *A. caudatus* is commonly called as Love-Lies-Bleeding and produces graceful long dropping catkins of purple crimson, pale green or white flower. *A. hybridus* produces crimson red plumes above light green foliage and looks very attractive. Another species *A. tricolour ruber* produces foliage of bright red whereas *A. tricolour splendens* produces rich scarlet foliage. *A salicifolius* is popular in the garden due to its undulating and drooping foliage of bright orange colour. A bicolour which is commonly called as Molten Fire produces deep maroon foliage and crimson top.

Balsam : *Impatiens balsamina* (Balsaminaceae)

It is very showy annual and is native of India. It attains the height of 25 to 75 cm and produces many branches alongwith the main stem. It needs proper training by romoving side shoots and only main shoot should be kept for display. Leaves are simple and margin is slightly serrated. The flowers are produced after 60-65 days of planting and flowering continues upto 20-30 days. The flowers are spurred, 5-6 cm across and like camellia are borne in leaf axils. The flowers are single or double in pure or variegated colours. Dominating colours are red, pink, violet, rose, and white. The important varieties which are grown in the garden are Double Camellia flowered mixed, Rose Flowered, Royal Balsams Mixed and Tall Double Mixed. The dwarf variety Tom Thumb is uniform, bushy and compact. There are two perennial species which have been used to evolve new types. *i.e., I. holstii* and *I. sultanii.* The important varieties are Holstii Hybrid, Bright Rose, Blaze Scarlet, pink Baby, Baby Mixed and Sultani Nana hybrid.

Seed pods burst and seed scatter when seed pods are ripened fully. They should be harvested before scattering.

Cock's Comb : *Celosia* sp. (Amarantaceae)

It is native of tropical Asia, Africa and America. The generic word has been derived from Greek word *Kelos* which means burnt and flowers of some species appear as thought burnt. It is very popular in the garden for their flowers. Leaves are simple and have variation in light red or crimson colour. There are tall as well as dwarf forms. Flower colour varies from deep crimson, deep ink cardinal red to deep orange flowers. The species which dominate in the garden are *C. argentea* var. *cristata, C. argenta* var. *plumosa* and *C. argenta* var. *childsii.* The sp. *C. argentea* var. *Cristata* is commonly known as Cock's comb which produces large compact head measuring about 25 or 30 cm. Plants are tall and dwarf and suitable for pot culture. The important varieties in tall group are Toreador Royal Velvet and Gilbert Celosia whereas dwarf varieties are Golden King, Lilliput, Jewel Box, Glasgow Prize, Aurea, Flame of Fire, and Thompson Magnifica.

C. argenta var. *childsii* (Chinese *woo-flower)* produces wooly globular flowers which are 7-10 cm across. *C. argenta* var. *plumosa* produces flowers like large plumes resembling Ostrich's plumes.

Gaillardia : *Gaillardia pulchella* (Compositeae)

It is also commonly called as blanket flower and is native of North America. Under Indian conditions it is grown throughout the year and seeds are sown in all the three sowing times and seedlings are transplanted accordingly. Plants are medium in height, bushy (30-45 cm) and produce profusely single to double flowers on long thin sterns measuring about 5-7 cm across. Flowers are produced profusely in many colours *viz.* yellow, orange, scarlet, copper colours or combinations of these. There are two common varieties of gaillardia *i.e.,* picta *(G. pulchella* var. Picta) and Lorenziana *(G. pulchella* var. *lorenziana*). The flowers of picta group are large and single. Important varieties are 'Indian Chief' red and 'Picta Mixed' whereas the flowers of lorenziana are double having quilled petals with split tips. Important varieties of this group are Lorenziana 'Sunshine'. 'Gaity double mixed', and 'Double Tetra Fiesta'—a tetraploid variety. The perennial species, *G. grandiflora* is also very popular in gardens which produces large flowers of yellow-orange and red flowers. Important variety in this group is 'Goblin'.

Gomphrena : *Gomphrena globosa* (Amarantaceae)

It is native of India and commonly called as globe amaranth. Plants are medium or dwarf. Round flowers about 2 cm across are produced of white, purple, magenta, pink and rose colour. Flowering starts after two months of transplanting of seedlings but profuse flowering occurs in September-November. The common varieties are var. Alba—White; var. Aurea Superba—Golden yellow; var. Nana Compacta Rubra—15 cm high purple flowers; var. Rosea—Pink; var. Rubra—Purple-red. The important varieties are Globosa mixed, Lilliput, buddy and Cissy (white).

Kochia : *Bassia scoparia* var. *trichophylla Syn. Kochia scoparia* var. *tricophylla* (Chenopodiaceae)

Its generic name has been given after Prof. W.D.J. Koch and is commonly called as summer cypress or burning bush. This is the only annual which provides lush green foliage during summer and looks very refreshing and cooling. It is also very popular due to its uniform ovoid conical or globular shape which looks naturally trimmed. It grows to a height of 90-120 cm. It requires sunny situation for its healthy growth. Plants do not produce colourful flowers.

Portulaca : *Portulaca grandiflora* (Portulaceae)

It is known as sun plant or Rose moss and is a native of South America. The flowers open in morning and close by afternoon. It is most ideal for beds, post and hanging baskets. Plants are dwarf and have trailing growth habit. Branches are round and succulent. Leaves are thick, fleshy, small and pointed. Flowers are about 3 cm across and are produced in white, yellow, red, orange, pink, lavender, crimson, purple, etc. In some varieties, flower petals have stripes of contrasting colours. Important varieties are Grandiflora single mixed, Double mixed and Magic carpet. The perennial form producing bright deep rose coloured single or double flowers is also commonly found growing.

Sun Flower : *Helianthus annus* (Compositae)

It is native of western parts of United States of America and its generic name is composed of two Greek words *i.e. Helios*—sun and *anthos*—flower. In tall varieties flower rotates with the sun movement. It is hardy, vigorous and produces magnificent flowers. The tall growing varieties (150-180 cm) are: Tall Single yellow—huge (20-25 cm) Yellow Single Flowers and Tall Double chrysanthemum—Yellow chrysanthemum like flowers which is dominating new in the gardens. Sun Gold and yellow Pigmy are double dwarf varieties which produce medium golden orange flowers.

Now-a-days, the hybrids of *H. annus* and *H. cucumerifolius* which are known as Sultan's Autumn beauty are available. They are well branched with small flowers on long stems of yellow, orange-yellow, chestnut-brown and maroon flowers.

The Russian sunflower is a tall (180-240 cm) while produces edible oil and cattle feed.

Tithonia : *Tithonia rotundifolia* syn. *T. speciosa* (Compositae)

It is commonly called as Mexican sunflower and is a native of Mexico. It is exceptionally tall growing about 120-240 cm, but its branches are brittle and break very easily. Flowers are large measuring about 5-7 cm having deep orange to scarlet colour with yellow centre.

Zinnia (Compositae)

It has become a very common flower in gardens during summer season flowering of which continues upto September-October. It is a native of Mexico and name after Johnan Gothifried Zinn—a professor of medicine. *Zinnia elegans* dominates in the garden due to wide variation in plant height (25-120 cm), flower size (2.5 to above 12 cm) and colour range varying from white, green, red, pink, yellow, orange, salmon, lavender, mauve and various intermediate shades. Flowers are single, semi-double and double. Ray florets are flat, quilled or ruffled. There are large number of varieties available in different groups and new varieties are being added every year. The important varieties of different groups are :

Giant Dahlia. Canary Bird (yellow); Dream (Lilac); Meteor (dark red); Oriole (Orange scarlet); Polar Bear (White).

Giant of California. Brightness (Pink), Cherry Queen (rose); orange Queen (golden orange), Lavender Gem (Lavender), Purity (white).

Giant Cactus. Empress (pink); Red Man (scarlet); Snow Man (white); Sun Gold (light yellow).

Lilliput or Pompon. Peach Blossom, Rose Gem, Canary yellow, white Gem.

Cupid. Flowers are like small button. Goblin (burnt orange), Pixie (yellow); Snow Drop (white); Pink Buttons (pink).

Other groups are Fantasy, Dwarf giant, Gaillardia flowered, Peppermint stick, Ortho Polka, Tom Thumb and Thumbelina.

F_1 hybrids are very common and are being marketed. Flowers of big size (12-13 cm) are produced on long sturdy flower stalks and are ideal for cut flower. These are available in Dahlia and Cactus group. In Dahlia group important varieties are : Blaze (orange scarlet), Eskimo (creamy white), Riverside Beauty. Latest addition in this a Peter Pan series which is available in different colours and Fruit Bowl. In cactus group varieties are Bonanza (golden orange), Fire Cracker (red), Princess (Salmon Pink), Sunny Boy (deep yellow), Cherry Time (Cherry rose).

Tetraploids are also very popular due to free flowering and vigorous growth.

Z. angustifolia (Z. haageana or *Z. mexicana)* is also common and important varieties are old Maxico and Gold Tip.

Another species *i.e., Z. linearis* is ideally suitable for edging, beds and mass effect. Flowers are small and profusely produced in September-November. Deep yellow, light yellow, and white colours are common.

Winter Annuals

Acroclinum : *Rhodanthe chlorocephala* subsp. *rosea* Syn. *Acroclinium roseum* (Compositae)

It is a native of Australia and is popularly known as paper flower and also ever lasting flower on account of its hard and paper like petals. Plant attains a height of 40-70 cm. Rose coloured flowers are produced profusely. Flowers are single or double and keep their form and colour on drying. *A. album* another species bears pretty white coloured flowers. *A. grandiflorum* produces very large flowers of pink colour. For drying, fully opened flowers are plucked and to keep the stem straight, thin steel wire is inserted carefully and hanged downward in a room. The important varieties are Double hybrid and Giant Flowered mixture.

Ageratum : *Ageratum houstonianum* (Compositae)

It is commonly called as floss flower and is very common in Indian gardens and is a native of central America. The garden varieties have been evolved from two species *i.e. A. conzoides* and *A. houstonianum.* Plant grows upto 15-60 cm and produce flowers of blue, pink and white. A tetraploid form having bigger flowers is also grown in the garden. It is very suitable for rock garden. The flowers remain fresh for long time. There are many varieties but important are : Midget Blue, Blue Cap, Tall blue, Fairy Pink, Violet blue, Blue Dip, Blue Mink Capri, Little Blue Star, White Cushion. F_1 hybrid Blue Angel is an other attractive variety.

Anchusa : *Anchusa capensis* (Boraginaceae)

It is a native of South Africa and is perennial but is grown as an annual flower in Indian gardens. It is highly suitable for pot. Plants attain a height of 45 cm. The flowers are bright blue and produced in clusters. The important varieties are: Blue Bird, blue with white centre; Bedding Bright Blue-azure. *A itailice* a perennial species that can be grown as biennial in the hills.

Annual Chrysanthemum : *Chrysanthemum* spp. and *Glebionis* spp. (Compositae)

It is a native of mediterranean, North Africa and Europe. There are three common annual species which are cultivated in Indian gardens. They are *Glebionis coronaria* Syn. *Chrysanthemum coronarium* commonly called as garland chrysanthemum or crown daisy in Europe whereas, Bhambiree or Guldhak in Punjab, market in Delhi and Gendi in U.P., *Ismelia carinata* Syn. *Chrysanthemum carinatum* (summer marguerite) and *Glebionis segetum* Syn. *Chrysanthemum segetum* (corn marigold). Amongst these species only *C. coronarium* is grown in large areas of western parts of U.P., Delhi and Punjab. In gardens it is ideally suited for shrubberies, beds and as cut flower. Plant grows about 100-120 cm tall. *G. coronarium nanum compactum* is a dwarf and plant grows upto only 30 cm. Leaves are deeply cut of shining green colour. Numerous single, semidouble and double flowers are produced of large size measuring 5-7.5 cm across weighing about 2.917 kg per plant which are plucked in 9-10 pluckings. The flower colour varies from white, creamish white, creamish yellow and yellow in *G. coronarium* and deep· yellow in *G. segetum*. In *G. carinatum* flower colour is white, yellow colour having band of purple or crimson colour and thus flower looks like tricolour and is good for garden display. There are several varieties available in all three species. The important varieties of *C. carinatum* are Court Jester, Rainbow mixed, Pole Star, Torch, Spectalite and Chelsea. In case of *C. segetum* good varieties are Eldorado, Evening Star, Golden Glow and Morning Star.

Antirrhinum : *Antirrhinum majus* (Scrophulariaceae)

It is a native of South America and due to curious shape of flowers which resemble to dog or rabbit, it is popularly called as snap dragon, dog flower, or bunny rabbit or bunny mouth. The height of plant varies from 30 cm to more than 90 cm. Leaves are simple. Magnificent flowers are produced on long spikes of many colours and various shades like white, cream, yellow, rose, salmon, pink, mauve, red, magenta, orange, bronze. Flowers have good keeping quality and flowers remain fresh for 6-7 days. It is most ideally suited for beds, pot, edging, window boxes, herbaceous border and rock garden.

There is a wide range of varieties suitable for various purposes. The important varieties of tall Giant or Grandiflora group are Tip Top Canary yellow, Tip Top apple blossom, Tip Top White, Tip Top Orange Rose, Triumph, Bright orange, Triumph Scarlet, Triumph Gold whereas the varieties of intermediate or semi-dwarf are Intermediate bright Pink, Intermediate Guardsman, Intermediate White, Intermediate Mauve, Bedding scarlet. Dwarf or Tom Thumb varieties hardly grow about 30 cm and important variety is Orange Scarlet. The crosses between *A. majus* and *A. molle* resulted into a new series which is known as miniature or magic carpet and important varieties are Magic Carpet and Floral Carpet. F_1 hybrids are getting very popular and two series are very common *i.e.* Rocket type with prominent varieties like F_1 Golden Rocket, F_1 White Rocket, F_1 Bronze Rocket. Super Giants are another series in which plants grow tall and produce standard blossoms. Different colours like red, yellow, white, gold, bronze, pink and rose are available. Other important varieties of this series are Candel Light, Sun Set, and Sky Scraper. Lollipop Nanum and F_2 hybrid are also popular due to their variable height and mixture flower colour. In double Azalea series of snapdragon, flowers are double like azalea and are bigger. Madame Butterfly is a good variety of double azalea series. Sweet Heart series, a tetraploid, is also getting popular because plants grow 30 cm tall and are double. The different colours such as bronze, pink, red, white and yellow are available. Super tetraploids are resistant to rust and important varieties are : High Moon, Glacier, Crimson Giant and Volcano.

Recently there are flowers in such varieties which do not snap and are called without snap or penstemon type. The common series is bright Butterflies in which each floret looks like a colourful butterfly. Plants grow 60-70 cm tall, produce branches freely. Flowers are available in a wide range of colours *i.e.* pink, red, bronze, rose white and yellow. There is another Pixie series available in different colours like Orange Pixie, Pink Pixie, White Pixie, Rose Pixie, Red Pixie, etc. There is scented antirrhinum also which have soft clove like fragrance. The scented varieties are Super Jet, Vanguard and Venus.

There is another introduction of Monarch strain which is available in different colours : crimson, scarlet, orange, and coral.

Arctotis : *Arctotis stoechadifolia* (Compositae)

It is commonly called as African daisy and is native to South Africa. It is suitable for pot, bed and herbaceous border. Plant grows tall and have deeply cut foliage. Daisy like flowers measuring about 4-5 cm of white colour having lilac centres are produced. Flowers are closed in the night and reopen in the morning.

Interspecific hybridization has resulted in large flowered hybrids. They are vigorous growing and produce bigger size of blooms of about 8-10 cm in different colours like white, orange, pink, yellow, and dark red.

Aster : *Callistephus chinensis* (Compositae)

It is native of China and Japan and is an excellent cut flower. The remarkable improvement has been made in its shape and colour range in Europe after its introduction in 18th century. Dwarf (15 cm) and tall (90 cm) forms are available. Flower size also varies from very small button like pompons to large size. Different shapes like, single, double, anemone, reflexed, paeony and quilled are available. There is a wide range of colour of flowers available varying from white, yellow, blue, lavender, red, salmon rose, magenta, and scarlet. Several varieties are available and some of them are wilt resistant. In India, Ostrich Plume Salmon Rose, Ostrich Plume Blue, Ostrich Feather are very common. Some of wilt resistant varieties are Early Wonder, Giant Comet, Queen of Market, Star-Dust, Unicene Record. Other good varieties are Super Giants, Giant Curved, Super Princess, Early Charm, Crego series and Giant perfection series.

Bells of Ireland : *Molucella laevis* (Labiatae)

It is native of mediterranean region and is very attractive when the plants dry up. Flowers are produced in leafy axils and calyces are big pale translucent, green and veined. These calyces appear like bells. The calyces are persistent and turn ivory white at maturity. Long branches are cut at green as well as at dry stage and are arranged in flower arrangement. Dry stems can be kept in rooms for long time.

Brachycome : *Brachycome iberidifolia* (Compositae)

It is popularly known as Swan River Daisy and in Greek language Brachycome means soft short hair. It is native to Australia. Plants are dwarf and grow upto 15-18 cm and have finely cut foliage. Flowers are small (2.5 cm across) and are produced in abundance which are white, blue and light pink. It is ideally suited for pot, bedding and rock garden. Important varieties are Azure, Fairy, Blue Gem, Mauve Beauty, Purple King and Summer Beauty.

Calendula : *Calendula officinalis* (Compositae)

It is also called as pot marigold and is native to South Europe. It is very popular amongst gardeners on account of its easy culture, early to bloom and profuse flowering habit. Plants are medium in height and grow upto 20-50 cm. Leaves are long, hairy and somewhat sticky. Flowers are single, semi double and double which are produced in light yellow, yellow and orange colours measuring about 5-7 cm. It is highly suitable for pot, bedding and window boxes. Flowers are also used as cut flowers. There are many varieties available in different colours. Important varieties in orange colour are Orange King, Grandiflora, Campfire whereas in yellow colour are Golden Emperor, Golden King, Sunshine, Yellow Colossal. Art shade is another variety of soft shade of light yellow to light orange coloured flowers. Other varieties are Pacific Beauty, Apricot Beauty, Flame Beauty, Lemon Beauty, Orange Coronet, Dwarf Family Circle mixed, F_1 Mandarin, Dwarf Double Fiesta, Giant Early, Geisha Girl.

California poppy : *Eschscholtzia californica* (Papaveraceae)

It is native of California (U.S.A.) and plants are suitable for bedding—alongwith walks and paths in the gardens. Plant grows to a height of 30-40 cm. Leaves are finally cut and smooth. Flowers are open cup shaped, single or semidouble or double in various colour shades *i.e. yellow*, orange, white, red, pink, crimson, etc. The varieties are grouped into three groups *i.e. (a)* Tall Single—Chrome Queen, Flambeau, Mandarin, Orange King; *(b)* Dwarf single—Aurora, Fireflame, Mikado and *(c)* Double flowered—Art Shades, Autumn Glory, Carmine Queen, Crocea, Enchantress. Scarlet Glow.

Candytuft : *Iberis* sp. (Cruciferae)

It is native of Europe and is a dwarf annual. It is suitable for pot, bedding, edging and herbaceous border. There are two distinct species grown in the garden. *I. umbellata* attains the height upto 45 cm. Leaves are long, small and having smooth surface. Flowers of white and purple colour are produced at the top of branches into flat umbel. Another species *i.e., I. amara* which is also called as hyacinth flower, produce flowers arranged on long branches about 15 cm spikes. The important varieties are 'White Spiral' and 'Empress'.

Cineraria : *Pericallis cruenta* Syn. *Senecio cruentus* (Compositae)

It is a native of Canary Islands and being shade loving plant, it is most ideally suited for pot growing, shady places in garden and conservatories. Plants attain the height of 30-60 cm. Leaves are large and green which are cut. Flowers are produced at the top of branches which are very attractive and eye catching. The flowers of self or contrast colour in a wide range varying from white, blue, lilac, purple, pink are available. There are various forms which are being grown in the garden like tall and compact, large flowered singles, stellata singles-looking like stars, open habit, intermediate and multiflora Nanaproducing dwarf plants. The important varieties are Copenhagen Market, Maxima Grandiflora, Hybrida Grandiflora, Maxima Nana, Nana Multiflora and Mammoth Rainbow mixed.

Clarkia : *Clarkia pulchella* Syn. *Clarkia elegans* (Onagraceae)

This is native of California (U.S.A.) and a showy flower. Plants are stately growing and attain the height of 60-90 cm. It is generally grown in semi shady situation. Leaves are oval and flowers are produced in axil of leaves. The flowers are single or double and produced in white, rose, pink, salmon, or purple colours. The double flowered varieties are very showy and preferred in the gardens. The important double flowered varieties are Apple Blossom, Salmon Queen, Scarlet Queen, Firefly, Salmon Perfection, Royal Bouquet, Lady Satin Rose and May Blossom.

Clianthus : *Swainsona formosa* Syn. *Clianthus dampieri* (Leguminosae)

It is a native of Australia and due to its peculiar shape of showy flowers resembling to parrots beak, it is also called as Parrot's Bill. It is most ideally suited for pot culture. Plant grows upto 60-70 cm in height. Flowers of dark scarlet colour are produced in clusters of 4-6 blooms. Flowers attract the attention of viewers in exhibition due to unusual shape and its pretty look. The important variety is Dampieri.

Coreopsis : *Coreopsis tinctoria* (Compositae)

It is native of California (U.S.A.) and commonly called as Tick Seed on account of seed colour and seed shape. It is a dwarf annual suitable for bedding purpose and along with walks, paths in the garden. The yellow flower heads with brown spots are produced in abundance. It comes in bloom quite early. The important varieties are Golden Crown, Dwarf mixed, Double Sunburst, Mayfield Giant (Deep Golden Yellow), Baby Gold, Sunray, Grandiflora Sun Burst, Gold Fink, Grandiflora Mayfield Giant.

Corn Flower : *Cyanus segetum* Syn. *Centaurea cyanus* (Compositae)

Delicacy is represented by corn flowers in the world of flowers. It is native of Europe and Britain and commonly called as corn flower because it grows as a weed in corn field. It is also called as 'Hurt Sickle', 'Blue Bottle' or 'Bachelor's Button' or 'Ragged Sailor'. It is a tall annual growing upto 90-120 cm and suitable for herbaceous border, beds, shrubberies and as cut flower. The common colours are Blue, White, Light Pink and Maroon. The important varieties are Blue Boy, Pinky, Red Boy, Snow Man, Black Boy and Blue Diadem. There are dwarf varieties which are also suitable for edging and beds. The dwarf varieties are Jubilee Gem, Polar Bear, Lilac lady, Polka Dot. Corn flower starts producing flowers after 90-105 days of transplanting which are produced in great profusion.

Cosmea : *Cosmos bipinnatus* (Compositae)

It is native of Mexico and is very common in the gardens. It is very good annual for bedding. Plants are tall, well branched and leaves are thin and deeply cut. The flowers are single, semi-double, double and are produced in many colours like white, mauve, purple, rose, crimson, etc. Flowers are 3-6 cm across, sometime 10 cm and produced on long and thin stems. The important varieties are Double White, Double Crimson, and Alipore Beauty. Dwarf variety is Sunset.

Another species *C. sulphurus* is very common which can be grown throughout the year though it is planted in summer and rainy season. The leaves are dark green. The orange flowers are produced in great abundance. Important varieties are Fiesta, Orange Flame, Orange Ruffle, Mandarin, Sensation, Early Mammoth (Separate colours), Bright Lights, Klondyke Gold Crest, Sensation Dazzler and Sensation Gloria.

Daisy : *Bellis perennis* (Compositae)

It is also called as 'English daisy' and native of Europe. It is mostly planted in pots, beds and rock gardens. Plants are dwarf and attain the height *i.e.,* about 20-30 cm only. Stem is thick and leaves are strap shaped. The flowers are single or double which are produced in white and pink colours. Flowers are small measuring about 2.0-5.0 cm. The showy varieties producing double flowers are Dwarf White, Snow Ball, Long Fellow, Ruby, Gigantea mixed.

Dimorphotheca : *Dimorphotheca sinuata* Syn. *Dimorphotheca aurantiaca* (Compositae)

It is native of South Africa and is popularly known as African Daisy. It is a very good for bedding and for pots. It is very sensitive to weather conditions and flower shuts in dark or cloudy weather. *D. sinuata* has been crossed with *D. pluvialis* to produce an attractive race of variously coloured hybrids. It is a bushy and medium plant giving about 30-60 cm. Leaves are narrow either toothed or entire. Flowers are white, yellow, orange and salmon in colour with dark brown disc in center measuring about 5-6 cm. Important varieties are Glistening White, Giant Orange, Salmon Beauty, White Sparkle and Buff Beauty. Tetra Goliath is a tetraploid variety with large orange flowers with dark green disc.

Gamolepis : *Steirodiscus tagetes* Syn. *Gamolepis tagetes* (Compositae)

It is highly suitable for bed, edging pots and rock garden. Plants are dwarf and grow about 20-25 cm tall. Leaves are fine and deeply cut. Flowers are small measuring 1-1.5 cm, yellow in colour and are produced profusely.

Gazania : *Gazania splendens* (Compositae)

It is native of South Africa. The plants are dwarf and grows about 25 cm. Leaves are long, narrow and silvery on underside. The flowers are very showy, large and are available in different colours varying from white, pink, orange, yellow, red or brown. Flowering continues for a long period, but flowers close in the afternoon. Sunshine hybrids are very common.

Gypsophila : *Gypsophila elegans* (Caryophyllaceae)

It is native of Caucasus region and popularly known as Baby breath. Plant grow 60-75 cm tall and produce numerous flowers of white, pink and crimson colour. Flowers are borne in mist like sprays and look very graceful. Flowers are highly suitable for bouquet arrangement and are being sold in the market along with other cut flowers. Important varieties are King of Market, London Market, Rosea, Convent Garden.

Helichrysum : *Xerochrysum bracteatum* Syn. *Helichrysum bracteatum* (Compositae)

It is native of Australia and is popularly called as everlasting flower or straw flower of immortal on account of good keeping quality of flowers after ripening. The generic name is comprised of two Greek words *i.e. Helios*—sun and *chrys*—gold. This is most suitable for bedding and rock garden purpose. This is also suitable for cutting and drying. Plant grows about 60-90 cm tall. Flowers are of yellow or orange colour measuring about 2.5-6 cm across. There are numerous varieties and important are Pure White, Bright Terracotta, Fireball, Golden, Salmon King, Silver Pink, Luteum, and Roseum. Dwarf varieties growing about 45 cm are Tom Thumb, Nanum and Hot Bikni.

Holly Hock : *Alcea rosea* Syn. *Althea rosea* (Malvaceae)

It is a native of China. Plants are tall, stately growing about 180-200 cm ar d, hence are more suitable for planting in the back row of herbaceous border or for screening purpose. Leaves are deep green, coarse and round in shape. Flowers are single, semi-double and double which are produced in the axil of leaves in various colours varying from white, cream, yellow pink, mauve, lilac, rose, salmon, red, scarlet crimson and many intermediate shades. Flowers are large about 7-10 cm. Highly ornamental varieties suitable for growing in the garden are Giant Double, Charter's Double, Powderpuff, Summer Carnival and Triumph Double mixed.

Ice Plant : *Stigmstoccarpum criniflorum* Syn. *Mesembryanthemum criniflorum* (Aizoaceae)

It is a native of South Africa and is commonly called as Living Stone Daisy or Fig Marigold. It is a very dwarf plant and grows about 20 cm. It is best suited for sunny situation, rockeries, shallow baskets, pots etc. It can also be planted to cover the space under standard roses. The branches and leaves are succulent. Leaves are elongated thick and fleshy and dew deposits on them into small droplets which look like snow crystals. The flowers open in sun which are very shining and attractive and close at night. Flowers are single measuring about 2.5 cm across in many colours like pink, white, salmon, orange, red, crimson, rose, apricot colours. Flowers are available in self coloured or white edged with contrasting colours.

Lady's lace : *Pimpinella monoica* (Umbelliferae)

It is a tall annual with finely cut foliage and is native of India. Flowers are small, white, delicate and are produced in umbles. Dry flowers maintain their shape and dried flowers are commonly used for dry arrangement. Queen Anne is the important variety.

Larkspur : *Delphinium elatum* Syn. *Delphinium hybridum* (Ranunculaceae)

It is native of Europe and other temperate regions. Other important species are *D. formosum, D. cardinale, D. belladonna, D. nudicaule.* It is a good cut flower and also suitable for mixed annual borders. Plants are much branched with finely cut colours are blue, white, pink, lilac, and mauve. Flowering occurs late in season *i.e.,* end of February-March. The double flowered strain known as 'hyacinth flowered' has been produced. Another strain, stock flowered Larkspur has been produced in several fine colour forms in dark blue, rose pink, lilac rose, rosy scarlet, white and soft salmon-rose. The plants of these varieties grow about 75 cm tall of compact erect habit. Recently selection and development of upright growing type known as 'Giant Imperial' have been done. It is characterized by plants of erect branching habit growing upto 120-130 cm tall with graceful spikes of double flowers in a wide range of delicate shades. The important varieties of this class are Blue Bell, Blue Spire, Brilliant Rose, Carmine King, Daintiness, Miss California, Rosalie and White King.

Linaria : *Linaria incarnata* Syn. *Linaria bipartita* (Scrophulariaceae)

It is also known as Toad Flax and is native of Spain, Portugal and Morocco. The shape of flower is similar of snap dragon but flowers are smaller in size. Plants are dwarf and attain a height of 30-40 cm. Branches are erect and are thin. The flowers are borne on these erect branches which are tiny and look peculiar due to their shape. Flowers are of self colour or bicolour and important colours are purple, blue, deep pink, yellow and white and combinations of these. The important varieties are Fairy Bouquet, Fairy, Brides Maid, Ruby King, White Pearl and Yellow Prince.

Linum : *Linum grandiflorum* var. *rubrum* (Linaceae)

It is a native of North Africa and Europe and is commonly called as flax. It is ideal for bedding purposes. Plants are medium and grow up 45-60 cm tall. Stems are slender and leaves are narrow and pointed. Flowers have five petals which are united and have double centre. Flowers measure about 2.5 cm. The common colour of flowers is red and white. Blue colour flax is also grown in the garden.

Lupin : *Lupinus hartwegii* (Leguminosae)

The name lupins has been originally derived from Greek word *i.e., lupe* which means grief. It is native of N. America specially of California and Virginia states. There are many species like *L. perenne, L. luteus, L. varius, L. hirustus.* Romans used to grow *L. albus* as food for man and beast. Modern lupins *i.e., L. hartwegii* have been derived from *L. polyphllus* and *L. arboreum.* Lupins are most ideally suited as cut flower, and as bedding, border and pot plants. The plants attain the height of 50-75 cm and have palm shaped foliage. Flowers are borne on a long spike and resemble pea flowers. The colour range varies from white, blue, pink and discoloured. The important varieties are Azure Blue, Roseus, Colestinus, Albus, Pixie Delight, Russel hybrids, dwarf Russel hybrids, Blue Jacket, Radiant, Day dream, Heather Glow, Freedom etc.

Mignonette : *Reseda ordorata* (Resedaceae)

It is native of North Africa. Although plants do not produce attractive flowers but they are valued for fragrance. Flowers produce nectar and lot of honey bee activity is observed around this flower. Plants are dwarf growing about 35-40 cm and have crawling habit. The flowers of white, light yellow, red or crimson are small in size and produced in loose cluster. It is suitable for pots, beds and window gardens. Machet is the variety which is being marketed in India.

Mimulus : *Mimulus tigrinus* (Scrophulariaceae)

It is native of Western parts of North and South America and also commonly called as Monkey flower due to its sported flowers. In Latin language it means mimick. Plant grows about 45-90 cm high and angular branches are produced. Leaves are toothed and heart shaped. The flowers are large, tubular and yellow, pink, red, generally bi-coloured with burnt and maroon spots. The plants are free flowering and numerous flowers are produced. It is most ideal for pot and for bedding purposes. The modern varieties are hybrids of *M. luteus, M. cupreus* and *M. guttata.* The common varieties are A.T. Johnson, Bee's Dazzler, Bonfire, Canary Bird and Queen's Prize.

Nasturtium : *Tropaeolum majus* (Tropaeolaceae)

It is native of Mexico and South America and is very common in Indian gardens. Plants have wiry stems and green round leaves. There are dwarf and trailing type nasturtium and in both groups flowers are single or double. Dwarf type hardly grows about 25-30 cm and most ideally suitable for pot, and bedding purposes, whereas trailing types can reach the height of about 2 m and need support for climbing. This type is most suitable for temporary screening, training in different forms and against wall. Flowers are large and have long spur. Flowers are of yellow, orange, scarlet and crimson colour and in some cases have spots and markings of contrasting colours. Flowers are generally hidden by foliage so production of leaves may be checked by restricting supply of nutrients and water. The common varieties in dwarf single are Empress of India, King of Tomb Thumbs (scarlet), Gold Kind (yellow). The dwarf double varieties are Orange Gem, Salmon Gem, Golden Globe, Scarlet Globe, Cherry Rose and Jewel mixed. The important trailing single varieties are Lucifers and Spot Fire (scarlet). Recently gleam hybrids have been introduced which are very attractive. The common varieties are Golden's Orange, Scarlet, Salmons Gleams and Gleam mixed. Now-a-days varieties like Cherry Rose and Golden Jewel, Golden Yellow, Scarlet Jewel are preferred because they produce flowers above foliage. Other important varieties are Peach Melba, Salmon Baby, Scarlet Express and Canariense.

Nemasia : *Nemasia strumosa* (Scrophuliaraceae)

Nemasia is a native of South Africa. It is semi hardy and growing about 20-45 cm in height. It is ideally suited for pot, bedding purpose and dwarf varieties are suitable for window boxes and rock gardens. Plants are bushy and compact with erect branches on which flowers are produced freely. These flowers are very pretty and attractive on account of their orchid like appearance. Wide range of flower colours and throat markings are available in nemasia. The flowers are of yellow, white, pink, purple, scarlet, crimson. Popular tall growing varieties are suttoni of separate colours as well as of mixed colour and Carnival mixture whereas amongst dwarf group, varieties are Triumph (mixture), Orange Prince (Orange), Fire Ball (Scarlet), Fire King and Blue Gem.

Nigella : *Nigella domascena* (Ranunculaceae)

It is popularly called as Love in a-mist or Devil-in-the bush and is native of North Africa and Europe. It is mostly liked in its dried stage when seed capsules become egg shape and can be arranged for dry decoration. It is also mostly suitable for pot as well as cut flowers. Plants attain the height of 45-60 cm and have finally cut thread like leaves which cover bloom. Flowers are semi double measuring about 3-4 cm of white light blue or rose in colour. The popular varieties are Persian Jewel and Miss Jekyll which produces sky blue flowers. Oxford Blue is also another good variety.

Pansy : *Viola wittorckiana* (Violaceae)

Pansy is one of magnificent flowers which is mostly sought in the garden due to its pretty shape and contrast colour markings or blotches on the flower. These markings or blotches resemble the face of cat. It is most ideally suitable for pot, edging, border, rockery, bedding, window boxes and makes good combination with roses when planted in rosery.

It is a native of Europe. It grows about 20-30 cm and makes bushy growth. Leaves are heart shaped. Flowers are produced in great profusion and size varies from 5 to 10-13 cm across. Flowers are of white, red, deep violet, blue, yellow, self coloured or have blotches or markings of contrast colours. Flowers are mildly sweet scented, particularly early in the morning.

The pansy varieties are grouped into two groups depending upon the size of flowers *i.e.* small flowered or large flowered pansy. Small flower varieties produce flowers into a great profusion which are most suitable for bedding and mass effects. The important varieties are Lord Beacon Field, Tormordeau mixed, Hiemalis mixed. For getting a colourful mixed population, seeds of separate colours are produced and different colours are mixed in a certain proportion. Under North Indian conditions, there is no problem of seed set in small flowering pansy.

The large flowered pansies are also known as Show pansies or Giant size and are specially preferred for pot culture and window boxes. These varieties produce less number of flowers but of good size around 10-13 cm. The important varieties are Sakata's Giants, Roggolis Swiss Giants, Aalsmeer Giants, Floridale Giants, Her Majesty, World's Seventh Wonder, Triumph of Giants, Fancy Queen, Fancy Giant etc. F_1 hybrid of Super Giants are also available. F_2 and F_3 hybrids are also available for mixed colourful planting though the size of flower is little less. Under North Indian conditions, seed set in these giant types is poor unless flowers are hand pollinated.

Petunia : *Petunia hybrida* (Solanaceae)

It is a native of South America and is very popular in Indian gardens due to its floriferous and hardy nature. Petunia is commonly grown as bedding, pots, hanging basket, window boxes, rockery and for mass planting. Plant grows about 30-40 cm and have spreading habit with trailing branches. Leaves are small, round and thick. The flowers have five petals which are joined and are trumpet shaped. The colour and form of petals differ widely. The flowers are self coloured of white, cream, yellow, pink, blue, purple mauve, salmon and bicoloured with star like patterns. The petals may be entire or frilled. The seeds of double petunia and fringed varieties do not produce the plants of true-to-type. For F_1 hybrid seed production, the single petunia is used as female parent which is fertilized with pollen of a double variety.

There are different types of petunia and large number of varieties available in different groups. The common varieties of dwarf and compact bedding are : Blue Bird (blue-violet), Cream Star, Peach Red, Rose of Heaven, Snow Ball, Butter Scotch, Fire Chief, Lady Bird. Smooth edged grandiflora varieties are Bingo, Dazzler, Pop Corn and Purple Prince. The important varieties of frilled type are Giants of California, Superbissima, Romona, Snow Storm, Theodosia, Defiance, Super Frills, etc. Balcony and Pendula's Petunia's are trailing type and produce small flowers and common varieties are Blue Wonder, Rose Wonder and Rose. The varieties of all double petunia are Sonnata, Prestos, and Alegro. Multiflora F_1 hybrids produce small flowers of smooth edged of petals and common varieties are Sugar Plum, Montana, Polaris, Coral Satin, Glitters, etc. whereas F_1 grandiflora petunia produce large flowers with ruffled petals. The important varieties are White Magic, Pink Magic, Lapalona, May Time, Spring Time etc. The varieties of F_2 hybrids of grandiflora are carnival and of multiflora are Colorona and Canfetti. There is a recent introduction of Single Joy Fest Series of F_1 hybrids multiflora and important are Blue, Rose Red, Rosy Star, Blue Star, Fiery Star, and White.

Phlox : *Phlox drummondii* (Polemoniaceae)

It is also known as Star flower and is native of U.S.A. specially the region of Texas. Phlox is very popular in all gardens due to its easy culture, free flowering and wide range of colours with long blooming period. It is highly suitable for pot culture, beddings, border, window boxes and is a good cut flower. Phlox is a dwarf annual and grows about 15-40 cm tall. Plants are bushy with medium sized, narrow and pointed leaves. The flowers are produced in clusters at the end of branches well above foliage and covers the whole plant. The individual flowers are small, about 2 cm across with round or pointed petals. The flower colour is wide which varies from white, cream, pale yellow, rose, pink, salmon, mauve, blue, red, scarlet, red, crimson etc.

There are numerous varieties in mixed or separate colours in both groups of phlox *i.e.* Grandiflora with large flowers and Nana Compacta which produces dwarf and compact plants. The important varieties of Grandiflora are Snowball, Vermilion, Coccinea, Ford Hook, Art Shades Brilliant, Yellow Beauty, Blue Beauty etc. Common varieties of Nana Compacta are Globe, Cecily, and Beauty Cuspidata and Twinkle are good varieties of star phlox. Tetraploid varieties like Tetra Road and Giant Tetra with stronger stems and larger blooms are also available and are quite popular.

Rudbeckia : *Rudbeckia biocolor* (Compositae)

It is also called as Cone flower and is native of North America. It is suitable for bedding purpose and border. Plants attain the height of 40-60 cm. Flowers are like daisy flower *i.e.* yellow with a crimson central zone. The important varieties are Hirta, Gloriosa, Daisies Mixed Golden Flame, Kelvendon Star and Summer Festival.

Cutleaf Coneflower : *Rudbeckia laciniata*

It is perennial herb producing loose clumps of tall and wiry stems which grows about 1-3 metres. Leaves are thin, pinnatisect each with 3-5 lobed and veins are prominent.

Daisy like flower heads measuring 7-15 cm across with reflexed yellow ray florets which are hemispherical to conical center. Top of the plant is laden with yellow flowers during mid of April to May. Golden Glow variety is very vigorous and well surviving in plains of North India. It is easily propagated division of rhizomes. However, it susceptible to aphids.

Saliva : *Saliva splendens* (Labiatae) **(Plate 1)**

It is also called as sage flower and is native of South America, specially the region of Mexico and Brazil. Saliva is ideally suited for shady situation and corners of the garden. It is an ideal plant for pots, bed and under the trees. Plant grows about 30-90 cm high and has many square, smooth branches of erect growth. Leaves are broad, pointed and their edge is slightly cut. The flowers are produced on long spikes which are 15-30 cm in length which are long, tubular with a long tip. The common colours are red, white, cream and purple. There are also pink, violet and salmon. There are several varieties and common dwarf varieties are Hussara (Red), St. John's Fire, Blaze of Fire, Scarlet Pygmy, Hot Shot whereas tall varieties are Scarlet Cream and Crimson King. *S. coccinea* is commonly grown in South India and important varieties are Red Indian, White Dove and Pink Pearl.

Saponaria : *Vaccaria hispanica* Syn. *Saponaria vaccaria* (Caryophyllaceae)

It is a native of Europe and is also popularly called as soapwart. It is tall plant growing about 75 to 90 cm. Plants make bushy growth and have straggling branches. Leaves are long. Numerous flowers of star shape are produced in loose clusters which look very attractive. Pink and White colours are common. Flower is suitable as cut flower. Another species, *S. calabrica* which is dwarf (20-30 cm) is suitable for edging or rock garden.

Schizanthus : *Schizanthus wisetonensis* (Solanaceae)

It is commonly called as Butterfly flower of Poor man's orchid and is native of Chile and Peru. This is excellent for growing in pots and cut flowers. The modern varieties have been developed by crossing two species *i.e., S. grahimi* and *S. pinnatus.* The plants are about 90 cm tall and foliage is deeply cut like ferns. Numerous small flowers about 2 cm across are produced which are attractively coloured and cover almost the entire plant. The common colours are white, apricot, yellow, rose purple, mauve, lilac, pink, salmon or cardinal red and marked beautifully and blotched in contrasting colours. The common tall varieties are Butter Giant, Monarch, Brilliant, Benery's Giant, Dr. Badger's Hybrids, Angelwings, Cattleya orchid, Excelsior while Dwarf Pansy and Dwarf Bouquet are dwarf varieties.

Shirley Poppy : *Papaver rhoeas* (Papaveraceae)

It is native of Europe and is also commonly called as corn poppy. Plants are medium to tall and grow about 60-90 cms. Stem are long slender. Leaves are bluish green. The flowers are cup shaped, single or double with crinkled petals. There is a wide range of colours available varying from dark crimson, salmon scarlet, pink, mauve, yellow to blue. The common varieties are Shirley Single, Double Paeony flowered, Oriental Mixed, Iceland Mixed, American Legion, Sweet Briar, and Mixed. Other species of poppy are *P. glaucum, P. nudicaule, P. somniferum, P. orientale.*

Statice : *Limonium sinuatum* (Plumbaginaceae)

It is also popularly known as Sea Lavendar or Pink because some species were found growing in salt marshes. It is native of Mediterranean region, and is suitable for planting in beds, borders and for flowers which keep colour after drying. Plant grows 60-70 cm in height. Branches are angular, leaves are leathery and deeply lobed. These branches carry dense cluster of numerous small flowers. These flowers have papery texture and are about 1.25 cm across, usually blue, lavender or rose and a tiny yellowish corolla hidden inside. There are several varieties available for commercial growing and important varieties are Mid-Night Blue, Twilight lavender, Blue Ronnet, Iceberg, Gold Coast, Blue Perfection, Charmois Rose, Lavender Queen, Market Growers Blue, Market Rose, Purple Monarch and White. There are many other species but *L. suworowii* is very ornamental *L. latifolium* is also highly suitable as cut flower and can be used as dry flowers which can be kept for several years. Blue Cloud and Violetta are good varieties.

Stock : *Matthiola incana* (Cruciferae)

Stock is an excellent cut flower ideally suited for pot growing, beds and herbaceous border. It is native of Mediterranean region. Plant grows 45 to 90 cm tall. Leaves are single and dark green.

The colour range is wide and available in white, lilac, rose, lavender, apricot, crimson to canary yellow. Flowers are fragrant, and single or double. Single flowers have 4 petals, 4 stamens and a pistil having 30-60 ovules whereas double flowers have 40-70 petals and are male and female sterile. There are four types of stock plants (according to production of single and double flowers).

1. Singles—which produce only single flowers.
2. Singles—which produce 75% single and 25% double.
3. Eversporting singles—which produce 54-56% double flowers and 44-46% single.
4. Doubles—which produce double flowers but do not have ability to produce seeds due to complete sterility.

There are several varieties in different groups and important varieties are *Ten week stocks*—Giant perfection, Princess of Alice. Snow Drift, Salmon Beauty, Almond Blossom, Crimson King, Fire Ball, Violet Queen, Celestial and Mauve Beauty; *Nice stock*—Beauty of Nice, All the year Round, Cote d' azure, Queen Alexandra, Empress Elizabeth, Christian Pink, Yellow Prince, Monte Carlo, Crimson King, Peach Blossom, Riveria Market; *Brompton stock*—Cottage White, Sun rise, Old English Scarlet, Improved Green, Cottager's purple;

Intermediate stock—Crystal White and Convent Garden. Trisomic seven weeks is also getting popular.

Sweet Alyssum : *Lobularia moschata* Syn. *Alyssum maritimum* (Cruciferae)

It is a native of Western Asia and Europe and is very common in Indian gardens. Sweet Alyssum is an excellent flower for edging, pot growing, hanging baskets, beds and window boxes. Plant grow about 30 cm high and have spreading habit. Leaves are narrow of light green colour and produce spike of tiny sweet scented flowers which cover whole plant and look a white carpet of flowers. Other common colours are purple, rose and lilac. The important varieties are Snow Carpet (white), Royal Carpet (violet), Pink Heather (rose), Violet Queen and Snowdrift.

Sweet Pea : *Lathyrus odoratus* (Leguminosae)

It is a native of Sicily and Greece. It has climbing habit and needs support to climb up. Therefore, it is planted for artificial screening or as temporary tall hedge, or trained on stumps. Seeds are sown directly at 6-9 cm apart, 3-5 cm deep and in double rows spaced 30-40 cm apart in deeply dug (5-7 cm deep) and well prepared land by adding organic manure. When the plants are about 5-7 cm tall it need support which can be of various materials like sarkanda, chicken wire netting, bamboo poles or galvanized pipes or strings. There is a wide range of flower colours varying from white, light blue, lilac, mauve, pink, rose, salmon, cherry red, red, and maroon to Crimson. There are numerous varieties which are being grown. The original hooded types and plain grandiflora varieties have been replaced by spencer type having frilled waved edges. There are six different strains of sweet pea. They are : Giant Frilled, Late Flowering Spencers; Early Flowering Frilled: The Cuthberston, Frilled : Multiflora or Zvolanik's multiflora (5-7 blooms to a stem); the Dwarf Cupid (without tendril-non trailing) and the Dwarf Cupid Frilled or the Little Sweet Heart. Important varieties are Early Multiflora Galaxy, Late Spring Galaxy, Giant Ruffle Mixed, Royal Family Mixed, Knee Hi Mixed, Vulcan, Monarch, Blue Wonder and Black Prince.

Sweet Sultan : *Amberboa moschata* Syn. *Centaurea moschata* (Compositae)

It is another important seasonal flower belonging to genera Centaurea and is native of Caucusus region, Middle East Asia and Eastern parts of mediterranean region. It's common name Sweet Sultan is due to the fact that it was brought from East which was under the rule of Ottoman Sultan in the 17th century. Like other seasonal flowers, it is also mostly planted in the Indian gardens and home landscapes due to its delicate and sweet scented flowers. It is most suitable for herbaceous border and for mass effect. Plant grows 90-120 cm in height. Flowers are soft and fluffy which are borne at the tip of them. Flowers are grown commercially around big cities and are sold as cut flowers. The flowers are available in white, mauve, rose and sulphur yellow colours. Important varieties are the Bride White, Flavo-Yellow and Rosea-Pink.

Sweet William : *Dianthus barbatus* (Caryophyllaceae)

It is a native of Northern France and is closely related with carnation and pink. Sweet William is biennial but is grown as an annual in North Indian plains. It is ideally suited for bedding, borders, cut flowers and pots. The dwarf varieties are mostly planted for edging, rock garden, window boxes and pots. The plants grow about 30-40 cm tall. Leaves are flat and tips are pointed and green. Flowers are produced in clusters. There is large variation in colour of flower and flowers of pleasing colour such as crimson, dark red, dark maroon, purple, salmon and white are produced. The flowers may be of self, mottled, or have markings of different colours. The common varieties and Diadem, Indian Carpet, Messenger Mixed, Pink Beauty and Scarlet Beauty.

Venidium : *Arctotis fastuosa* Syn. *Venidium fastuosm* (Compositae)

It is a native of South Africa and is a popular flower for bedding purpose. Plant attains the height of about 60 cm and makes a bushy appearance. Leaves are irregular and lobed. Leaves and flowering stalk have silvery white hairs which give a wooly or fuzzy appearance. Flowers are large measuring about 10 cm across of brilliant orange, ivory, cream, yellow colour with a purple black centre and are produced on long stem. The important variety is Monarch. Dwarf hybrids which grows about 35-50 cm high are also available.

Verbena : *Verbena hybrida* (Verbanaceae)

It is native of South America and is a very common in Indian gardens during winter. It is a dwarf annual which grows about 25-30 cm in height and is ideally suitable for bedding, edging, window boxes, pots and rock garden. Leaves are of dark green colour and are wrinkled, small and deeply cut. Flowers are produced in loose clusters above the foliage. These clusters consist of several star like small flowers which give an appearance of mass of flowers. There is a large variation of colours which vary from white, cream, blue, lilac, violet, rose, mauve, and red to intermediate shades. The important varieties in large flowered are Candissima (white), Lavender Glory, Defiance (Scarlet) and Grandiflora mixed. There are dwarf compact varieties like Crystal (White), Dannebrog (scarlet), Salmon Queen (Salmon), Sparkle (purple with a white eye) and Star Light. Double flowered variety Miss Susie is liked for its ruffled petals.

Wall flower : *Erysimum cheiri* Syn. *Cheiranthes cheiri* (Cruciferae)

It is native of Europe. It is medium annual growing about 45-60 cm and has long green leaves. The flowers are produced on a long spike. The flowers are single and common colour is golden orange. Other colours are orange brown, scarlet, wine, dark blood red, cream, yellow and flame red. Early annual varieties and Persian Wall flowers, Early Wonder, Early Flowering Phoenix and Early Flowering Yellow Phoenix. There are number of biennial varieties which can be successfully grown in hills.

DESCRIPTION OF IMPORTANT ANNUAL FLOWERS

S. No.	Botanical Name	Common Name	Height (cm.)	Colour range	Remarks
			WINTER ANNUALS		
1.	*Rhodanthe chlorocephala* subsp. *Rosea* Syn. *Acroclinum roseum*	Paper flower	60-75	Pink and white	Flowers keep their form and colour, even they dry up; so these can be used for interior decoration.
2.	*Agertaum houstonianum*	Floss flower	40-50	Blue white or pink	Grown for bedding purposes.
3.	*Alcea rosea* Syn. *Althaea rosea*	Hollyhock	120-180	White, crimson, pink, yellow and red	Can be used for screening. Purpose, background of border.
4.	*Lobularia maritima* Syn. *Alyssum maritimum*	Sweet Alyssum	10-30	White, lilac	Bedding purpose, pot growing.
5.	*Antirrhinum majus*	Snap dragon	15-120	White, yellow, pink, rose, mauve	Bedding and pot growing good cut flower.
6.	*Arctotis stoechadifolia*	—	60-70	White	Bedding purpose.
7.	*Brachycome iberidifolia*	Swan River Daisy	15-25	White, blue, light pink	Bedding purpose.
8.	*Bellis perennis*	English daisy	15-25	White, light pink	For pot cultivation.
9.	*Calendula officinalis*	Pot marigold	40-50	Yellow, orange	Bedding and pot growing, good cut flower.
10.	*Callistephus chinensis*	Aster	40-50	White, pink, blue	Bedding and pot growing, good cut flower.
11.	*Cyanus segetum* Syn. *Centaurea cyanus*	Corn flower	90-100	White, pink, blue	Bedding purpose.
12.	*Amberboa moschata* Syn. *Centaurea moschata*	Sweet sultan	90-100	Mauve, white, yellow	Bedding and pot growing, scanted and good cut flower.
13.	*Erysimum cheiri* Syn. *Cheiranthes cheiri*	Wall flower	30-40	Yellow, burnt orange	Bedding and pot growing
14.	*Glebionis coronaria* Syn. *Chrysanthemum coronarium*	Annual Chrysanthemum	90-120	White, yellow	Bedding purpose, good loose flower.
15.	*Clarkia pulchella* Syn. *Clarkia elegans*	—	70-100	White, rose, pink and purple	For bedding purpose.

16.	*Coreopsis stillmanii*	—	30-45	Yellow	For bedding purpose.
17.	*Cosmos bipinnatus*	Cosmea	90-150	White, pink, yellow, mauve, magenta	Can be grown throughout the year.
18.	*Dahlia pinnata* Syn. *Dahlia variabilis*	Dahlia	90-150	Yellow, red, white, mauve, crimson	Bedding and pot growing.
19.	*Consolida ajacis* Syn. *Delphinium ajacis*	Larkspur	45-150	White, pink, crimson, mauve, yellow, violet	Bedding and pot growing.
20.	*Dianthus barbatus*	Sweet William	30-40	White, pink, crimson, mauve, yellow, violet and red	Bedding purpose with scented flowers.
21.	*Dianthus caryophyllus*	Carnation	45-75	White, pink, crimson, mauve, yellow, violet, and red	Bedding, pot growing good cut flower with long lasting vase life and scented flower.
22.	*Dimorphotheca sinuata* Syn. *Dimorphotheca auriantiaca*	African daisy	50-60	White	Bedding, pot growing.
23.	*Gypsophila elegans*	—	60-90	White, pink, yellow	Bedding purpose, good cut flowers used as filler.
24.	*Helichrysum bracteatum*	Straw flower	70-90	Yellow	Bedding purpose and good cut flower.
25.	*Iberis amara*	Candytuft	30-40	White, pink	Can be grown in pots and for edging of herbaceous border.
26.	*Lathyrus odoratus*	Sweet pea	90-150	White, pink, orange, blue scarlet, crimson mauve, violet	Mainly grown as background of herbaceous border and screening purpose. Sown directly in the permanent places, it requires support, good cut flowers, scented.
27.	*Limonium sinuatum*	Statice	45-60	White, pink, yellow, purple	Fairly drought resistant, very good cut flower for dry arrangement.
28.	*Linaria incarnata* Syn. *Linaria bipartita*	Linaria	30-40	Yellow, white, pink, blue, red and purple self and double colour	Can be grown in semi-shady places, good for bedding and pot growing.
29.	*Matthiola incana*	Stock	70-100	Red, white, yellow	Good cut flower.

S. No.	Botanical Name	Common Name	Height (cm.)	Colour range	Remarks
30.	*Stigmatocarpum criniflorum* Syn. *Mesembryanthemum criniflorum*	Ice plant	20-30	Pink, white, yellow	Leaves are fleshy and the angular branches have small dew like natural eruptions. Grown for edging. pots hang ing baskets and rockery.
31.	*Mimulus tigrinus*	Monkey flower	30-45	Large, yellow, tubular flower, have irregular brown spots	Partially shady places, suitable, good for pot growing.
32.	*Nemesia strumosa*	—	15-40	Mixed colour white, pink purple	Bedding and pot growing.
33.	*Nigella demascena*	Love in mist	45-60	Blue, white purple	Sunny places are good for growth. Very good dry flower.
34.	*Petunia hybrida*	—	50-75	Red, blue, white, pink, violet	Good for pot and bedding.
35.	*Phlox drummondii*	—	30-50	White, red, blue, violet, pink	Dwarf plants are very showy when grown in flower border, baskets and pots.
36.	*Papaver rhoeas*	Shirley poppy	60-75	Scarlet, pink, orange, white, yellow	Early flowering, good for bedding.
37.	*Saliva splendens*	—	45-60	Scarlet, red white, purple	Easily grown in shady places and rockery.
38.	*Pericallis cruenta* Syn. *Senecio cruentus*	Cineraria	25-45	Blue, purple, salmon, crimson, red, pink, white	Grown in shady places, good for pot growing.
39.	*Tagetes erecta*	African Marigold	60-120	Yellow, orange	Easily grown, good for bedding, grown for loose flower.
40.	*Tagetes patula*	French Marigold	30-40	Red, orange, yellow	For bedding and pot growing.
41.	*Tropaeolum majus*	Nasturtium	30-45	Yellow, orange, red	Mostly grown in rockeries, it has trail-ing habit so used as climbers also.
42.	*Verbena hybrida*	—	25-45	Purple, pink, blue, red, white	Good for bedding, pots, hanging baskets, window boxes and rockery.
43.	*Viola wittorckiana*	Pansy	25-30	Yellow, blue, red, purple and violet, and bicolour	Sunny site is preferred, good for bedding and pot.

		SUMMER AND RAINY SEASON ANNUALS			
44.	*Amaranthus caudatus*	—	60-75	Pink & white	Due to varigated leaves it is also grown in pots for indoor decoration.
45.	*Gaillardia pulchella*	Blanket flower	45-60	Yellow, orange, scarlet, brown	It can resist dry conditions and can be grown as perennial also. Good cut flower.
46.	*Gomphrena globosa*	—	60-75	Pink, white, purple	It can resist dry conditions and can be grown as perennial also. Good cut flower.
47.	*Portulaca grandiflora*	—	20-30	Orange, red, yellow, pink, white, purple	Easily propagated by cutting, can be grown in pots, hanging baskets and small boxes.
48.	*Bassia scoparia* Syn. *Kochia scoparia*	—	60-75	—	Mostly grown for foliage beauty due to uniform, well shaped and bushy growth.
49.	*Zinnia elegans*	—	75-90	White, orange, red, pink, yellow violet, cream	Very hardy and grown easily for longer duration.
50.	*Impatiens balsamina*	Balsam	60-75	White, orange, red, pink, violet, cream	Early flowering and for short duration. Very delicate and needs much care.
51.	*Celosia argentea* var. *cristata* Syn. *Celosia cristata*	Cock' comb	45-75	Pink, yellow, orange,	Very hardy and longer flowering period.
52.	*Tithonia rotundifolia* Syn. *Tithonia speciosa*	Mexican Sunflower	70-150	Orange, scarlet	Sown directly from seeds.
53.	*Catharanthus roseus Syn. Vinca rosea*	Sada Bahar	60-75	Purple, white	Can be grown throughout the year.
54.	*Helianthus annuus*	Sunflower	60-150	Yellow, orange	Can be grown throughout the year and sown directly from seeds into permanent places.

4

Chrysanthemum

It is commonly called as Guldaudi or Autumn Queen and belongs to family compositae. It is native of China and was growing before 500 B.C. This flower was highly respected by Chinese and its export was not allowed to other countries. However, few plants trickled to Korea and to Japan around 386 A.D. In Japan it became very popular. Its cultural practices were standardised and lot of new types were evolved. It is clear from the ancient records that from 900 A.D. onwards chrysanthemum festivals were used to be organised and prizes of highest orders like Grand Cross of the order Chrysanthemum (G.C.O.C.) were used to be given to gardeners by Kings. From Japan it reached to England in 1789 via France and Holland. Good improvement work was done and many exhibition varieties were produced. It reached America, India and other countries in early 19th century. In America more emphasis was given on the development of spray type of chrysanthemum representing a wide range of form such as pompon, spider, decorative and anemone. In India also, only chrysanthemum shows and competitions are held every year in many cities and prizes are distributed.

Chrysanthemum plant is not very attractive but it produces most showy flowers. It has a wide range of flower shape, size and colour and is highly suitable for pot culture and bedding purposes. Flowers of standard varieties are produced on long, sturdy stems and have good keeping quality. These characters make it highly suitable for flower arrangements. Flowers of spray type varieties are highly suitable as loose flowers for making garlands, veni, bracelets. In India it is commercially grown in Maharashtra, Tamil Nadu, Rajasthan, Karnataka and M.P.

Now-a-days large number of varieties are available and natural blooming period of November-December has been extended from September-December by just selecting suitable types and planting on different dates. In America, Europe and Japan, it is commercially grown under artificially controlled conditions (photoperiod, temperature, shading, use of growth regulators, etc.) for the production of bloom around the year.

IMPORTANT GROUPS AND VARIETIES

Broadly, chrysanthemum is classified into two categories on size of flowers *i.e.*, large flowered and small flowered. Large flowered varieties have the genetic potential for production for large flowers if proper cultural practices are followed whereas in the latter group the varieties cannot produce large flowers despite of best cultural practices adopted. These two main categories are further divided into following groups according to the shape and floret arrangement. **(Plate 2)**

Large Flowered Groups

(*i*) Incurved. In such varieties, ray florets are incurved and overlapping smooth and regular. Florets are not twisted and blooms are compact and globular when fully developed. Examples : Snow Ball—White; Sonar Bangla—Yellow; Chandrama—Yellow; Kiku Biori—Yellow; Alfred Simpson—Crimson with Bronze, Grape Bowl—Rosy purple, Pink Cloud—Coral Pink, Eva Turner—Silver Pink; Ghenghis Khan—Copper, bronze and orange, Red West Field—Chestnut Red, etc.

(*ii*) Reflexed. This is just opposite to incurved and ray florets reflex outside and overlap one other. Blooms are globular or flattened. Important varieties are : Cresta—White; Dorothan—White; Beatrice May—Cream white; City Beauty—Yellow; Golden Rule—Yellow, Day Dream—Pink; Peach Blossom—Pink; Picardy—Pink; Sweet Heart—Pink; Regalia—Purple, etc.

(*iii*) Intermediate. This class is intermediate between incurved and reflex. The inner florets are in curved ones whereas outer florets are reflexed type. *e.g.*, Mrs. W.A. Reid, Sun Flight, General Petain.

(*iv*) Quilled. Ray florets are elongated and tubular with tips open or closed. *e.g.*, Green Sensation, Tribhuban.

(*v*) Spider. Ray florets are large and tubular and are usually curved. Tips of these florets may be open, for example Rupasi Bangla—White; Bidhan's Best—Mauve, Coronation—Bronze, Mahatma Gandhi—Cream.

(*vi*) Ball. In this class flowers have ray florets straight and radiated in all directions to give a compact ball shape. *e.g.*, Pride of Madford, Nigeria, Red Jack.

(*vii*) Japanese or irregular. Twisted ligulate petiols are its distinguishing character which give the bloom an open airy appearance with irregular overlapping petals, *e.g.*, Mountaineer, J.H. Salisbury, Hommand Philip.

Small Flowered (Spray-type)

(*i*) Korean (single). In this type ray florets are flat, strap like and blooms are flat. Number of whorls of ray floret are five or less than five and disc is open *e.g.*, Cardinal—red; Charmain—bright yellow; Gul-e-Sahir—yellow.

***(ii)* Korean Double.** These are like Korean singles but the number of whorls of ray florets are more than five and the disc is open. Jante Wells—yellow; Flirt—maroon; Man Bhawan—yellow and red turn yellow at later stages.

***(iii)* Decorative.** These are similar to Korean double, except that flower is completely double and centre of blooms is not visible. Examples : Arctic—White; Blue Chip—Purple; Dolly—Purple; Flame Blair—Bronze; Hurricane—Yellow; Lemon Polaris—Light yellow; Jubilee—Yellow; Yellow Divinity—Yellow; Elegance—White; Golden Elegance—Yellow; Alankar—Orange.

***(iv)* Anemone.** Ray florets are flat or may be twisted or quilled but disc florets are well developed and prominent *e.g.*, Caleb Cox—light bronze; Grace Land—White with yellow cushion; Long Island Beauty—White with yellow centre; Cloud Bank—White; Golden Sands—yellow, White sands—White. Tan Vedova—Light bronze; Vedova—Purple.

***(v)* Pompon.** They are getting popular and produce large crops of small bloom. Pinching once or twice is very helpful. Ray florets are short, broad, regularly arranged to give bloom of compact hemispherical shape and florets are incurved or reflexed. Disc florets are covered or inconspicuously open *e.g.*, Cameo—White; White Bouquet—White; Dandy—Bronze; Eve—Purple; Lustre—Coppery bronze; Baggi—Glistening White, Fred Yule—Burnt orange; Otome Zakura—Pink; Beeswing—Bronze, Nanako—Yellow, Cotton Ball—White, Birbal Sahni—White. **(Plate 3)**

***(vi)* Button.** Blooms are compact, small and hemi-spherical measuring 2-3 cm in diameter. For example : Gold Dust—Yellow, Liliput, King Fisher.

***(vii)* Quilled.** The florets are tubular. For example Golden Crystal yellow; Yellow Hector—Yellow, Snow Crystal—White.

***(viii)* Spoon.** These are slightly different than quilled in which ends are open to the tubes showing inner petal colour *e.g.*, Anokha—White, Modipon—Deep pink.

***(ix)* Semi-quilled.** The florets are tubular upto some length and then open. Open portion of the florets may be flat, reflexed or incurved. Disc is open.

***(x)* Cineraria.** Now-a-days this type is getting very popular specially for pot culture. Ray florets are flat, Korean type and are smaller in size, not more than 3 cm. *e.g.*, Jassie, Bindya, Charmis.

***(xi)* Stellate.** Ray florets are regular with sides reflexed and may or may not be twisted. Disc florets are short. *e.g.*, Laura, Red Star, Stella.

***(xii)* No Stake and No Pinching.** Appu-mauve, Sharad Kumar—yellow.

VARIETIES SUITABLE FOR EXPORT

Chrysanthemum in International Market

Standard

White : Bounty, Dignity, Nobhill, May Shoesmith, Hartmann's Dignity.
Pink : Indian polis, Pink, Escapade, Promenade, Streamer.

Yellow : Golden Bounty, Bright Yellow, May Shoesmith, Yellow Nobhill, Golden Shoesmith, Yellow Dignity.

Bronze : Detroit News, Onward, Topaz, Wild Fire, Dark Indian Polis Bronze.

Spray

White : Divinity, Iceberg, Ping Pong, Hurricane, White Marble, Spider yellow White.

Yellow : Florida Marble, Cavalier, Yellow, Iceberg, Statesman, Early yellow.

Pink : Pink Marble, Westland Pink, Parliament, Deep Jelstar.

Bronze : Bronze Dolly, Bronze Marble, Dazzler.

Pot mums

White : Alpine, Free Spirit, Ruffled Spirit, Joy, Surf.

Yellow : Gettys Burg, Sunburst Spirit, Yellow Spin wheel, Ritz.

Pink : Fantasy, Anthem, Loyality, Albert, Chic.

Bronze : Aglow, Copper Hostess, Minuteman, Mandarin.

Propagation

Chrysanthemum is propagated vegetatively either through rootsuckers or terminal. Latter method is preferred over former one for neat and sturdy plant whereas plants raised through root suckers are shaggy and heggard. The details of these methods are as follows :

Terminal cuttings. These cuttings are taken from healthy stock plants from middle to end of June. 5-7 cm long cuttings are made by shearing basal leaves and cutting half of open leaves. For enhancing rooting, these cuttings can also be treated with Seradix-l powder or 25 ppm NAA. Then these cuttings are planted in sand in pots or beds and are kept in partial shade. Water is sprayed 4-5 times a day. Rooting takes place in 2-3 weeks and these cuttings, are ready for transplanting. To avoid the rotting of cuttings, Captan (0.3%) or Brasicol (0.2%) should be applied in irrigation water once or twice.

Root suckers. After flowering, the plants are planted in partial shade and ample of irrigation and fertilizer are applied for encouragement of root suckers. When these root suckers are 10-15 cm high, they are separated in February or March and planted in small pots and later on planted in bigger pots.

Seeds. This method of propagation is used by breeders to evolve new varieties. In majority of varieties seed set is less due to self incompatibility and lack of pollination in winter by honey-bees. By artificial crossing seeds are obtained and sown for creating new variations.

Pinching and Disbudding. These are important operations which should be done carefully at appropriate time. The objective of pinching is to encourage the side branches and it should be repeated to encourage more number of branches depending upon the number of blooms to be retained. On the contrary, disbudding is done to remove the side branches which arises from axillary buds so that number of flowers is limited and blooms of better size are obtained. If disbudding is not followed, many branches will be produced and bear the flowers which will affect the quality of blooms ultimately.

Standard chrysanthemum. In this case single bloom on a branch is allowed to produce. The standard chrysanthemum varieties have the genetic potential to produce single bigger sized bloom on a branch if disbudding and proper feeding is done. The pinching is not done if only one central bloom is desired on the main branch and only disbudding is done regularly in such cases. Single pinching is done, if two flowers are desired whereas double pinching is done for four flowers. Similarly more number of pinching can be done to produce more standard flowers per plant. Regular disbudding is done to produce single flowers on single branch. First pinching is done in July and second in August and finally in September.

Spray chrysanthemum. These types produce small to medium sized numerous flowers and do not have genetic potential to produce bigger size bloom irrespective of disbudding and best fertilization. These are planted for mass effect in beds or for commercial production. In such varieties two pinchings are required to encourage lateral growth. First pinching is done at 4 weeks after planting and second after 7 weeks of planting. No or little disbudding is done. These spray chrysanthemum are also trained as cascades or in different shapes.

Staking. Staking is very necessary to provide the support whether plants are grown in pots or in field. There are only a few flowering varieties which neither require pinching nor staking and hence known as no stake no pinch varieties. For standard varieties, number of stakes will depend upon the number of main branches which have been allowed to produce bloom. In spray type, 3-4 stakes are inserted on the band of the pot and are tied with string from bottom to top which gives good support to developing branches and finally flowers. When the plants are grown for cascade, the framework of steel or wooden strips is prepared first and plants are trained according to shape.

Growing of Chrysanthemum in pots. Chrysanthemums are grown in pots to decorate verandahs, window gardens and to create colourful islands in the lawn. It provides a great pleasure in growing chrysanthemum in pots. Mostly earthen pots of different sizes are used. Now-a-days use of plastic pots is also on the increase. The common pot mixture used is comprised of 3 parts garden soil : 1 part sand, 1 part leaf mould and 1 part of well rotten farmyard manure. Pots are filled by putting concave side of 3-5 pieces of crocks over the drain hole. For growing chrysanthemums in pots, plants are propagated by root suckers or terminal cuttings but latter method is preferred. The terminal cuttings are got rooted by the method as described earlier. Rooted cuttings are transplanted in mid to end of July in small pots measuring 10 cm. The pots are watered twice depending upon the requirements. The plant grows sufficiently in a month and it needs repotting in bigger size pot measuring about 15 cm.

Final repotting is done in end of September when plants are transferred tc 25-30 cm pot. Staking is done according to type of variety and method of training. Pinching and disbudding is also done according to type of variety and method of training as described earlier. Regular feeding with nitrogenous, phosphoric and potassic fertilizers in small doses is helpful in keeping plants healthy and to obtain good size blooms. Urea should not be applied as this is known to cause phytotoxicity. At the time of appearance of buds, feeding with dilute solutions of cakes and superphosphate is highly beneficial in obtaining good size and shine of bloom. The use of growth retardants like phosphon or Alar is very effective in producing better size of blooms on dwarf plants. The soil drenching is done with Alar @ 33 g/100 L or phosphon at 10% liquid formulations normally at the rate of 100 ml of solution for 15 cm pot has been found effective.

Growing for Loose Flowers. The commercial cultivation of small flowered varieties of pompon or decorative group like Cameo, Baggi, Santi, Birbal Sahni etc. around the big cities has been found profitable. For raising commercial crop, a fresh crop is to be raised and grow every year. Soil is prepared in end of June by cultivating 2-3 times and adding of 10-15 cartloads of farmyard manure alongwith 100 kg single superphosphate and 133 kg of Muriate of Potash/acre before planting as basal dose. The rooted cuttings prepared by the method described earlier are transplanted at the distance 30 × 20 cm in the evening hours of the day in middle of July. Irrigation is done immediately. In order to encourage the branching and ultimate flowering, pinching of apical portion of shoots should be done 4 and 7 weeks after transplanting. The application of 320 kg CAN/acre should be split into three doses and should be applied at the interval for 30 days.

Flowering occurs in November if planting has been done in July which continues upto six weeks. Flowering time can be extended by delay in planting upto end of August. Fully opened flowers are picked in late hours of day when dew has dried up. Soon after harvest, flowers are packed in baskets containing 3-4 kg of flowers. These varieties produce about 6000 kg of flowers/acre and are sold at a good price.

IMPORTANT DISEASES AND INSECT PESTS

A number of diseases and insect pests have been found damaging chrysanthemum. If these are not controlled at right time, these can cause severe damage to the crop. Important disease are as follows :

Diseases

Black leaf spot. It is caused by *Septoria obesa* and *S. Chrysanthemella* and is a serious disease. Greyish brown circular spots appear on the leaves. Leaves turn yellow and die. It can be controlled by spraying Mancozeb 75% WP @ 2 g/l or Propiconazole 250 EC @ 1 ml/l of water.

Powdery mildew. It is caused by a fungus (*Oidium chrysanthemi*). The disease is characterised by the white powdery mass on leaves and stems. To control this disease spraying with 0.5% Kerathane (40 E.C.) has been found useful.

Flower disease : Flower rot. It is caused by grey mould (*Botrytus cinerea*) which appears as brown water soaked spots on the petals. The infected flowers show tan or light brown specks at the base of petals which may result complete rotting. It can be contr olled by spraying Mancozeb 75% WP @ 2 g/l or Propiconazole 250 EC @ 1 ml of water.

Insect-Pests

Aphids (Myzus persicae). Black aphids appear at the time of flowering and suck the cell sap of stem, stalk, bud, flowers, etc. Flowering is affected drastically. It can be controlled by spraying Imidacloprid 200 SL 0.4 @ ml/l or Thiamethoxam 25 WG @ 0.4 g/l of water.

Red hairly caterpillar (Amasacta moorie). The larvae eat away the foliage leaving only the skeleton of veins. At later stages larvae migrates to other plants. It can be controlled by spraying Quinalphos 25 EC @ 0.4 g/l of water.

Gram caterpillar (Heliothis armigera). The caterpillar which are of green, black and yellow colour attack the flower buds and eat them away. Their presence is indicated by very fine holes. It can be controlled as in case of red hairy caterpillar.

***

5 Rose

Roses are symbol of beauty, fragrance and are used to convey the message of love. Without roses gardens are not considered complete. Gardens exclusively for roses have been made in various parts of the world for showing the respect to this flower. Great diversity in plant growth, colour of flowers, flower shape, fragrance, slow opening of flowers and good keeping quality made roses so popular that it is grown commercially to meet the demand of cut blooms. It is top ranking cut flower, in the flower trade on the basis of average, production and consumption. In India roses are grown for cut flowers, making essential oil, rose water and *gulkand.* Dry petals of roses are also used for making incense sticks and milk sweets.

Roses are native of Himalaya regions, West Asia, China, Japan, Europe and North America. There are about 150 species but very few species have played a major role in evolving modern roses. These species are *Rosa gallica, R. damascena, R. chinensis, R. foetida, R. gigantea, R. moschata, R. multiflora,* and *R. wichuriana.* Modern roses are grouped into the following classes.

A. Hybrid Tea's (H.T.)

As the name indicates, the varieties of this group have been obtained by crossing between hybrid perpetuals and Tea roses of China. The first variety 'La France' of this class was produced by French breeder Guillot (1867). They have characteristics of both parents. These are winter hardy than the tea roses and produce blooms for longer period than Hybrid perpetuals. Main characteristics of H.T. roses are production of blooms on long canes, elongated buds, slow opening of flowers and good keeping quality. The continuous breeding programme has produced varieties having all possible good characters in self colour and striped coloured varieties. Every year there is an addition of new varieties. There is a wide choice to select the varieties for different purposes in different colours. Following are some suitable varieties for different purposes.

I. Bedding varieties. These varieties are vigorous growing and free flowering and produce elegant flowers. In different colours, following varieties have been found suitable :

Yellow—McGredy's Sunset, Golden Giant, Buccaneer, Vasant, Aalsmeer Gold, Dutch Gold, Fragrant Gold, King's Ransom.

Orange, Scarlet, Coral and Vermilion—Montezuma, Super Star, Fragrant Cloud, Queen Fabiola.

Orange Yellow—Vienna Charm, Chantre, Raja Ram Mohan Roy, Bettina, Kana Kangi.

Pink—Confidence, Picture, South Seas, Michele Meilland, Mischief, First Prize, Sonia.

Red, Dark Crimson—Christian Dior, Ena Harkness, Avon, Crimson Glory, Baccara, Oklahama, Happiness.

Apricot/Bronze, Thais—Sunset Song, Valencia, Whisky.

Mauve or Lavendar—Africa Star, Blue Moon, Sterling Silver, Paradise, Anuraag.

White—June Bride, Pascali, Tushar.

Bicolour—Bajazzo, Flaming Sunset.

Colour Blend—American Heritage, Careless Love, Kiss of Fire, Tahiti, Double Delight, Sea Shell.

Striped and Hand Painted—Anvil sparks, Careless Love, Siddhartha, Madhosh.

II. Exhibition varieties. The blooms of these varieties are large, properly shaped and are at the peak of floral perfection. Following are important varieties :

Anvil Sparks, Avon, Bel Ange, Christian Dior, Eiffel Tower, First Prize, Garden Party, Golden Giant, King's Ransom, Michele Meilland, Mischief, Montezuma, Pusa Sonia, Show Girl, Sterling Silver, Super Star, Thais, Vienna Charm.

III. Scented varieties. These varieties are prized also for their sweet fragrance : Avon, Blue Moon, Crimson Glory, Heart Throb, Kana Kangi, La France, Seventh Heaven, Sugandha, Hadley, General Mac Arthur, The Doctor.

IV. Commercial varieties. These varieties are grown for cut flowers *e.g.*, Super Star, Montezuma, Happiness, Ilona, Sonia, Mercedes, Pasadena. The variety Gladiator is very popular among rose growers of Nasik and Pune.

B. Floribundas

These have been produced by crossing Hybrid Tea and dwarf polyantha by Danish breeder Poulsen in 1912 and first floribunda was named 'Rödhatte'. It was in 1924 when impressive varieties like 'Else Poulsen, and 'Kirsten Poulsen' were released and this group got popularity. After that number of varieties have been added every year. Floribundas possess the characteristics of both parents and produce flowers in clusters like polyantha of better shape and bigger size than the flowers of polyantha but the flowers are smaller than Hybrid Teas. They are produced in great profusion appearing like bouquets. Following are the important varieties in different colours.

Red—Marlena, Gabriella, Rob Roy, Week Jock, Jantar Mantar.

Orange-Scarlet—Flamenco, Independence, Orangeade, Shola, Zambra, Zorina.

Yellow—All Gold, Fugitive, Ballena, Sunflare, Gold Bunny.

Pink—Queen Elizabeth, Else Poulsen, Junior Miss, Loree, Cherish Baby Talk, Mercedes, Belinda.

White—Iceberg, Himangini, Margret Merril.

Lavendar—Angel Face, Lavendar Prinçess, Shocking Blue.

Bicolour—Red Gold, Sear Pearl, Court Jester, Fantasia, Sabre, Dance.

Colour blend/Multi Colour—Charleston, Banjaran, Bicentennial, Charisma, Madhura, Parfait, Vitul Spark.

C. Climbers

These produce long arching canes and need some support to keep them upright. Some of climbers are natural bud sports of hybrid tea roses, so these have similar flowering habit of their parents. These climbers are most suitable for arches, against wall, pillar and produce magnificent effect. Important varieties are :

Alec's Red clg—red; Sympathie—red; America—Salmon orange; Fragrant cloud clg—Coral red; Breath of Life—apricot; Compassion—pink; Sonia clg—salmon pink; Marechal Niel—yellow; Golden shower—yellow; Royal Gold—deep yellow; Iceberg clg—white; Swan lake—white; Delhi White Pearl—white; Prosperity—white; Lamarque—white; Angel Face clg—deep Lavendar; Paradise clg—deep Lavendar with ruby petal edge.

Rambler is another class of roses which are different from climbers but have climbing habit. These have been produced by crossing Hybrid perpetuals with *R. wichuriana* and *R. multiflora*. The flowers in ramblers are produced only in one flush during the year but it last for several weeks. Important varieties are : American Pillar—red with yellow centre; Excelsa—Tyrian rose to bright light crimson; Sander's white—white; Dorothy Perkins—pink; Albertine—coppery pink.

D. Miniature Roses

These are also called as Baby or Fairy roses. They are neat and compact but dwarf plants. They have long blooming period and least affected by cold weather and diseases. These were introduced from China as Pigmy or Lawrence rose (*R. chininsis* var. minima syn. *R. rouletti*). By hybridization, single to double flowers in various colours have been developed.

These are most ideally suited for edging, pots, rockeries or window gardens. These are easily propagated by semihard to hard wood cuttings. Important varieties in different colours are; *Red*—Red Flush, Little Red Devil, Desert Charm; *Pink*—Cupcake, Rosemarin, Cuddles; *Orange/vermilion*—Sun Blaze, Mary Marshal, Starina, Chattem Centennial; *Yellow*—Yellow Doll, Summer Butter; *White*—Cindrella; Green Ice, Misty Dawn; Mauve K. Lavendar—Jewel; Multi colour—Party Girl, Puppy Love, Sun Maid, Rose Window; *Bicolour*—Dreamglo, Gloriglo, Sassy Lassy; *Cascading miniatures*—Orange Cascade, Pink Cascade, Red Cascade, Snow Carpet.

E. Shrub Roses

These perpetual flowering roses are taller than bush types but shorter than climbers and provide continuous flowering. These are excellent for an informal layout, specimen, hedges and very useful for the shrubbery. The important varieties are : *White*—Delhi White Pearl, Butterfly Wings; *Pink*—Carefree Beauty, Pearl Drift; *Red*—Cocktail, Fountain; *Yellow*—Lucia.

Performance of Rose Varieties under Green House Conditions

Growing of rose under green houses has become a commercial practice in many countries including India. Many export-oriented projects have come up around Delhi, Pune, Bangalore, Hyderabad, etc. Technical know-how alongwith varieties have been imported either from Holland or Israel. Roses are being grown at 25°C (day) and 16°C (night) temperature and are fertigated with nutrients. Various nutrient compositions are being tried by various units. However, 150 ppm of N, P, and K of each at 5.5 pH is most ideal. The performance of different varieties which are being grown in India has so far not been reported, however, their performance as reported by breeders is mentioned hereby.

PERFORMANCE OF ROSE VARIETIES UNDER GREEN HOUSE
(Production under Holland conditions)

Name of variety	*Colour*	*Length of stem (cm)*	*Vase life (days)*	*Production m²/year*	*Name of breeder*
(1)	(2)	(3)	(4)	(5)	(6)
H.T. Type					
Bacarole	Dark Red	100	7	80	Tantau
Black Magic	Dark Red	80	12	150	Tantau
First Red	Red	70	10	130	NIRP
Grand Prix	Red	70	11	140	Terra Nigra
Grand Gala	Red	90	8	130	Meilland
Sacha	Red	50	12	350	Kordes
Sangira	Red	60	12	275	—

Femma	Salmon	65	18	250	Schreur
Noblesse	Pink	60	8	150	Tantau
Osiana	Pink	100	8	105	Tantau
Pavarotti	Pink	90	10	120	De Ruiter
Ravel	Purple	90	10	120	De Ruiter
Prophyta	Pink	50	14	220	De Ruiter
Vivaldi	Pink	90	10	120	De Ruiter
Kiss	Pink	55	10	300	Kordes
Lamabada	Orange	55	12	250	Kordes
Madelon	Orange Red	70	12	180	—
Pareo	Orange	65	10	170	Meilland
Frisco	Yellow	45	14	350	Kordes
Konfetti	Bicolour	80	8	150	Tantau
Sunny Prophyta	Yellow	50	14	240	De Ruiter
Starlite	Yellow	50	12	240	Meilland
Yellow River	Yellow	45	12	300	Meilland
Bianca	White	65	9	200	Preesman
Spray Type					
Evelien	Pink	55	11	170	Interplant
Majolika	Pink	50	8	150	Tantau
Red Ace	Red	40	10	175	De Ruiter
Rumba	Yellow	50	10	120	De Ruiter
White Evelien	White	55	11	170	Inter Plant

Growing of Roses

Situation. The ideal position for planting roses is sunny, open and protected by hedge, fence or building against strong winds. It should be quite away from shade or from the underground roots of trees.

Soil. Roses require fertile and clay loam and loam soils. Soil should be deep having good water holding capacity with proper drainage. Roses do well in soils having pH between 6.0 to 7.5 but it can also grow satisfactorily in alkaline soil with pH upto 8.4. The soil pH can be brought in safe limits by adding gypsum or other acidifying agents in alkaline soil whereas pH of acidic soil can be raised by adding well ground dolomite lime stone.

Layout. Rose is a beautiful flower, hence, it should be displayed in an attractive manner. The beds can be made of formal designs or informal designs keeping the style of layout of the garden in view. Formal shapes of beds like square, rectangular, circular, U-shape or L-shape can be prepared.

For informal designs, the beds of informal shapes suiting to the design should be prepared.

Planting. The place selected for planting roses should be dug thoroughly to a depth of 90-120 cm and kept open for few days. The soil should be dried and refilled with 10-15 kg/sq. m. well rotten farm yard manure and good garden soil at the top. The spacing between plants varies with the vigour of the variety but generally H.T. varieties can be planted at the distance of 75 cm from each other while for the varieties of floribundas which are used for massing, a distance of 60 cm can be kept.

For planting roses, best time is from end of September to middle of October but it can be extended upto November. At the time of planting roses, the soil of the size of earth ball should be removed from the bed and plant should be placed in this pit. Soil should be refilled and well pressed. Care should be taken that bud union is just above the ground. Light pruning *i.e.,* tipping back of the branches should be done. After planting, frequent irrigation, removal of root suckers and manuring should be done.

Propagation. Roses are commonly propagated by 'T' or shield budding on the rootstock. The common rootstock used is Edouard rose (*R. bourboniana*) or *R. multiflora*. Recently *R. indica odorata* has been found better than the former rootstocks. Thornless rootstock is also getting popular. These rootstocks are easily propagated by hard wood and semi hardwood cuttings which are prepared in September-October at the time of annual pruning. When these cutting attain pencil thickness which takes about a year, they are ready for budding.

Roses are grouped into different classes according to the height at which budding is done on the rootstock. For Bush roses budding is done at 7.5-10 cm above the ground whereas for standard rose budding is done at the height of 100 cm. These are more useful in the garden because flowers are produced 1 m above the ground and are suitable for both formal and informal layouts. Sometimes *Half standard* is also produced by budding at 50 cm. *Weeping Standard* roses are prepared by budding of climbing roses at 100 cm above the ground. In Weeping roses branches of climbing roses droop down and create artistic effect.

These rootstocks are easily budded by 'T' method and there is a high percentage (about 90-95%) of success. The best budding time is from November to February and after union of buds, budding starts growing. The upper portion of rootstock is cut into instalments. By the time plants are ready for transplanting it completes about two years.

The new method of cuttage buddage is followed to reduce this long period. In this method, cuttings are budded immediately and planted in sand or burnt rice husk media for rooting under polythene cover in the month of December or January. It takes about 3-4 weeks for cutting to strike roots and bud to grow. High success can be obtained by treating the cuttings and buds with 1000 ppm of IBA so that in one year plants are ready for planting. The miniature roses are propagated through cuttings.

Manuring. Many arbitrary recommendations are available about feeding of roses. Roses should be fed with both organic and inorganic sources. One hundred gram of mixture containing groundnut cake—5 kg, bonemeal—5 kg, ammophos (11 : 48)—2 kg, Ammonium Sulphate—1 kg, super phosphate (single)—2 kg and Potassium Sulphate—1 kg should be applied per bush for better results. Results obtained at PAU reveal that addition of 60 g N, 20 g of P_2O_5 and K_2O should be applied per sq. m. containing nine plants. These fertilizers should be applied in two splits *i.e.*, half amount of N, full dose of P and K at the time of prunning and remaining half one month after the first application. In the market many ready-made rose mixtures are being sold which can also be applied.

For better effects foliar feeding of roses has also been found very useful. Foliar sprays of 30 g of a mixture of 2 parts urea, 1 part Dihydrogen ammonium phosphate, 1 part potassium nitrate and 1 part of potassium phosphate in 10 L of water at weekly or fortnightly interval, improve growth and flowering of roses. A weekly spray of 30 g urea/10 L of water, with an insecticide is also recommended.

Irrigation. Water requirement of roses depend upon soil type and seasons. Light soils require more frequent irrigation than heavy soils. During summer, water requirement is more than winter. Therefore, irrigation is adjusted in a way that soil is moist but not wet. During rainy season, watering is generally not done except during drought period. During winter, irrigation is done at about 7-10 days interval whereas during summer it should be done at an interval of 5-6 days.

Pruning. It is most important and technical aspect of rose growing. Root suckers should be removed whenever they appear. Every year rose plants produce some shoots which become weak by the young shoots growing out and become bushy in appearance. If not pruned properly, plants produce poor quality of flowers and become unproductive. A judicious pruning is done which removes those plant which would become ultimately useless, rob and endanger the healthy wood. Pruning also helps in securing healthy and clean canes and production of quality blooms and availability of light and air to central parts, thus maintaining the health of plants.

The ideal time of rose pruning is when rains are completely over and winter is approaching. This condition prevails under North Indian plains from end of September to middle of October. Most of the varieties take about 60-65 days for blooming after pruning. Therefore, to secure flowers for particular occasion, pruning can be adjusted accordingly.

Pruning should be done systematically by cutting out all dead wood and removing all weak and spindly growth. Overcrowding of shoots at the centre should be thinned out to keep the centre open. If too many shoots are left after thinning out process, reduce their number to few strong ones. The canes are shortened according to the type of which it belongs. Pruning should always be done to an eye pointing outward and cut should be covered with fungicidal paint to check the infection of die-back pathogens.

Hybrid Teas and Teas are pruned to 5-6 eyes but for exhibition purpose severe pruning is done to 3-4 eyes.

In case of Floribundas, dwarf polyanthus and miniatures, pruning consists of removal of all weak and diseased growth and also the ends of branches should be tipped back.

In climbers, inlaced stems and all weak growth should be removed. Old stems may be cut after 3-4 years.

IMPORTANT INSECT—PESTS AND DISEASES

Roses are attacked by many insect-pests and diseases and to keep them healthy, suitable control measures should be taken timely failing which great loss can result.

Insect-Pests

White ants. They attack the plants and damage the root system and slowly plants die completely. White ants can be controlled by applying Chlorpyrifos 20 EC @ 4 ml/pit or Imidacloprid 200 SL 0.5 @ ml/pit at time of planting.

Red scale. This is very serious pest of roses which attacks mostly in August and September. The branches are covered with a reddish-brown incrustation under which the insect sucks the juice of the plants. The plants may dry up under severe attack. This pest can be controlled by spraying Chlorpyrifos 20 EC @ 3.75 ml/of water or Methyl Parathion @ 0.05%.

Aphids. These appear in winter months on leaves and flower buds. They suck the cell sap, discolour the leaves and affect flower buds which fail to open. This can be effectively controlled by spraying Imidacloprid 200 SL 0.4 @ ml/l or Thiamethoxam 25 WG @ 0.4 g/l of water.

Digger Wasp. This insect has been found damaging freshly pruned rose plants. They dig burrows into the pith through the cut ends and nest there. Later on fungal attack may be there and stems may dry up. These can be effectively controlled by applying Chlorpyrifos 20 EC @ 3.75 ml/l of water.

Chafer Beetles. The adults of these beetles appear in August-September and cut away the leaves. These can be effectively controlled by applying Chlorpyrifos 1.5%.

Diseases

Die-back (*Diplodia roseum*). This is very serious disease of roses and appears after pruning. The drying up and blackening of pruned shoots start from the downwards. If it is not properly controlled it may spread on large scale. For its effective control, the pruned cut end should be painted with a fungicidal paint, prepared of 4 parts copper carbonate, 4 parts of red lead and 5 parts of linseed oil.

Black spot (*Diplocarpon rosae*). This disease appears in November and continues till the end of March. Conspicuous circular black spots (less than 1 cm) with fringed margins appear on either side of leaf. Leaves become chlorotic, dry up and prematurely drop. It can be easily controlled by spraying 0.2% Captan or Ferbam fungicides at fortnightly intervals.

Powdery mildew (*Sphaerotheca pannosa var. rosae*). It is a serious disease in warm, humid and cool weather conditions. Young growing shoots and leaves are coverd with white powdery growth. Infected leaves turn purplish and drop. Flower buds may fail to open. It can be checked by dusting 80% sulphur or spraying Karathane 350 EC or Propiconazole 250 EC @ 1 ml/l.

***

6

Gladiolus

It is a leading cut flower of India as well as of world. Gladiolus is very much liked for its majestic spikes which contain attractive, elegant and delicate florets. These florets open in sequence over longer duration and hence has a good keeping quality of cut spikes. There is a wide range of colours *i.e.,* self or bicolour with or without central mark varying from white to darkest crimson. These spikes of gladiolus are mainly used for garden and interior decoration and for making bouquets.

It belongs to the family *Iridaceae* and native of Africa and Asia minor. It was introduced in India in 16-17th century by the Britishers and it has been naturalized in different parts of the country. First it was thought that only hilly tracts of the country were suitable but later on plains were also found suitable for growing it commercially.

Climate

It can successfully be grown in wide range of climatic conditions and planting is to be adjusted in such a way that gladiolus crop enjoys favourable weather conditions throughout its growing, flowering, corm and cormel production period. The favourable of lowest temperature is about 16°C and higher temperature range is 25-30°C. There should not be rains especially during flowering period.

Gladiolus requires open sunny situation, therefore, site for planting should be selected carefully. The longer day length improves the spike quality. About 12 hours day length with high light intensities are preferred over short days with low light intensities. The site should also be well protected against strong winds by planting hedge or wind-break.

Soil

For growing gladiolus successfully well fertile and well drained soil is better than heavier soils. The texture of heavy soils can be improved by mixing organic manure and sand for better production of gladiolus. The soil reaction should be around neutral. If it is acidic or alkaline, it should be corrected.

Varietal Wealth (Plates 4 & 5)

Gladiolus is very rich in its varietal wealth. The commercial life of a variety is about 10-15 years. Every year new varieties are introduced which replace the old varieties. This is possible due to easy hybridization, high pollen fertility and good seed set and germination. Therefore, a gladiolus grower should keep constant vigil in the market for varietal changes. Before introduction of a new variety, grower should ensure its adaptability under local conditions.

On the basis of trials conducted at Punjab Agricultural University for the last several years, promising varieties have been found, are : Snow Princess, Melody, Sylvia, Suchitra, Mayur, Friendship, Pusa Suhagin, Jacksonville Gold, Sancerre, Rose Supreme, Yellow Stone. The brief description of these varieties and other varieties is as follows :

***(i)* Melody.** It requires about 90-100 days for flowering; spike length 87 cm; florets 15 per spike; florets colour Peach with red throat; cormel production 67/plant.

***(ii)* Snow Princess.** It requires about 80-90 days for flowering; spike length 65 cm; 11-14 florets per spike; floret colour white with pale throat; cormel production 15 per plant.

***(iii)* Sylvia.** It requires about 120 days for flowering; spike length 75.5 cm; florets 13-15 per spike; colour claret rose; cormel production 15 per plant.

***(iv)* Suchitra.** It requires about 74 days for flowering; spike length 84 cm; 17 flowers per spike; floret colour light pink with dark pink blotches; cormel production 98 per plant.

***(v)* Mayur.** It requires about 95 days for flowering; spike length 80 cm; 18 florets per spike; floret colour mauve with purple blotch; cormel production 112 per plant.

***(vi)* Friendship.** It takes about 86 days for flowering; spike length 78 cm; 12 floret per spike; floret colour carmine rose with yellow throats; cormel production 15 per plant.

***(vii)* Jacksonville Gold.** It requires about 87 days for flowering; spike length 69 cm; 11 florets per spike; floret colour deep yellow; cormel production 37 per plant.

***(viii)* Happy End.** It requires about 82 days for flowering; spike length 65 cm; 10 florets per spike; floret colour salmon carmine rose with yellow throat, cormel production 37 per plant.

***(ix)* Sancerre.** It takes about 85 days for flowering; spike length 94 cm; 19 florets per spike; floret colour snowy white; cormel production 48 per plant.

***(x)* Rose Supreme.** It takes about 103 days for flowering; spike length 93 cm; 15 florets per spike; floret colour light peach; cormel production 24 per plant.

***(xi)* Yellow Stone.** It takes about 99 days for flowering; spike length 68 cm; 14 florets per spike; floret colour golden yellow and ruffled; cormel production 4 per plant.

***(xii)* White Prosperity.** Florets white in colour (white group 155 B), ruffled, spikes 90-110 cm long each with 18-20 florets, florets 9.5-10.5 cm in diameter. Cormel production about 30-40 per corm.

***(xiii)* True Yellow.** Florets yellow in colour; spikes stout, 100-110 cm long each with 13-15 florets; nearly 10-11 cm in diameter, ruffled, each corm produces about 8-10 cormels.

***(xiv)* Nova Lux.** Florets yellow with deep yellow throat and purplish base; spikes 90-100 cm long with 16-18 florets; florets 9-10 cm in size and ruffled; each corm produces 15-20 cormels.

***(xv)* Hunting Song.** Florets scarlet red with maroon throat having yellow markings; spikes 95-100 cm long; each with 17-19 florets; florets about 10 cm diameter; cormel production about 20 per corm.

***(xvi)* Rose Spire.** Florets light pinkish mauve with creamish white throat; spikes 95-110 cm long each having 17-18 florets of 10.5-11.0 cm diameter; each corm produces about 25-30 cormels per corm.

***(xvii)* American Beauty.** Florets reddish pink in colour with whitish throat and whitish blue anthers; spikes 70-75 cm long; each with 15-16 florets; florets 9.5-10.5 cm in size; each corm produces 70-80 cormels.

***(xviii)* Praha.** Florets white with creamish throat; spikes 100-105 cm long each with 18-20 floret size 10.5-11 cm; each corm produces 15-20 cormels.

***(xix)* Bonaire.** Florets peach coloured with deep pink and light yellow throat; spikes 90-100 cm each having 15-17 florets of about 10.0 cm diameter; cormel production about 40 per corm.

***(xx)* Trader Horn.** Florets bright red with creamish throat having prominent stripes. Spikes 90-100 cm long each having 18-20 florets; florets 10-10.5 cm in diameter; cormel production 70-80 per corm.

***(xxi)* Priscilla.** Spikes pinkish mauve with prominent light yellowish white bases; spikes 60-70 cm long with 14-15 florets; 8.0-9.0 cm in diameter; each corm produces 20-25 cormels.

***(xxii)* Her Majesty.** Florets deep violet in colour, spikes 65-70 cm long each with 13-14 florets; florets 8.0-8.5 cm in size; cormel production on about 20 per corm.

***(xxiii)* Eurovision.** Florets bright red in colour spikes 75-80 cm long each having 18-20 florets; florets 10-10.5 cm in size; cormel production 25-30 per corm.

***(xxiv)* Jester.** Florets deep yellow with red throat; spikes 90-100 cm long with 16-18 florets; florets 9.5-10.0 cm in diameter; each corm produces 15-20 cormels.

***(xxv)* Applause.** Florets deep pink with red throat; spikes 90-100 cm long having 15-16 florets; 8.0-8.5 cm in size, each corm produces about 20 cormels.

***(xxvi)* Regency.** Florets deep maroon with velvetty appearance and highly ruffled. Spikes 75-80 cm long each having 14-15 florets. Florets 8.5-9.0 cm in diameter, cormel production 10-20 per corm.

***(xxvii)* Windsong.** Florets purplish mauve with prominent light yellow throat. Spikes 90-95 cm long each bearing 15-16 florets. Florets 10-11 cm in diameter; each corm produces 30-40 cormels.

***(xxviii)* Oscar.** Florets maroon in colour; spikes 90-100 cm long with 15-18 florets. Florets about 10.0 cm in diameter. Each corm produces 25-30 cormels. This variety is highly susceptible to fusarial wilt.

***(xxix)* Peter Pears.** Florets orange red. Lip petal scarlet with light stripes; Florets 10.5 cm in diameter; each corm produces about 25-30 cormels.

***(xxx)* Mascagani.** Florets bright red; 9-10 cm in diameter; spikes 80-90 cm long each having 15-17 florets. Each corm produces 10-15 cormels.

***(xxxi)* Anglia.** Florets yellow; spikes 60-70 cm long each having 13-14 florets; florets 9.5-10 cm in diameter; each corm produces about 10-15 cormels.

***(xxxii)* Green spire.** Florets yellowish green in colour; spikes 75-80 cm long each with 12-13 florets; florets 10-10.5 cm in diameter. Cormel production 40-50 per corm.

***(xxxiii)* Blue Sky.** Florets violet blue in colour; Spikes 80-90 cm long each having 15-16 florets; Florets about 10 cm in diameter. Cormel production 25-30 per corm.

Timing Production

There is a demand of flowers round the year. Therefore, efforts are done to produce the flowers for longer possible period. Experiments conducted revealed that quality spikes of gladiolus can be produced from end of October to April by adjusting the planting from July to December in the plains of North India. The interval of planting may be 15 days or a month. The selection of varieties should be done carefully because some varieties may not produce the spikes at all or quality spikes under high temperature conditions. In the hilly areas, planting is done in March-April and flowers are produced from June to September.

Land Preparation and Fertilization

Soil should be exposed to sun during May and June. It is advised that soil should be sterilized by covering with black polythene during these months. This will kill all harmful fungi of the soil. Twenty tonnes of well rotten farmyard manure should be mixed per acre at least one month before planting gladiolus corms. 40 kg of P_2O_5 (250 kg single super phosphate) and 40 kg K_2O (66 kg. Muriate of potash) per acre should be applied.

Planting. For planting gladiolus, well rested corms measuring more than 5 cm are used. This stage can be easily identified by the presence of rootlets at the basal disc. The corms after taking out from cold storage are kept under shade and they are descaled and treated with 0.2% Bavistin solution for 30 minutes. Corms raised through cormels produce better spikes than corms raised through corms.

It is generally planted in flat beds at the distance of 30 × 20 cm at the depth of 7 cm. At this distance about 60,000 corms can be accommodated per acre. This distance between row to row or plant to plant can be increased or decreased depending upon the machines used for cultivation. Higher plant population *i.e.*, 80,000 or 1,00,000 corms per acre can also be accommodated. When application of nitrogen into two splits followed by earthing up is done, ridges are formed automatically. Whenever there is danger of water logging conditions even for few days, planting should be done on ridges which are made 30 cm apart. Every year field should be changed. After 3-4 years, the same field may be used again.

Application of Nitrogen. The application of N @ 160 kg N/acre should be done into two splits. Half dose of N. *i.e.* @ 80 kg N/acre should be applied at 3-leaf stage and remaining half nitrogen and 40 kg K_2O (66 kg of Muriate of potash) should be applied at 6-leaf stage.

Irrigation. Irrigation schedule for gladiolus depends upon weather conditions, soil types and rainfall. Normally in sandy loam soils irrigation is required at the interval of 7-10 days.

Weeding and Hoeing. Weeding and hoeing are important operations to raise a crop successfully. Manual labour is used to carry out this operation in India, so far. Therefore, the cost of cultivation enhances considerably. Weeds should be removed whenever they appear. Otherwise weeds will compete with the main crop for nutrition and moisture. Four to five times weeding is required to raise a crop successfully.

Experiments conducted here vividly revealed that the application of 2.5 L Basalin/ha as preplanting treatment kept field considerably clear for 70 days without any harmful effect on corm and cormel production.

Yield of spikes. Normally single vegetative bud sprouts per corm which inhibits the sprouting of other buds resulting into single spike formation of a corm. However, in many cases more than one bud sprout and depending on number of bud sprouts, number of spikes are produced. Thus the yield of spikes is increased than the number of corms planted. It has been observed that about 10-15% more spikes are produced than actually corm planted. The yield of spike also depends upon the density at which planting has been done. Thus 70,000-75,000 spikes are harvested when planting is done at the density of 60,000 corms per acre.

Harvesting of spikes. Spikes of gladiolus have better vase life and can be transported for long distances when properly packed. The stage of harvesting of spikes depends upon the distance to be travelled. For distant market, spikes are harvested when basal floret of the spike shows colour whereas for local market spikes are harvested when basal floret is fully opened.

For enhancing the vase life of spikes, pulsing with 20% sucrose and 200 ppm hydrocxyquinoline citrate for 24 hours has shown very encouraging results. Cut spikes also can be successfully stored for 2 weeks at 1-2°C.

Packing of flowers. Cut flowers should be packed in cardboard boxes made for this purpose. Generally card board measuring 100 cm × 25 cm × 10 cm are used. Boxes of this size are costly, therefore, other old suitable boxes like tube light boxes or of other industries can also be used.

Corm harvesting. Corms and cormels are ready for lifting from the ground after 6-8 weeks of harvesting of spikes. Irrigation is withheld atleast 2-3 weeks before lifting the corms. Crops show the sign of maturity *i.e.*, yellowing of leaves start. After lifting the corms from the ground, they are kept in shade alongwith the leaves for drying for a week. Leaves are removed, corms and cormels are cleaned. These cleaned corms and cormels are then dipped in 0.2% Bavistin solution for 30 minutes. Then these are dried in shade for 2-3 weeks, packed in shallow wooden crates or gunny bags and kept in cold storage till the next planting season.

Diseases

1. Wilt or Collar rot. It is caused by *Fusarium oxysporum* f. sp. *Gladioli* and is very devasting disease. Most typical symptom of this disease is the appearance of sickel shaped leaves. The plants exhibit yellowing of leaves and reddish brown cankers may also appear on collar region. For controlling this disease the aim should be for the prevention of the disease because it is not possible to cure a diseased plant after the symptoms are noticed. To control this disease, plant apparently healthy corms after treating with 0.2% Bavistin solution. Do not plant the gladiolus crop in the same field for more than 2 or 3 years. Use corms for planting which are raised from cormels. Drenching the soil with 0.2% Bavistin and turning of the soil for 2-3 times has been found useful in controlling the disease. Prestorage treatment with 0.2% Bavistin has also been found useful in controlling this disease. An integrated use of hot water treatment of corms (57°C for 30 minutes) and soil drenching with captan (0.3%) was found to be effective in minimising the disease.

2. Storage rot of corms. It is also caused by *Fusarium oxysporum* and by species of many saprophytic fungi like *Alternaria, Penicillium, Aspergillus*, etc. During storage corms become soft or corky with foul smell. Black spots appear on the surface of the corms. To control this disease, proper storage of corms in thin layers and prestorage treatment with 0.3% Captan or 0.2% Bavistin should be followed.

***

7 Carnation

Carnation is one of the most important cut flower of the world. It is commercially grown in many countries like Columbia, Kenya, Canary Island, Italy, Spain, Holland, U.S.A. etc. In India, earlier marguerite type of carnations were being grown as winter annual but with the introduction of Sim carnation, now these are being cultivated in many cities like Nasik, Pune, Bengaluru, Ludhiana, Solan and Shimla etc. It is popular as cut flower on account of its exquisite shape, wide range of colours, good vase life and light in weight which makes it ideal for distant market. It is being cultivated more than 2000 years. Its common name carnation has been taken from Latin word carnation—carnis which means flesh colour on account of similarity of original colour of carnation and it is also derived from Greek word coronation as the carnation flowers which were being used in decorating the crown of Greek athelets. It's generic word Dianthus is from Greek word. Dios which means Divine and anthos means flower. The species name caryophyllus has been given after generic name of the clove tree—*Caryophyllus aromaticus* which means caryon-nuts; phyllon-leaf, due to resemblance of fragrance of carnation flowers to clove.

It is native of mediterranean region. There are 280 species of carnation and are found from Siberia to Arctic America, Japan and in Himalayas, where blooms spring when long days and congenial temperature is found. The centre of origin of carnation is considered to be spain. Some species are native to Hungary and Alps region.

Types of Carnation

The important species which took part in the development of modern carnation are : *Dianthus caryophyllus, D. chinensis, D. barbatus, D. knappii, D. plumarius, D. winteri, D. heddewigii, D. lacinatus, D. superbus, D. arenarius, D. nobilis, D. gratianopolitanum,* etc.

Development of Perpetual Flowering Carnation

Perpetual flowering carnation was a hybrid between *D. caryophyllus* × *D. chinensis* which have originated in the city of Lyons in France around 1830. This race was called "Oeillets Mayonnais" or "Oeillets de Mahon". James Mayor bred 'Duchess of Lancaster' first laced pink. In this, the edge of petals are tipped with same colour as central blotch. Montague Allwood raised a race "Allwoodii" by crossing pink with Perpetual (*D. plumarius* × *D. gratianopolitanum*) × (*D. caryophyllus* × *D. chinesis*): Another race of "Allwoodii alpinus" was developed by crossing Allwoodii with dwarf species *i.e. D. alpinus, D. arenarius* and *D. superbus.* It is a race of dwarf plants producing pink or purple flowers. Other colours, however, lacked. C.H. Fielden raised a race of pinks which was called as lancing pinks. It was a result of crossing between Allwoodii with *D. winteri.* The plants were having longer flowering period. A french breeder M. Dalmais crossed an "Oeillet de Mahon" with 'Bisbon'-border carnation and the resultant was true perpetual. Stocks of these type were introduced in America in 1852. An early breeder Donati developed 'Victor Emmanuel' in 1860 which was a yellow ground Picottee and became a parent of subsequent varieties of this colour. Another breeder of America, Peter Fisher developed an important variety "Mrs. Thomas Lawson" in 1895 by crossing 'Day Break' with 'Van Leeuwen'. This was commonly used as parents of later varieties. Another important cultivar of Peter Fisher's was the 'Enchantress'—large pale pink, handsome flowers but male and female sterile.

In due course of time stocks from France and America were introduced in England and 'Winter Cheer', 'Miss-Jolliffee, important cultivars were released by James Veitch nursery. 'Winter Cheer' despite of its name was not practically floriferous in the winter month but was compact with good scarlet flower. It was further crossed with 'Mrs. T.W. Lawson' in 1904 and hybrid released was 'Britannia', a scarlet.

The cultivar 'William Sim' was produced in 1938 or 1939 by William Sim of North Berwick, Maine, USA, was the greatest contribution to the carnation industry. Many periclinal chimeras have been isolated in white, red, pink, orange, yellow, salmon, tinted etc. Important varieties are Arthur Sim—white with red tinted : White Sim—White; William Sim—red; Scania—red; Crowley Sim—Pink; Pink Sim—Pink; Harvest Moon—Yellow; Laddie—salmon pink; Aleck's Spark—red; Dusty Sim—salmon pink; Tangerine—orange; Yellow Sim—yellow; Joker—crimson. In England also, number of good cultivars have been developed. Important cvs are : Bailey's Splendour—pink; Bailey's Supreme—salmon pink; Bailey's Masterpiece—Crimson; Rose Splendour—rose cerise; La Royale—magenta; Bailey's Amarillo—pastel shade of cream. These sim type carnations are standard type excellent cut flower and dominating in commercial cultivation in the world. However, the Sim strains have low disease tolerance and do not produce profuse branches.

Another class of spray type was being developed simultaneously to save the labour employed for pinching and disbudding in standard carnation. The raiser was Mr. W. Thompson in Connecticut, America who developed Exquisite variety as first spray type carnation and Elegance was another good sport. Number of other spray types were produced in course of time.

Development of Marguerite Type Carnation

Marguerite type carnations were developed by crossing *D. chinensis* × *D. caryophyllus,* show a remonant quality, have broader and green leaves. These are raised through seeds as winter annual in India. Wide range of colours are available.

Development of Malmaison Type Carnation

In 1857 a Monsier Lassie was raising seeds of perpetual carnations and out of them, few seedlings bore flowers which were markedly larger and more handsome than previous cultivars. It was named 'Souvenir de La Malmaison'. The pollen production this cultivar was very scanty and was not viable. Three more bud sports were selected. 'Princess of Wales' malmaison was an excellent. 'Marmion' another white with red edge malmaison was produced by Burnett in 1906. Martin Smith crossed original malmaison with border carnation and produced 'Mrs. Martin Smith' which was five feet high and had flowers of spectacular size. They were called tree carnation. However, flowers were produced very sparingly.

Development of Royal Carnation

Mr. Allwood was successful in combining the malmaison with perpetual carnation in 1951 and race was called as 'Royal carnation'. An important cv. 'Royal Fancy' was developed which gave rise to a number of large perpetual flowering. In U.S.A. 'White Perfection' another important cultivar was raised. It was further crossed with 'Brittania', which gave rise to 'Wivel's field'—White, which became very popular.

New Trend of Modern Carnation

To save the cost of labour required for pinching, disbudding, staking, tieing etc., miniature carnations are being preferred. Such varieties should be fast cycling and more productive. Stem may be shorter and flower somewhat smaller but are produced in great profusion so that these types could be used for mass market bouquets. Demmink (1982) raised a new group of glass house carnations. These are characterized by compact growth and smaller flowers. *D. chinensis*, *D. pencicii* and *D. pontederae* were the parents which were used. Recently F_1 knight series have been developed by Gold Smith. These are compact plants having 25-30 cm stem length but produces 15-20 flowers per stem. Dwarf carnation will be a future crop for pots. Different race of carnation *i.e.*, micro carnation has been developed which produce button like flowers in large quantity *i.e.*, about 1000/sq. m. Very recently there is a development of 'Pico' variety.

Varieties

Carnation is very rich in its varietal wealth. Every year large number of superior varieties than existing ones are being introduced in the trade by the breeders. Hence, carnation grower has to be very vigilant about new promising introductions in the market.

Before 1970, in Indian gardens mostly Marguerite, Riveria Giant and Royal type of carnation were grown as winter annual. These flowers hardly find their utility in the market as cut flower. After introduction of Sim carnation, earlier types have been replaced by them completely and these are grown for cut flower production because these are rich in colour, long stem, bigger size and good vase life.

During 1970-90, the common varieties which were being grown by carnation growers were : Scania, Arthur Sim, White Sim, William Sim, Shocking Pink, Tangerine, etc.

Performance of Sim Cultivars

Ten sim cvs. showed varying response to the growing environments under the open field conditions of Punjab. Periods from planting to peak flower production varied from 160 days to 175 days. cv. Shocking pink was the earliest to reach peak production while cv. Scania took two weeks more for its peak production. Average stem length and diameter of cut flowers observed at full anthesis stage ranged 46-48 cm and 6.7 to 8.4 cm, respectively. Yield of cut flowers per plant for a single cropping from October to March varied from 5.3 to 6.7. Calyx splitting was common in all the cultivars and varied from 10.3 per cent to 15.7 per cent. These differences might be due to interaction between environment and genotypes (Table 7.1).

TABLE 7.1
Performance of ten cv. of Sim carnation

Cultivars	*Days taken for peak flowering*	*Average stem (cm)*	*No. of flowers per plant*	*Bloom size (cm)*	*Calyx splitting (%)*
White Sim	170	48.3	5.3	7.6	12.3
Arthur Sim	170	48.2	5.7	7.6	13.6
Scania	175	48.5	5.7	8.4	15.7
William Sim	170	45.3	5.3	7.5	14.8
Laddie	165	45.7	5.4	7.6	12.3
Can Can	165	44.8	6.7	6.7	10.7
Shocking Pink	160	46.9	5.2	7.5	12.5
Tangerine	165	46.8	5.4	7.2	12.3
Harvest Moon	165	47.3	5.3	7.3	10.7
Clear Yellow	165	47.9	5.4	7.5	10.3

During 1990-1996, the new varieties were introduced which started replacing earlier varieties. These varieties are Red Corso, Limara, Espana, Manner, Razziode Sole, Alec's Spark. The performance of these cvs. is as follows :

Variety	Average height of the plant (cm)	Number of flowers per plant	Length of flowering shoot (cm)	Number of branches per plant	Flower size diameter (cm)
Espana	59.33	5.33	35.67	13.33	5.33
Alec's Spark	47.17	3.33	32.67	12.67	5.33
Mannon	56.00	4.00	47.33	9.00	5.61
Limara	56.00	5.33	39.67	16.00	5.70
Razziode sole	51.67	3.00	38.67	12.67	5.61
Red Corso	48.33	5.33	37.67	12.00	5.42

During 1996-97, the varieties which have been introduced are : Master (Red), Laurella (Pink), Venesse (Pink), Gelatea (yellow), Corella (Magenta), Kelley (White), etc. and are performing well.

Climate and Soil

Carnation does not like extreme of temperature and has been classified as cool crop. The ideal range of temperature for carnation is considered to be 10-20°C and exposure of plant to 7°C favours development of branches. Sun light for 12 hours and humidity in the range of 50-60% is considered very favourable for growth of plants as well as flower development. High cost of green house production of carnation has forced to grow it in natural areas which are located near 4° N or south of equator at higher altitude (2000 m).

Root system of carnation is very shallow and fibrous and hence it prefers porus loose and friable soil. It should be rich in organic matter, high in fertility and slightly acidic in nature, having 6-7 pH. Therefore, sandy loam soils, high in fertility is mostly preferred for its successful growing.

Propagation

Annual carnation is propagated through seeds whereas perennial carnation is propagated through terminal cuttings. These cuttings should be healthy, vigorous, free from diseases and should be weighing about 16 g. After rooting, such cuttings produce healthy plants. Weak cuttings do not make healthy plants. Stock plants established during October starts yielding cuttings from January and yield about 20-24 cuttings till April. For better rooting, carnation cuttings need hormonal treatment with 500 ppm NAA for five minutes and high humidity. High percentage of rooting is observed in cool months than summer months.

River sand mixed with burnt rice husk mixed in equal proportion is better than alone sand or burnt rice husk or vermiculite.

Effect of Time for Rooting of cuttings and their establishment

Month	No. of days taken for rooting	Per cent rooting of cuttings
April	22	51
June	40	45
August	25	58
October	16	83
December	30	88
February	20	89

Unrooted cuttings of carnation can be stored from April to September in the cold storage when stored in unsealed plastic bags in card boxes at 0-1°C. Successful carnation plants are raised from such stored cuttings, carnation is also multiplied through micro propagation technique on large scale. A well developed tissue culture laboratory is required for this purpose.

Preparation of Land and Planting of Carnation

Commercial production of carnation is done in container system of planting. For this raised beds (30 cms) above the ground are prepared of the size of 1 m × 10 m with work aisels of 60 cm supported by bricks on all sides. Soil media having 3 parts sandy loam and 1 part well rotten farm yard manure is considered good. After mixing of desired fertilizers, these container should be sterilized with formalin or through solarization by using black polythelene. The rooted cuttings planted in sterilized beds give better performance than unsterilized one. Carnation plants should not be planted either too shallow or too deep and it should be planted in the same depth as it was in nursery. Carnation plants require optimum spacing for better growth and flowering. Wider spacings or lesser density per unit area forces growers to manage larger area and thus increasing the cost of production. It has been experienced that closer spacing *i.e.* 15 × 15 cm or 20 × 15 cm produces taller plants and highest flower production than wider spacing.

Manuring. Balanced and regular supply of nutrients to carnation plants helps in better growth and flower production because carnation plants make a good reserve for N during tufting stage which is utilized during flowering. Tissue analysis reflects the uptake and distribution of minerals by plants and a relationship between absorbed nutrients and growth exists. Fifth pair of leaves from terminal end of primary shoots are considered best for sampling until the flower bud visible on these shoots. At the time of flowering sampling of 5th pair is done from side shoots. The N, P, K, content of carnation plants throughout the cycle approximates 1 : 0.5 : 1.5. Relationship between leaf N, P and K and flower production shows that highest number of flowers are produced when 5.27 : 0.31 : 3.8% N : P : K, present in leaves, respectively. However, the range for N : P : K varies 3.2-5.2 : 0.2-0.3 : 2.5-6.0, respectively.

Results of a fertilizer trial on carnation cv. Marguerite Scarlet in which three levels of N (0, 20 and 40 g/m^2), P_2O_5 (0, 10 and 20 g/m^2) and K_2O (0, 30 and 60 g/m^2) alongwith 5 kg well rotten farm yard manure show that application of 10, 10 and 30 g/m^2 of N, P_2O_5 and K_2O, respectively proved beneficial in terms of growth and flower production. Application of N into three split doses *i.e.,* 1/3 before transplanting, 1/3 one month after transplanting and 1/3 after one month of second application. By many specialized carnation growers, carnation crop is being fertigated with potassium nitrate, calcium ammonium nitrate, superphosphate and magnesium sulphate @ 765, 156, 848 g/1000 liter of water, respectively.

Irrigation, weeding and hoeing. Carnation needs regular supply of moisture due to shallow root system. Thus, soil is kept wet throughout the cropping period. However, common practice is to give heavy irrigation followed by dryness before next irrigation. Frequency of irrigation is determined by soil texture and growing environment of the plants. Generally two irrigations per week from August to October and thereafter once in 10 days interval till end of February. Irrigation once a week or at an interval of 3-4 days during March to July proved sufficient to keep the bed moist. Carnation plantation should be kept clean and weed free. In early stage of plantation, unchecked weeds compete both nutrient and moisture affecting growth. Application of fluchloralin @ 0.96 kg/ha or pendimethalin 30% @ 0.90 kg/ha is very effective in controlling most of the weeds. These weedicides are used as pre plant and their incorporating in ground will keep plantation free upto 80 days. After 80 days carnation plants grow sufficiently leaving no place for weed growth. Hoeing is followed in early stage of plantation to provide good aeration to the root system.

Staking, Pinching and Disbudding

Carnation is a leggy plant and it breaks very easily from swollen nodes. After planting, it grows straight about 25-30 cm high. Branches are not produced for some time due to inhibitory effect of apical bud. The plants tend to lodge and may easily be broken from nodal region, if suitable support is not provided. Carnation blooms are valued for their upright long stems and hence support is required for their entire growth period. Depending upon duration of crop 2-3 grid support structure is provided. First grid is provided 20 cm and second at 35 cm above the ground. Three heavy supports in the end of both sides of bed are provided. Wire and strings are commonly used to make a support network for the plants. Wire is run through length of bed and string are run across the bed besides each row of plants. Flowering stems exert lot of pressure being heavy, hence, these should be stretched tightly. These days ready-made plastic nets especially made for carnation are available and are being used by specialized carnation growers. These plastic nets are raised to desirable height as the plants attain height.

The pinching of carnation plants is a regular practice which is done to increase the number of branches, flower and ultimately profit to grower. Due to pinching flowering get delayed so it helps in regulating the crop also. Pinching is done leaving 6 pairs of basal leaves which are inductive. There are many kinds of pinching which are being followed by growers keeping advantages and disadvantages of each in view. In pull type pinching apical bud is pulled whereas in single pinch, apical portion measuring 5-7 cm is pinched off. In single and half pinch, after following single pinch of main shoot half of the side branches are pinched off. In double pinching, after following pinching of main shoot, all the secondary branches are pinched off.

It is the habit of carnation plant when apical bud on shoot develops into flower bud, basal axillary buds and shoots start developing into flower buds. Thus many flower buds develop and if they are allowed to develop all, the flower size of main bud will get reduced. Therefore, in standard carnation only main bud is retained and others are removed when they are large enough (pea size). Lot of labour is involved in disbudding and thus cost of production of standard carnation get increased. In spray or miniature carnations disbudding is not followed and shoots look like readymade bouquets. Therefore, these are preferred by many florists/growers.

Harvesting, Grading and Packing

Carnation plants produce cut blooms after 16 to 20 weeks and it varies according to pinching method followed. But harvesting when 1.25 to 2.5 cm of petal show colour with stem as long as possible is considered ideal stage for distant markets whereas for local retailers, flowers are harvested when they reach 3/4 stage of anthesis. Harvesting is done by hand as flowers are easily broken from the node. These flowers are immediately placed in clear water to a depth of half to two-third in length of the stem. Foliage at the bottom half of the stem are removed. Preconditioning of cut flowers in solution of silver thiosulphate (mM) is an important practice to avoid ethylene injury and prolonging life. Standard carnations are graded according to length, strength of stem and flower size. According to Society of American Florists', flowers were graded as follows :

Factors	*Blue (Fancy)*	*Red (Standard)*	*Green (short)*
Minimum flower (mm)	Tight-50 Fairly Tight-62 Open-75	Tight-44 Fairly Tight-56 Open-69	Tight (none) diam. Fairly Tight-none Open-none
Minimum length (cm) overall stem length	55	43	30
Stem strength is determined by holding the stem horizontally at a point of 25 cm above the minimum length for the grade and the deviation of the flower head shall not be more than 30° below the horizontal place the natural curvature of the stem down.			

The grades and standard developed by the European Economic Community are as follows :

Code	*Stem length (cm) 5 or less*
0–5 or less	50–50-60
5–5-10	60–60-80
10–10-15	80–80-100
15–15-20	100–100-120
20–20-30	120–120 or more
30–30-40	
40–40-50	

Carnation flowers being light in weight are mostly preferred for distant market and it is being marketed by air. After harvesting, it is maintained under cool chain throughout till retailed. Standard carnations are bunched into 25 stems per bunch with 24, 28, 32 bunches per box according to grade. Spray carnations are bunched into 5 stems per bunch and placed in plastic sleeve with 100 sleeves per telescopic corrugated board carton with lid of 120 × 46 × 25 cm size. Precooled cartons for 2 hrs. at 0°C and atleast 95% humidity are shipped for destinations.

IMPORTANT INSECT-PESTS AND DISEASES

Insect-Pests

Aphids (*Myzus persicae*). They suck the sap from the leaves, and leave sticky deposits on leaves and disfigure the young growth. It can be successfully controlled by spraying Imidacloprid 200 SL 0.4 @ ml/l or Thiamethoxam 25 WG @ 0.4 g/l of water.

Thrips (*Thrips tabaci*). These are sucking insects, suck the sap, causing leaves turn to yellow and patchy often with black specks and slight crinkling. It can be controlled by spraying Ethion 50 EC @ 2 ml/l of water or Triazophos 40 EC @ 0.5 ml/l of water.

Red spider mites (*Tetranychus uriticae*). The mites feed on the underside of the leaves, suck the sap and finally leaves turn yellow and have a dusty coating and fine webs. Plants become stunted and flowers are also disfigured under heavy infestation. Its attack is very common during hot months and dry season. It can be also be controlled by spraying Ethion 50 EC @ 2 ml/l or Triazophos 40 EC @ 0.5 ml/l of water.

Gram hairy caterpillar (*Helicoverpa armigera*). Eggs are laid on carnation plants. Larvae bore flower bud of carnation and eat away petals and reproductive parts. Buds may develop into full flower from damaged flower. It can be controlled by spraying Quinalphos @ 1.5 ml/l of water.

Diseases

Fusarium wilt (*Fusarium oxysporum dianthii*). It is very devastating disease and appear during hot weather associated with high humidity. If not controlled, cause huge losses. Main symptoms are comprised of yellowing of leaves and wilting of entire plants in very short period. Fungus is soil borne, therefore, sterilization of soil is very essential. Wilt can be contolled by soil application of Trichoderma. 2 kg talc based formulation of Trichoderma should be mixed with 50 kg well rotten farm yard manure and broadcasted on moist beds before planting. This application should be continued at least for three years for effective control. Spot application of 0.2% Bavistin also helps in checking the spread of disease. Fusarium wilt tolerant varieties should be selected.

Alternaria leaf spot (*Alternaria dianthii*). It is very common foliage disease. The fungus causes black spots on leaves and stems and finally leaves wilt and die prematurely. It can be controlled by spraying Propiconazole 250 EC or Tabuconazole 430 SC @ 1 ml/l of water.

Bacterial wilt (*Pseudomonas caryophylli*). Its symptoms are similar to fusarium wilt, however in this disease, a sticky substance oozes out from stem and roots. The control measure are similar to fusarium wilt.

Physiological Disorder

Calyx splitting. It is very common in double carnation and determines the quality. The petals during opening exerts pressure on calyx tube which support the base of petals and cause splitting of calyx. There are various causes like genetical, nutritional, environmental and cultural practices. It is generally believed that irregular temperature during flowering is mainly responsible for this malady.

***

8 Orchids

It belongs to important monocotyledons series microspermae, order—orchidales and family orchidaceae. This family includes about 800 genera and 35,000 species and is the largest family amongst flowering plants. Orchids are wide in origin and are available in abundance in tropics and temperate regions, however, main centres are : Indo—Malayan and tropical American regions like Brazil, Mexico and other areas, New Guinea and Australia. Main regions in India are Himalaya—Sikkim, Meghalaya, Tripura, Assam, Bengal, Western Ghats of North Kanara, Malabar and Travancore. The important genera which have been found are Eria, Dendrobium, Cymbidium, Vanda, Perisylus, Calanthe, Orchis, Bulbophyllium, Liparis, Coelogyne, Paphiopedila, Vanilla, Rynchostyllus, Aerides, Microstylis, Spiranthes, Epipactis, Hemepelia, Goodyera, etc. Brazilian catteleyas. Mexican laelia and Indian dendrobium, cymbidium Vanda have played a major role in the development of orchid industry.

Orchids are perennial, terrestrial, epiphytic, saprophytic and intermediate herbs with rhizomes or pseudobulbs, or tuberous roots or assimilating roots or aerial epiphytic. The growth habit may be either monopodium or sympodium. In monopodium main axis continues to grow year after year and bears flower on lateral branches *e.g.* Vanda, Angaecum, Polyrhiza, whereas in sympodium when main axis comprises of annual portions of successive axis each of which bears scale leaves and terminal flowers is called Acranthos sympodium *e.g.* Eulphia, Dendrobium and Liparis. Another habit in which inflorescence is lateral and main axis continues to grow for current year and stop growing at the end of season is called Pleuranthos sympodium *e.g.* Cymbidium, Phaius, Calanthe, Bulbophyllum etc.

According to habitat, orchids are grouped into four main groups :

***(i)* Epiphytes.** These orchids grow on the trees and fasten their support with the help of clinging roots which arise from rhizomes and form network between orchid and support and form a reservoir for humus. These orchids have absorbing roots which are adventitious and project into reservoir containing humus. Aerial roots hang down and are green inside and have capacity to store water and photosynthesised the food. During dry period they shed the leaves but they bear flowers and perennate with the help of pseudo bulbs *e.g.* Dendrobium, Vanda, Bulbophyllum, Obernia etc.

***(ii)* Terrestrial Orchids.** These are sympodium and usually have a rhizome or tuberous roots or pseudobulbs. These are xerophytic or mesophytic and are perennial in nature which store food and water in modified plant parts, enable them to overcome the dry period. Each annual shoot sprouts into leaf shoot and inflorescence.

***(iii)* Lithophytes.** These orchids grow on moist and shaded rocks and cervices of stony walls *e.g. Cymbidium munroniamum, Diplomeris hirsuta, Geodoram durpureanum* and *Hernanium josephii.*

***(iv)* Saprophytes.** These orchids lack green leaves and possess a fleshy underground rhizome with (Neotha) or without roots (Epipogium). The rhizome is usually which branched and absorb moisture from humus soil. Endotrophic mycorrhiza occurs in almost all saprophic orchids.

Orchid Flower

Most showy part of orchid is its flower which may be produced as solitary terminal (*Cyperipedium, Pogonia*) or usually a spike sometimes in raceme. Flowers are usually trimerous and vary in form and size. In *Angraecum sesquipedale* from Madagascar the white flowers are of 25 cm in diameter and 45 cm long spur. Perianth is usually six in two whorls, of three each and both whorls may be alike or outer whorl may be like calyx and inner may be corolla like. Many modifications are there. The posterior tepal or inner whorl is curiously modified and may be slipper like spurred, butterfly like, tubular, broad, strap shaped, shoe shaped, lobed or lacerated, sac like, spider like, etc. This modified tapel is called lip or labium and acts as a landing place for insect. The lateral tepals of the inner series may remain separate and form wing like structure or may fuse like hood, or sachete structure and are usually similar. The parianth lobes are curiously coloured and bear diverse and variegated patterns.

The androecium is made of one or two or three fertile structure and a variable number of staminoids arranged in two whorls. Stamens and style fuses to form a column or gynodrium or gynostemium. The column is present opposite the labellum. The anthers are bithectus, dehiscence, longitudinal, pollen numerous, granular normally agglutinated into mealy, waxy or bony masses called pollenia and the number of which varies 1-4 in each anther. Gynaecium consists of stigma of three lobes, out of which two are fertile and one is fertile which is called rostellum. Ovary is tricarpellary, unilocular, with marginal dehisces, ovules are minute. Pollination is done by insects and there is large variation in orchid flowers for pollination for different kind of insects. The flowers have nectary in the spur of labellum or base of column. The insect lands on labellum and tries to reach nectary. The sticky mass of pollenia glued to its body while it is drilling for honey. Some orchid species attract males of many hymonoptrous insects by emitting feromones and their most unusual method of discharging their pollinia. Fruit is a capsule which dehisces by means of three to six lateral longitudinal slits. Seeds are non-endospermic or endospermic or endosper nucleus and are produced in large numbers extremely light and with undifferentiated embryo. As many as 40,00,000 seeds/capsule are produced. The seeds are dispersed by wind.

Seed Germination

The orchid seeds being non endospermic in nature, can not utilize their own reserves and hydrolyse larger molecules of starch or cellulose. Hence, under natural conditions seeds germinate after being affected by a fungus—the orchid mycorrhizia which supplies sugar to germinating seeds of orchids. As a result asymbiotic germination in the absence of sugar proceeds only upto the early protocorm stage. After analysing orchid salep which contained starch, proteins, sugars, minerals and germinated seeds of Cattelya, Laelia, Epidendrum or artificial media. Lewis Knudson (1966) advocated that fungus was not necessary for seed germination. Later on many workers used different media for growing seeds of orchids, however, Kundson and Vacin and Went medium are used commonly which are available in ready form in the market. The composition of important media are given below :

Chemical	*Kundson C. (1946) (g)*	*Vacin and went (g)*	*Moral (1965) (g)*	*Wimber (1963) (g)*
Ammonium sulphate	0.50	0.50	1.00	0.50
Ferrous sulphate	0.025	—	0.025	—
Magnesium sulphate	0.25	0.25	0.25	0.25
Manganese sulphate	0.0075	0.0075	0.0075	—
Ammonium nitrate	—	—	0.50	—
Calcium nitrate	1.0	—	0.50	—
Potassium nitrate	—	—	0.50	—
Monopotassium Acid	0.25	0.525	—	0.525
Phosphate	—	0.25	0.25	0.25
Dicalcium phosphate	—	0.20	—	—
Tricalcium phosphate	—	0.20	—	—
Potassium chloride	—	—	0.25	—
Ferric tartarate	—	0.028	—	—
Tryptone	—	—	—	0.20
Sucrose	20.0	20.0	20.0	20.0
Agar	17.50	16.00	17.50	12.00
Distilled water	1000.00 CC	1000.00 CC	1000.00 CC	1000.00 CC

Seeds from green pods are sown on the autoclaved media under aseptic conditions. After 15-20 days, the embryo start swelling and after 30-35 days, 2 leaf stage is achieved. At 4-leaf stage the seedlings are taken out from flasks and after thorough washing are planted in community pots in 1 : 1 mixture of shredded tree fern fibre and charcoal.

The orchids are propagated through tissue culture technique for producing virus free seedlings of rare plants or hybrids which otherwise could not be propagated. Commercially nodal sections and in certain cases lateral bud meristem, or explants from shoots growing on pseudobulbs or flower stalk cutting or leaf tips have been used. First mericlones are raised and on transplanting them orchid plants are produced. There are commercial laboratories which propagate orchid plants on commercial basis and earn good profits. For tissue culture solid or liquid media are used.

Growing of Orchids

The orchid houses are made for cultivating orchids. According to the orchid types proper conditions and rooting media are provided. Thus, orchid houses are made for tropical orchids or temperate orchids. Former, grow in shade or under indirect light, under high humidity and do not like sudden change in temperature, whereas latter type are grown in cool orchid houses. These houses are made running north-south from split bamboo, glass, fibre glass, etc. The optimum requirement of light varies according to species grown. Cyperipedium and Phalaenopsis required only 2200-3200 Lux whereas genera like Vanda and Aranda require about 8500 Lux. It has been observed that filtered sunlight stimulates formation of male flowers and direct sunlight favours female flowers. Majority of orchids, are day neutral and are not influenced by day length. But in Catteleya both short day and long day plants are found.

There are also orchids which are grown in open sun in trenches filled with brick piece, charcoal. Orchids are commercial being grown in this way in Trivandrum, Ceylon, Thailand, Singapore. The species of Vanda, Archanis, Aranda, Rananthera, etc. are being grown.

Humid and warm atmosphere is an ideal for the growth of most of the tropical orchids which do not have well established root system. The humidity in the range of 30-80% during night and day, respectively, should be maintained. For this a central tank of water is very helpful in orchid house. The plants in pots, baskets and hanging pots need watering two-three times a day with a fine spray so that plants are not hit directly.

Pots/Containers

Terrestrial orchids are grown in pots and its size should be used according to the size of the plant. Normally 25 cm pot is used. Orchids are not disturbed frequently until very necessary and, hence repotting is done in 2-3 years. Orchids flower better when under potted. Pots are filled with the mixture of leaf mould. F.Y.M. and sand in the ratio of 1 : 1 : 1. Addition of charcoal or fern fibre or coconut husk or over burnt brick improves the aeration and ultimately growth and flowering. Requirement of N and P depends upon the stage of growth *e.g.* during vegetative phase more of N is required than productive stage when P is required. Micro nutrients like Cu, Mg, Mn, B, Fe, Zn are applied extra. Many formulations are available, however, Ohio W.P. is most commonly used. The composition is as follows.

The use of readymade preparations like NPK 20 : 20 : 20 or NPK 10 : 30 : 20 with trace elements alongwith coconut water (20-25%) has also been recommended.

COMPOSITION OF OHIO W.P. SOLUTION FERTILIZER (g)

Potassium nitrate	2.63
Ammonium sulphate	0.44
Magnesium sulphate	2.04
Monocalcium phosphate	1.09
Calcium sulphate	4.86
Ferrous sulphate	0.005
Manganese sulphate	Make 15% solution and 2.5 ml.
Water	4 litres
Use the above solution in half strength. pH is maintained slightly acidic but not alkaline.	

Breeding of Orchids

In nature orchids are pollinated by insects and birds and there is free crossing in two species and two generas. There is no genetic barriers in species and genera. There also exists polyploidy and introgressive hybridization. The important genera which have given maximum number of man-made hybrids are Cattleye, Cymbium, Paphiopedium, Vanda, Dendrobium, etc. Important generic hybrids are : Ascocenda (Ascocentrum × Vanda), Aranda (Arachnis × Vanda), Aeridovanda (Aerides × Vanda), Brassocattleya (Brassovola × Cattleya), Saphrocattleya (Cattleya × Sophronities).

Some important hybrids evolved for cut flower production on commercial scale are : Archianis—'Maggie Oei', Aranthera—James Storie; Dendrobium—Pampadour, Walter Oume, Tomie, Spell Bound, Caesar; Vanda—Miss Jo aquim, Rothschildiana AscocendaYeh Sum Wah; Oncidium—Cymbidium—Babette (Mini), Ballerian (Mini), Broadmoor (Mini), Capella (Mini), Cassak (Mini), Cradle mont (Mini), Drumm (Mini), Fantasy (Mini), Green Queen (Mini), Lime-light, Mary Pincess, Mem Rosl Green, Miranda, Mem Rosl Green Moonglow, Pink Perfection, Princess Rose, Sabrina, Sir Cotton, Towiantot, Wendy Brown, Molly, Micke, Vanguard. Paphiopedium—King Arthur 'Bougone', Amer. Hybr. Phalaenopsis—Hybrids.

Pests and Diseases

Orchids are attacked by various kinds of pests and most common are aphids, thrips, mealy bugs, orchid weevil, snail and slugs. They attack younger shoots, leaves and buds. However, these are effectively controlled by the use of common insecticides like Imidachloprid, Qunialphos etc.

Orchids are also prone to many fungal, bacterial and viral diseases. The common fungal diseases are leaf spot, caused by *colletotrichum*, sp. and *Gleosporium*, sp. leaf blight caused by *pythium*, sp. collar blotch—caused by Pencilliuthomii, and orchid wilt caused by *Sclerotium rolfsii* which can be controlled by spraying fungicides like Captaf 80 WDG or Equation Pro 52.5 WDG.

***

9 Marigold

Introduction

In India, marigold is one of the most commonly grown flowers and used extensively on religious and social functions in different forms. It was introduced in India during the 16th century and since then it has been naturalized in different agro-climatic regions of India in such a way that it now appears to be native of this country. It has gained popularity amongst gardeners and flower dealers on account of its easy culture and wide acceptability. Further it's habit of profuse flowering, short duration to produce marketable flowers, wide spectrum of attractive colours, shape, size and good keeping quality attracted the attention of producers and traders most. Therefore, the commercial cultivation of this crop found its way in the vicinity of different cities and towns.

Flowers are sold in the market as loose or as garlands. Due to its variable height and colour, marigold is especially used for decoration and included in landscape plans. Besides it is also ideal for newly planted shrubberies to fill in the spaces with attractive colours. French marigold because of dwarfness and profuse flowering is most ideal for rockery, edging, hanging baskets and window boxes.

In the survey carried out in Bombay, Bengaluru, Calcutta, Madras and Delhi, the annual production of marigold has been reported to be 1985.7 tonnes. The price of marigold like any other produce, depends upon demand and supply. During rainy and winter season under the north Indian conditions, it is sold at ₹ 10-15 per kg whereas during summer season it fetches ₹ 15-25 per kg. However, the study conducted at Bengaluru, revealed that a net profit varying from ₹ 50,000 to ₹ 65,000 per ha is not uncommon even when it is sold at ₹ 5 per kg.

Varieties

There are 33 species of marigold and numerous varieties.

Most of the varieties are planted for garden decoration. The important varieties are :

T. erecta (African marigold)—Giant Double African Orange (**Plate 6**), Giant Double African Yellow, Cracker Jack, Climax, Dubloon, Golden Age, Chrysanthemum Charm, Crown of Gold, Spun Gold, Pusa Narangi, Pusa Besanti.

T. patula. (French marigold)—Red Brocade, Rusty Red, Butter Scotch, Valencia, Sussana.

T. tenuifolia—Golden Gem, Lulu, Pumila, Ursula.

However, in the market mostly orange coloured varieties are preferred and the variety which is dominating is African Giant Double Orange.

Soil and Climate Requirements

Soil. Marigold can be successfully cultivated on a wide variety of soils. However, a soil that is deep fertile, friable having good water holding capacity, well drained and near to neutral in reaction *viz.* pH 7.0-7.5 is most desirable.

Climate. Marigold requires mild climate for luxuriant growth and profuse flowering. It ceases to grow at high temperature thereby flower quantity and quality are adversely affected. During severe winter including frost plants and flowers are killed and blackened. However, plants if allowed to over winter, sprout during spring season and produce some flowers. Sowing and planting is carried out during rainy season, winter and summer season. Hence, flowers of marigold can be had almost throughout the year.

Under north Indian conditions in the plains the best flowering has been observed during winter months *i.e.,* from October to April. During summer which is characterized by high temperature and long days, plants remain mostly vegetative with sparse flowering. Under such conditions, flower size is reduced considerably but length of flower stalk is increased. On the contrary in the hills flowering commences from May onwards which continues upto October.

Preparation of soil. Land should be well prepared by ploughing it 2-3 times and 50 tons of well rotten farmyard manure should be well mixed/ha. Beds of convenient size be made to facilitate irrigation and other cultural operations.

Planting. There are two common methods of propagation of marigold *i.e.,* (*i*) by seeds, (*ii*) by cuttings. Crop raised from seeds is tall, vigorous and heavy bloomer; thus it is preferred over cuttings.

Seed rate and Nursery raising. Seeds of wide range of varieties of common species *i.e., T. erecta, T. patula, T. tenuifolia* are easily available and germinate quickly. Therefore, propagation through seed is advised. For better seed germination, optimum temperature range is between 18 to 30°C. For raising seedlings for one hectare, about 1.5 kg. seed is required. Seeds of marigold can be sown in pots, seed boxes or on flat or raised nursery beds. Nursery beds of 3 × 1 m size are thoroughly prepared and mixed with 10 kg of well rotten farmyard manure per sq m. About 8 to 10 such beds are needed to raise seedlings for one hectare. Before sowing the seeds, D.D.T. or B.H.C. should be dusted on outer side of nursery beds to avoid removal of seeds by ants. Seeds can be sown preferably in lines or by broadcast method. In case of broadcasting care should be taken for proper distribution of seeds so as to have healthy seedlings. For this, thinning is an important operation in both the methods. Seeds need to be covered with light soil or sand or strained leaf mould and watering should be done with rose can. For entire period, nursery bed should be kept moist but not wet and thus watered accordingly.

Sowing time. Marigold Crop Can be raised three times a year *i.e.*, rainy, winter and summer season. Sowing and planting times for each season are as under :

Season	***Sowing time***	***Transplanting time***
Rainy	End of June to 1st week of July	First fortnight of August
Winter	Mid of September	Mid of October
Summer	First week of January (Under glass house or plastic)	First week of February

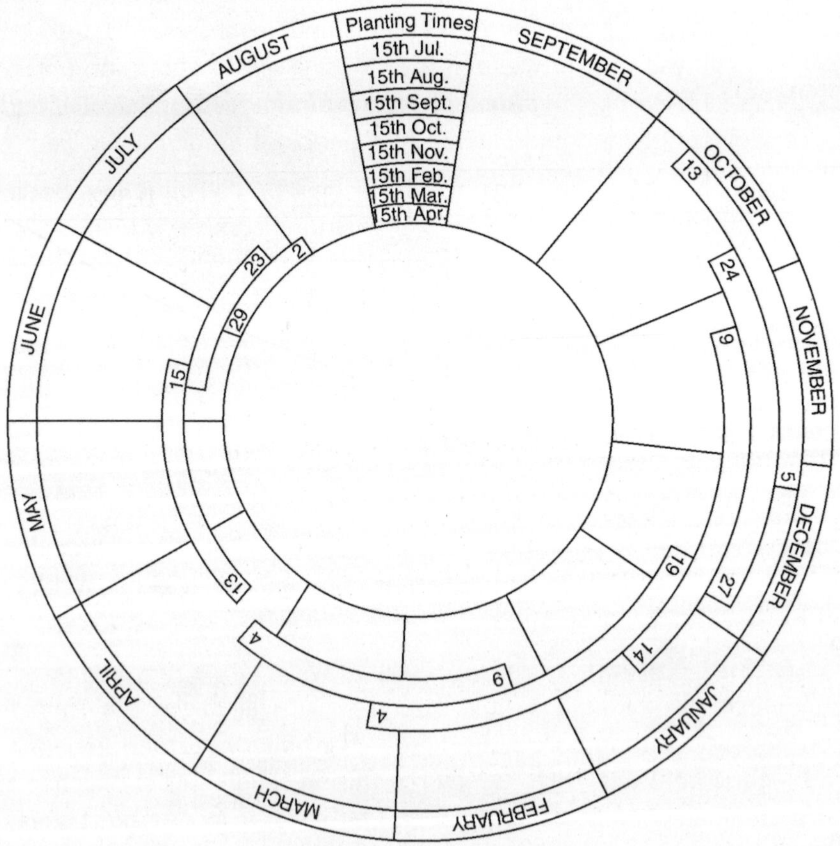

Fig. 9.1. Showing the Period of Availability of Flowers as affected by Planting Times.

A scheme has been developed to produce marigold flowers from October to May-June (Fig. 9.1). By following the growers around the cities can produce flowers for longer duration and good profit can be earned.

Performance of a variety varies from season to season which is reflected in the data given in Table 9.1 for Giant Double African Orange (Research conducted at P.A.U.)

Size of flowers is likely to vary within a season whether flowers plucked or not.

Cuttings. This method is commonly followed for maintaining the purity of varieties. Normally, the presence of adventitious roots along the stem helps in the establishment of cuttings. About 10 cm long cutting are made and treated with seradix No. 1. The cuttings are planted in the sand to strike roots easily and plants thus raised are used for bedding and pot planting.

TABLE 9.1
Effect of planting seasons on growth and flowering marigold (*T. erecta*) cv. Giant Double African Orange

Season	*Height (cm)*	*Days required for first flowering*	*No. of flowers/ plant (not plucked)*	*No. of flowers/plant (plucked regularly)*	*Yield of flower (kg/plant) (plucked regularly)*	*Flower size (cm)*
Rainy	98.5	58	35	54	0.492	8.20
Winter	59.9	53	21	32	0.382	9.05
Summer	35.5	58	6	12	0.042	4.55

Transplanting of Seedlings. Marigold seedlings are easily establish after transplanting in the field without much mortality. Fast growing root system present in this species enable the seedlings to establish better. At the time of transplanting, seedlings normally of about one month old should have attained 3-4 true leaves and must be stocky. Closely sown seeds give rise to thin, long and weak plants which are leggy and do not make a good plant. Very old seedlings are also not desirable because they have lost their juvenile phase in the nursery itself. Seedlings should be transplanted in well prepared land in the evening especially during rainy and summer seasons to avoid exposure to harsh weather, transplanting shock and to allow better establishment in cool hours of night. Soil should be pressed well around root zone so that there are no air spaces left. After transplanting, a light irrigation is essential. Heavy irrigation if done immediately after transplanting seedlings bent down and leaves stick with the soil which delay proper development of plant. On light sandy soil irrigation a day before transplanting is considered beneficial for the better establishment of seedlings.

Spacing. Proper spacing between plants is required for growth, development and flower production. Two marigold species which are commonly grown for commercial production of flowers require different spacings, for example *Tagetes erecta* requires wider spacings than *T. patula*. Experiments conducted at P.A.U. reveal that high flower production in Giant Double African Orange (*T. erecta*) was obtained by planting at a distance of 40 × 30 cm whereas in case of *T. patula* cv. Red Brocade higher number and weight of flowers per plant was recorded in spacings 20 × 20 cm; however, per unit area, highest yield obtained was from 20 × 10 cm spacing.

Manures and Fertilizers

Although marigold is one of the commercial flowers but much research work has not been done to determine its requirements for macro and micro nutrients for growth and flower production. The rate of nutrient utilization from soil has been reported to be 57.5 per cent of available N, 20.1 per cent P, 94.4 per cent K, 8.6 per cent Ca, 23.0 per cent Mg and 6.4 per cent Na.

Field trials were conducted at P.A.U. for two years to find out optimum dose of N and P of marigold (*Tagetes erecta*) cv. Giant Double African Orange for growth and flower production. The indication was that with the increase of N from 0 to 40 g/sq m, the number of flowers per plant and flower yield increased with the increase in dose whereas plant height responded only upto 30 g/sq m. The response to P was inconsistent on growth and flower production. Therefore, to get highest flower yield, 100 kg N (4 Q CAN) 100 kg P_2O_5 (6.25 Q single superphosphate) and 100 kg K_2O per ha should be mixed at the time of preparation of land. Remaining 100 kg N (4 Q CAN) per ha should be applied one month after seedlings are transplanted.

Weeding and Hoeing. Weeds are a problem in marigold especially in rainy season crop. After transplanting of marigold seedlings in the field, weeds grow faster than marigold in initial stages and cover large area in a few days. If these weeds are not removed in time, a great loss would occur in terms of growth and productivity of marigold. In India, weeding is done manually. It has been observed that 3-4 times weedings are required during the entire growth period.

Irrigation. Marigold is a herbaceous plant and puts up rapid vegetative growth during initial stages. It takes about 55-60 days to complete vegetative growth and to enter into reproductive phase. At this stage terminal flower buds appear which break the apical dominance and encourage secondary branches. At all stages of vegetative growth and during flowering period sufficient amount of moisture in soil is essential. Moisture stress at any stage of growth and development may hamper the normal growth and productivity of flowers. The frequency and quantity of water mainly depends upon soil and climatic conditions. In lighter soil more frequent irrigation is required than that in heavy soil. Season of planting determines the frequency of irrigation. If rainfall is normal and well distributed, irrigation is not frequently required, but if the rain is scanty, irrigation is needed frequently. It has been experienced under sandy loam conditions of Punjab that from September to March, weekly irrigation is enough. Though plants tolerate dry weather upto 10 days without irrigation but growth and flower production is affected adversely. From April to June, frequent irrigation at the interval of 4-5 days is required.

Pinching of Marigold Plants. In tall varieties of *Tagetes erecta* emergence of side branches and their flowering is influenced by the presence of apical dominance. It has been observed that the plants of marigold grow straight upwards to their final height and develop into terminal flower bud. By production of terminal flower bud, side buds become free from correlative inhibition of apical dominance and these buds develop into branches to produce flowers. However, if the terminal portion of shoot is removed early, emergence of side branches starts earlier and more number of flowers of good quality and uniform size are produced. Results of studies on pinching revealed that pinching the plants 40 days after transplanting enabled the plants to yield more flowers. However, the plants remained dwarf and flowering was delayed in comparison to late pinching *i.e.*, 50 or 60 days after transplanting.

Harvesting, packing and Transportation

Harvesting of flowers. Marigold flowers should be plucked when they attain the full size depending upon the variety. Plucking of flowers should be done in cool hours of the day *i.e.*, either in the morning or evening. Field should be irrigated before so that flowers keep well of longer period after plucking. Plucking is done by hand because flower stalk is a hollow structure which breaks easily when twisted between thumb and finger. Productivity of plants is increased considerably by regular plucking of flowers. It has been observed that in an acre flowers can be plucked by 6-8 labourers in a day. Flowers plucked should be covered with moist gunny bags if kept over night before taking to market.

Packing. Marigold flowers are primarily used for making garlands and hence plucked flowers are collected in gunny bags or bamboo baskets for carrying to the market. For the local market marigold flowers are taken into gunny bags whereas for distant market bamboo baskets are used.

Transportation. Different means of transportation including rickshaws, buses and trains are used to carry the flowers to market depending upon the distance.

Yield of flowers. Flower yield depends upon season of planting and cultural practices adopted. On an average a fresh flower yield of 200-225 q per ha during rainy season, 150 to 175 q per ha in winter and 100-120 q/ha in summer can be obtained.

Seed production. Marigold is a cross pollinated crop, hence, proper isolation distance of 1-1.5 km should be given amongst varieties. However, natural cross pollination amongst species is absent. The seed crop is ready in about six months. Under Indian conditions, winter crop of marigold does well for flower as well as for seed crop. Seed production studies carried out revealed that seed yield of 312 to 375 kg per ha in *T.*

erecta cv. Giant Double African Orange and 1000 to 1250 kg/ha in *T. patula* could be obtained.

IMPORTANT DISEASES AND INSECT PESTS

Marigold is relatively free from diseases and insect-pests. However, occasionally the following diseases and insect-pests have been observed.

Diseases

1. Damping off. It is caused by *Rhizoctonia solani* and appears as brown necrotic spots, girdling the radicle which later on extend to plumule and causes pre-emergence mortality. Post-emergence symptoms appear on lower part of hypocotyl as water soaked, brown, necrotic ring, leading to collapse of seeding. When infected seedlings are pulled, the root system appears partially or fully decayed. To control the disease soil drenching with Captaf 80 WDG @ 4 ml/l or Pencycuron 25 EC @ 2.5 ml/l of water should be followed. Proper drainage should be provided in the nursery bed.

2. Leaf Spots and Blight. Various species of *Alternaria, Cercospora* and *Septoria* are known to cause leaf spot or blight of marigold. *Alternaria tagetica* and *A. tenuissima* damage the foliage of the plants. The symptoms appear as minute brown circular spots on lower leaves and enlarge at later stage of infection leading to premature defoliation and ultimate death of the plants. To keep the disease under check the marigold crop should be sprayed with Mancozeb 75 WP @ 2 ml/l or Propiconazole 250 EC @ 1 ml/l of water at fortnightly intervals starting from the first appearance of disease symptoms.

3. Inflorescence Blight. This disease is caused by *Alternaria zinnae.* Elongated lesions are formed on inflorescence. Symptoms may appear as large irregular blotches on the leaves also which is light tan to dark brown with zonations. This can be effectively controlled by spraying the crop with Mancozeb 75 WP @ 2 ml/l or Propiconazole 250 EC @ 1 ml/l of water.

4. Flower Bud Rot. It is caused by *Alternaria dianthi.* The disease appears mainly on young flower buds and results into their dry rotting with brown scroched, necrotic discoloration of sepals and stalk. The ray and disc florets also turn brown. At later stages buds become shrivelled, turn deep brown and dry up. Symptoms are less prominent on mature buds but these buds fail to open. On older leaves few deep brown necrotic spots develop. To control this disease regular spraying of the crop with Mancozeb 75 WP @ 2 ml/l or Propiconazole 250 EC @ 1 ml/l of water should be followed.

5. Powdery Mildew. *Oidium* sp. and *Leveillula taurica* have been reported to cause powdery mildew of marigold. The whitish, tiny superficial spots appear on leaves which later on result in the coverage of whole aerial parts of plant with whitish powder. The disease can be controlled by following spraying with Karathane (40 E.C.) @ 0.5% or dusting with sulphur powder a fortnightly intervals.

Insect-Pests

1. Red Spider Mite. (*Tetranychus* sp.) These mites sometime appear on the plants near flowering time. Plants give dusty appearance. It can be controlled by spraying Ethion 50 EC @ 2 ml/l or Triazophos 40 EC @ 0.5 ml/l of water.

2. Hairy Caterpillar (*Diacrizia obliqua*, Walker). It is a polyphagus insect and caterpillar eats away foliage. This catepillar can be controlled by sprays of Quinalphos @ 1.5 ml/l of water.

***

10 Gerbera

Gerbera ranks fifth amongst cut flowers after rose, carnation, chrysanthemum and tulip. On an average 241 million blooms are produced every year. The reason for its popularity is its floral beauty, attractive colour, good keeping quality and suitable for long distance travelling. It is an important flower grown through out the world in a wide range of climatic conditions under open conditions or under heated or unheated glass house, plastic houses or tunnels.

Gerbera is commonly called as Transvaal Daisy or African Daisy. Botanically, it is *Gebera jamesonii* and belongs to family compositae. This genus has been named after German botanist Traugott Gerber. This is native of South Africa (Natal *G. jamesonii* and cape—*G. virdifolia* were crossed with each other to produce modern types) and Asia. In India gerbera is widely found in temperate Hamalayas from Kashmir to Nepal at the altitudes of 1300 to 3200 mts.

It is herbaceous, stem less plant and makes a rosette growth. Leaves are petioled, lanceolate and deeply lobed and are arranged at the base. Flower heads are solitary, composite (many flowered) having conspicuous ray floret in 1 or 2 or more rows. Inner row is very short and tubular and 2 lipped. Flowers are like daisy in many clear or bicolours *e.g.* White, Yellow, Pink, Orange, brick red, scarlet, salmon, terracotta, etc.

On the basis of doubleness, flowers are grouped into single, semi double and double cvs. The double cultivars are highly attractive due to contrast colour of disc *i.e.* yellow or black colour. The flower stem are long, thin and leafless. Seeds are beaked achenes with pappus (hairs).

Genetics and Breeding

There are about 40 species which are half hardy and perennial in nature. Important species are *G. jamesonii, G. virdifolia, G. asplenifolia, G. aurantiaca, G. kunzeana; G.*

aberdaria, G. abyssinica, G. ambigua, G. anadria, G. bojeri, G. bracteata, G. chiliensis, G. eineraria, G. elliptica, G. krausii, G. tuberosa, G. tomentosa, etc. Diploid and tetraploids forms exist. The basic no. of chromosome is 25. Increase in number of chromosomes have shown useful effects on plant and flower size. Different level of sterility exists and plants could be divided into (*i*) Completely self sterile plants, (*ii*) plants with distinct differences in seed set when pollination takes place within inflorescence, and when pollination takes place between neighbouring inflorescence on the same plant and (*iii*) plants in which these differences were less pronounced. For the production of hybrid seed first two types are suitable whereas third type was less favourable due to frequent selfing.

In gerbera, productivity, is due to additive gene effects. High heritability in *G. jamesonii* for inflorescence diameter and number of ray florets and relatively low for disc diameter and stem length has been reported. The numerous cultivars of gerbera have been developed through mostly hybridization and through mutation. The development of gerbera plants from seed till flowering takes 3-4 years. The seeds are to be sown immediately after harvest otherwise they loose viability. Seedling stage continues for about one month, the juvenile stage continues upto 4-5 months whereas the reproductive stage continuous 3-4 or more years.

Varietal Wealth

Gerbera is very rich in its varietal wealth due to ease in hybridization and survival of seedlings. Thousand of cultivars have been developed which are highly suitable for commercial production of cut flowers under open as well as under green house conditions. Several varieties suitable for pot culture which are characterized by their bushy nature, dwarf, bearing large number of flowers, may be, somewhat little smaller in size have also been developed.

According to flower size also cultivars are grouped into large and mini types. Similarly according to flower stem length, varieties are grouped as follows.

Tall varieties—when flower stem length is more than 50 cm. *e.g.* Funda, Diablo, Sunanda, Dark Red ; medium—(40-50 cm)—Ornella, Sun set, Thalassa, Tavila, Kolika, Ruby Red, Pink Elegance, Rose bella; Dwarf—(30-40 cm)—J.S. Lal, Alemasa, V. Record, Janson Hybrid, Verasace, Indu Kumari, Yellow Queen, Gen Khaiser, Nebulusa, Red Monarch, Sangria. The performance of varieties greatly vary from region to region and flower quality differs area wise.

Under Indian conditions, the popular gerbera varieties amongst commercial growers are : Jaffa, Sangria, Rosula, Oprat, Romona, Salina, Tecora, Starlight, Aida, Venturi, Ornella, Aalasmeer, Goliath, Pink Elegance, Rosa lina, etc.

Other important varieties of gerbera from Terra Nigra, Holland are being propagated and marketed by M/s Germini Agro Pvt. Ltd., Pune. These varities are Cariba (Red),

Lamborshini (Red), Bliss (Red), Dino (Yellow), Gucci (Yellow), Flow Basic (Pink), Grizzly (Pink), Cauca (Orange), Virginia (White), White Grizzly (White), Aligator (Violet), Crossroad (Yellow with red center), Solemio (Yellow with orange center), etc.

The characteristics of different large and mini varieties as reported by nurserymen are given in following tables.

Important Varieties of Gerbera and Their Brief Characteristics

Name	*Colour*	*Type*	*Flower Diameter (cm)*	*Length*	*Production/m² (7 plants/m²) flower/year in green house*
Beauty	Red	Double	10-12	Long	140-150
Red Favourite	Red	Double	10-12	Medium	130-140
Sangria	Red	Double	10-12	Long	130-140
Sunny Boy	Red	Double	10-12	Medium	140-150
Darobel	Pink	Double	10-12	Medium	140-150
Fiona	Pink	Dorbla	10-12	Long	140-150
Pinkas	Pink	Double	10-12	Long	140-150
Rosabella	Pink	Double	10-12	Long	140-150
Ornella	Orange	Double	12-13	Long	160-170
Optima	Orange	Double	10-12	Medium	140-150
Sunset	Orange	Double	10-12	Long	140-150
Tara	Orange	Double	10-12	Medium	130-140
Junkfrau	White	Double	10-12	Medium	140-150
Olivia	White	Double	10-11	Medium	130-140
White Sun	White	Single	10-11	Long	140-150
Lila bella	Violet	Double	12-13	Long	130-140
Blue bell	Violet	Double	10-12	Long	140-150
Aruba	Yellow	Semi Double	10-12	Long	130-140
Golden Fever	Yellow	Semi Double	10-12	Long	140-150
Golden Gate	Yellow	Semi Double	10-12	Medium	140-150
Gold Spot	Yellow	Semi Double	10-12	Medium	130-140
Sun Dance	Yellow/Orange	Semi Double	10-11	Long	130-140
Molina	Purple/Pink	Double	12-13	Long	150-160

Important Mini Varieties and Their Characteristics

Name	Colour	Type	Flower Diameter (cm)	Flower stem length	Production/m²/ year/plants (Under green house conditions)
Cheri	Red	Single	7	Long	250-275
Fancy	Red	Semi double	6	Medium	275-300
Northstar	Red	Single	8-9	Long	300-325
Ponsy	Red	Single	7	Long	250-275
Noki	Pink	Semi Double	7	Medium	250-275
Rose may	Pink	Semi Double	7	Long	250-275
Rusty	Pink	Semi Double	7	Long	250-275
Lily	Orange	Semi Double	6	Medium	275-300
Orny	Orange	Semi Double	7	Long	250-275
Sun Burn	Orange	Single	7	Medium	250-275
Bomby	Yellow	Semi Double	7	Long	275-300
Missy	Yellow	Single	7	Long	250-275
Nini	Yellow	Semi Double	7	Medium	275-300
Elsi	Violet	Single	7	Long	250-275
Lauri	Violet	Single	7	Medium	225-250
Sissi	Violet	Semi Double	7	Medium	275-300
Spiders					
Kontiki	Pink	—	7-8	Medium	200-225
Pinky	Red	—	6-7	Medium	225-250
Spanky	Orange	—	6-7	Medium	275-300
Spikey	Yellow	—	6-7	Medium	225-250
Videy	White	—	6-7	Medium	225-250

Climate. Gerbera needs mild climate for its vegetative growth and flowering. Extreme of cold or hot temperatures are not desirable. Therefore, under tropical and sub tropical climate gerbera is grown in open or in unheated plastic houses. In temperate climate, gerbera plants need protection from frost and are grown in green houses. Gerbera plants require sunny situation. In temperate climate, poor light conditions during winter adversely affects the flower quality and hence, artificial lighting is done. Night temperature plays an important role in production and quality of flowers. It has been observed that a night temperature of 12°C has been found optimum.

During entire growing and flowering period high humidity is to be maintained. If the air humidity drops below 50%, the young plants suffer. The humidity can be maintained more easily in closed structures.

Soil. For growing gerbera successfully, soil should be porus and well drained. Thus light soils with good organic contents are more suitable than heavy soils. The pH of the soil becomes a limiting factor when it is not in desirable limit. By nature gerbera loves acidic media than alkaline soil. The ideal pH for gerbera growing is 6.0-6.5 but it can tolerate little higher pH *i.e.* upto 7.2. More high pH of soil needs its correction to the desirable limits. On the other hand, too acidic soil can be made sweeter by adding lime. Most of the problems associated with its growth and productivity can be solved by adjusting pH and drainage system.

Before starting a cultivation of gerbera, soil sterilization is absolutely necessary specially against fungus like phytophothera, cryptogera and nematodes which causes serious wilt disease and may destroy a crop partly or even completely. For this, formaldehyde or methyl bromide gas can be used and soil is covered with polythene for 48 hrs.

Gerbera also responds well to artificial media if it satisfies the conditions of porosity, drainage and pH. Artificial media like coco peat, peaty soils, vermiculite, perlites, etc. do find favour for gerbera growing. Better grades of peat like mixes can be used. Gerbera will not grow happily in soil mixes with poor aeration.

Modification of Environmental Conditions

Area suitable for gerbera production should have mild weather conditions and temperature ranging from 16-20°C which should be in day whereas 12°C should be during night. This is the temperature which should be maintained in green houses. As gerbera plants can not tolerate extreme of weather conditions, there is a need to modify the environment and bringing it in limits so that optimum flower yield can be obtained. Mild weather conditions are available in places like Bengaluru, Pune, Nasik, Hyderabad, Kodai Kanal, etc. In these places gerbera is cultivated with less efforts. On the contrary under north Indian conditions due to extreme of weather conditions during summer and winter, it is not possible to grow gerbera under open conditions with optimum yield. During summer, lowering of temperature can be achieved by shading and having evaporative cooling system. Different materials like shading net (green or white), saran cloth, aluminium coated nets are available and can be selected judiciously. In the winter raising the temperature by heating or mere by covering with single or double polythene layers can be achieved depending upon the local conditions. Adoption of such expensive technology will depend upon it economics and profitability (**Plate 20**).

Propagation

Both sexual and asexual methods are employed in propagation of gerbera. Propagation through seed is used to breed new varieties by breeders. Gerbera seedlings exhibit variation when obtained from open pollinated seeds or in a planned programme useful variations are released as new varieties. It is a long processes and it takes 3-4 years to bloom a seedling. Care should be taken that seeds should be sown immediately after harvest otherwise they loose viability quickly.

Vegetative methods like division of clump, cuttings, and micro propagation are very commonly used.

Division

Large clumps are divided into seedling plants and are set in the field. Before planting, roots and leaves are trimmed keeping central axis intact. While setting the suckers care should be taken that soil does not cover up the central axis. Suitable time of separation will depend upon local conditions. At the time of separation environmental weather conditions should be mild and much ahead of winter so that winter crop is in the time.

Gerbera is also propagated through cuttings. Healthy plants are kept without water for 3-4 weeks, roots are pruned, planting in the peat are kept at 25-30°C with 80% humidity. The buds in the axil of leaves are detached and rooted in rooting medium. Plants become ready for transplanting in 2-3 months. Young stem cuttings produce roots and shoots much easily and quickly under misty conditions.

Micropropagation

Normally, gerbera is propagated through seeds and division, but with the availability of modern techniques of tissue culture of plants which allows the production of disease free planting material in unlimited number from single explant. It offers a very good opportunity for large scale multiplication of gerbera. It involves rapid multiplication of explants by repeated subculturing and preparation of divisions for transfer to the soil. Various explants in gerbera like, shoot tips, inflorescence buds, capitulum, and mid ribs have been employed. Various culture media for different explants have been found effective in obtaining the highest number of plantlets. Overall, Murashige and Skoog's medium with certain modification have been found successful in gerbera.

Planting

In green houses or in open field conditions, two rows or four rows planting system are generally followed. Hence, accordingly bed width is kept. For large flowering gerberas 8-10 plants/m^2 are planted. Close planting is followed for small flower cultivars. The spacing in rows ranges from 30-40 cms. Plants are planted at 30-37.5 cms apart. While planting, care should be taken that crowns should be well above the soil surface. Planting time varies with place of growing, its climatic conditions and method of growing.

Planting can be done round the year under Pune, Bengaluru, Hyderabad, conditions but under North Indian conditions optimum time for planting is September and October. The plants should be left in the field undisturbed for two years for flower production.

Manuring and Fertilization

Gerbera prefers organically rich medium for proper growth and flowering. Therefore, addition of organic matter in medium results in healthy plants and excellent flower production of better quality of flowers. It is advised before addition of fertilizers, leaf blade analysis should be done. Leaf blade analysis has been found better than petioles as it is better index of nutrient contents. It has been observed that good flowering yield was recorded when the leaves contained 2.7-3.13% N, 0.19-0.35% P, 3.06-3.64% K, 1.66-2.18% Ca and 0.3-0.4% Mg.

Soil analysis will decide the quantity of organic matter and other fertilizers to be added. Generally in fairly sandy soil, addition of 7.5 kg/m^2 well rotten farm yard manure gives desirable results. Gerbera responds well to the applications of NPK which improves the numbers of leaves, suckers and large number of blooms of big size. Excess of N reduces of flower quality and flower production. Generally, N @ 15 g/m^2 was found to be optimum for better flower production of quality bloom. N, P & K in the ratio of 3:1:1 produced highest flower yield.

Gerbera being sensitive to pH of the soil, if pH is not in ideal range, plants can exhibit the deficiency of micro nutrients. Under high pH of conditions, deficiency of Fe or Mn or both is very common. Hence, it is recommended that gerbera plants should be sprayed with micro nutrients time to time even if plants do not exhibit micro nutrient deficiencies to avoid development of deficiency symptoms of these micro nutrients and harm to occur to the plants.

Irrigation

Gerbera is a moisture loving plant and hence, it needs thorough irrigation at regular interval. However, wet feet is not desirable so avoid water logging conditions. Water requirement vary with the varying environmental conditions. It has been observed that there was an increase in flower number when the water regime was higher in summer and low in winter when it was compared with constant water supply. Hence, watering should be adjusted according to crop requirement. Water quality is also important and different varieties responds to various levels of salts. Irrigation water should be of good quality and low in total soluble salts.

Watering methods like flood irrigation, sprinkler and drip irrigation can be employed according to the facilities and finances available.

Weeding and hoeing

Depending upon season and method of growing, and weeds types will appear in gerbera field. Gerbera being perennial crop and remains in the field for 2-3 years and its rosette nature of growth, weed appear only in the early stages of crop. Later on gerbera plants cover most of the field except irrigation channels or furrows. Two to three manual weedings are required. Efforts should be done to keep the soil loose and friable by regular hoeing and free from weeds.

Flower Regulation

After planting, normally it takes 2-3 months to produce flower by the plants. If plants take longer period for flower production. Plants can be sprayed with 100 ppm of GA_3. This will result in early flowering with longer stems.

Flowering Yield

Once the gerbera plants enter into a reproductive stage, continuously plants keep on producing flowers. Under green house modern varieties are very heavy yielder and produces about 250-300 flowers/m^2 from 8-10 plants. Under open conditions different varieties show variation in yield depending on types of media, pH, climatic conditions and seasonal changes.

Storage of Flowers

Optimum storage temp. (wet) for gerbera is 4°C. The flowers can be stored efficiently upto 4 weeks. The cut flowers should be kept in clean water containing preservative and bactericide without delay after harvesting. Removal of heel of stem improves water uptake, Addition of HQS or chrysal is generally used.

Harvesting and Packing of Flowers

The harvesting of gerbera flowers should be done at optimum stage. This stage can be identified very easily by observing outer two whirls of disc florets which should be perpendicular to the stem and show pollen grains otherwise flowers will wilt and close at night. After harvesting individual flower head is provided with thin layer of plastic cup which protects it during transport and keeps fresh and do not allow any damage to occur to flower head. The gerbera cups of following three sizes are available.

Type	*Size*
Mini	112-22-85 mm
Standard	136-30-100 mm
Giant	150-30-120 mm

All the flowers are packed in insulated boxes individually for safe transport. Good quality gerbera should have atleast 40 cm stem length and should be firm and straight. Stems showing bending are not desirable in the trade. The flower should be uniform in size and should not be less than 7 cm in diameter. Flowers of gerbera are packed in flat boxes having paper shreds and holes for individual flowers.

Main problems observed in field or green house

(*i*) If flower stems are too long or foliage is too large, increase light intensity or use B-9.

(*ii*) If flowers are too short or hidden in foliage, this condition could result from excessive fertilizer (nitrogen or ammonia).

(*iii*) If plants are dying out too frequently, too much growth regulator or average growing temperature is too low.

(*iv*) If flowers are distorted, it is caused by mites or thrips.

(*v*) If plants fail to grow, usually caused by poor drainage, packing soil too tightly or low soil temp.

(*vi*) If plant wilts or dies, it might be planted too deep and eventually was killed by crown rot.

IMPORTANT DISEASES, INSECTS AND PESTS

Diseases

1. **Alternaria.** It is a common disease caused by *Alterneria* sp. *Phyllostica gerbera* and other fungus. It can be easily controlled by Difenconazole 250 EC @ 0.5 ml/l or Propiconazole 250 EC @ 1.5 ml/l of water.

2. **Root Rot and Foot Rot.** It is soil borne disease caused by Pythium irregularae and *Rhizoctonia solani*. Soil sterilization is through formaline is good solution. Drenching of soil with Captaf 80 WDG @ 4 ml of water gives the good control of disease.

3. **Powdery Mildew.** It is caused by fungus *Oidium erysiphoides* sp. gerbera and produces a white coating on the foliage. Good control can be achieved by spraying Karathane 40 EC @ 0.5% (50 ml in 100 L of water).

Insects and Pests

(***i***) **Leaf miners.** Gerbera is often attacked by leaf miners (Liriomyza trifolii). Adults of the insect lay eggs on the leaves. The larvae bore into the leaf and make irregular shaped tunnels which are easily identified by the presence of yellowish tan coloured marking on the leaves. On maturity, larvae fold leaves on inner surface. It can be controlled by spraying Imidacloprid 200 SL 0.4 @ ml/l or Thiamenthoxam 25 WG @ 0.4 g/l of water.

(***ii***) **Mites.** It is commonly found in green house. The development of leaves and buds was adversely affected and flowers became malformed and unsaleable. These can be controlled by spraying Ethion 50 EC @ 2 ml/l of water or Tirazophos @ 0.5 ml/l of water.

(***iii***) **Thrips.** These are small insects and suck the cell sap from leaves and flower stalk. It can be also controlled by spraying Ethion 50 EC @ 2 ml/l of water or Triazophos 40 EC @ 0.5 ml/l of water.

***

11 Jasmine

Whenever, we think of fragrance, flowers like rose, jasmine and tube rose are immediately come to our mind and sweet smell of these flowers start filling in our nostrils and give a magical relief to mind and provide solace. Though there are many more other fragrant flowers but these three flowers are top ranking in area and production either for loose, cut flowers or for extraction of oil. Amongst these, jasmine is a very important group of plants which is cultivated extensively in tropical and sub-tropical areas of southeast Asia and other parts of world for garden decoration and commercial purposes. The white and fragrant flowers of Jasmine are symbol of purity, eternal love, nobility and also symbolizes the beauty of girl. It is offered to the God and used for worshiping and hence, these are preferably planted in Hindu temple gardens and Mogul gardens.

The genus Jasminum belongs to family Oleaceae and is of about 200 species which are mainly shrubs and climbers. Mostly, the plants of jasminum species are ever green but winter deciduous species *viz. Jasminum nudiflorum* is also found. Leaves are dark green in colour, opposite, and varying from simple (*J. sambac*) to tri foliate or pinnate with upto 9 leaflets (*J. grandiflorum, J. Officinale, J. multiformum*, etc.). Flowers are waxy white in most of the species but yellow colour is also found in *J. humile, J. nudiflorum, J. mesnyi.* Similarly, there is a quite variation in flowering time. Earliest flowers are produced during spring only in *J. humile* whereas in most of the species flowers are produced in spring through summer with peak flowering in rainy season till autumn. Some species also produce flower during winter season *i.e. J. multiflorum, J. nudiflorum* and *J. mesnyi.*

Most common species are : *J. sambac, J. grandiflorum, J. auriculatum, J. humile, J. multiflorum, J. officinale, J. nudiflorum, J. mesnyi, J. angulare, J. angustifolium, J. nitidum, J. parkeri, J. polyanthum, J. pubescens, J. rex*, etc. Jasmines are widely cultivated for their flowers which are used for making veni.

Jasmines are native to tropical and warm, temperate regions of south east Asia, west Asia, and Europe. *Jasminum sambac* is the national flower of Philippines which was adopted in 1937. It is also national flower of Indonesia which has been adopted recently in 1990 by Indonesian Government alongwith Moon orchid (Phalaenopsis amabilis) and *Rafflesia arnoldii*.

Origin, Species and Varietal Wealth

There are about 200 species which are widely distributed in the countries of Asian and European continents. The important species are described hereby.

J. sambac. It is national flower of Philippines and Indonesia and commonly called as Arabian Jasmine, or motia or mughra. It is evergreen, makes bushy growth and produces long scandent branches in growing season which starts from spring continue to Summer and rainy season. It produces white fragrant flowers of varying degree of doubleness, single and semidobule and double flowers.

Flowers are mostly used in making garlands and for extracting perfume for cosmetic industry. Flowers of *J. sambac* are used for making green tea which is considered effective in preventing cancer when taken daily. Varities of *J. sambac* has been classified and described by Tamilnadu Agricultural University Coimbatore on the bases of number of whorl, length of pedicel, corolla tube length and shape of flower bud. The varieties Mandan ban, Gundumalli, Rambanam, have been considered better than other vartities in terms of yield of flowers and other characters.

J. grandiflorum. It is commonly called as chameli or Spanish jasmine. It is ever green with dark green imperipinnate leaves producing long branches. Flowers of bigger size of pure white colour are produced from April to November and flowers are used in making veni, garland, bracelet, etc. and for extracting oil.

Six varieties of *J. grandiflorum* were selected at T.N.A.U., Coimbatore depending upon floral characters and ploidy level. Out of six, five are pink and one is of white colour. One of pink is triploid (39) and rest are diploid (26). The triploid are highly sterile. An improved variety 'COI Pitchi' with an average yield of flowers 10.15 tonnes/ha and with recovery of 29.5 kg/ha of jasmine concrete has been released and has become very popular amongst growers in south India.

J. auriculatum. It is commonly called as juhi. It is evergreen shrub and produces white and highly fragrant flowers. Flower are sold in the market as loose flower for making veni, garland, bracelet, etc. and for extraction of essential oil.

Based on flower bud characteristics, at T.N.A.U. Coimbatore five variants were selected and named as long point, long round, medium point, short point, and short round. These clones exhibited variation in susceptibility to common gall mite pest, however, medium point was found to be highly resistance to the pest which was named as 'Parimullai'. Latter on another selection of long round bud *i.e.* 'Co Mullai was released which produced 8800 kg/ha of flowers with good market potential.

J. multiflorum* syn. *J. pubescenms. It is commonly called as kund or dela in Punjab. It is ever green medium in height producing white flowers during winter months *i.e.* Dec.-Jan. It's bushy growth makes it highly suitable for planting in gardens and shrubbery border.

J. humile. It is commonly called as pili-chameli on account of its yellow colour flowers. Leaves are dark green and makes luxuriant vegetative growth and hence, it is highly suitable for various situations in garden as ground cover, screening, against wall, or hiding unwanted sites. It produces semi double, deep yellow flowers in spring but for short duration *i.e.* 20-25 days.

J. mesnyi. It is called as primrose Jasmine. It is ever green shrub with long slender arching branches. It grows about 1.5-3.0 m without any support. The bright yellow, semi double (6-10 petals), fragrant and trumpet shaped flowers are produced in spring and sporadically during summer. It is ideally suitable as a specimen or to trail over fences or walls.

J. nudiflorum. It is commonly called as winter jasmine. It is winter hardy and produces yellow flowers in leafless condition during winter months.

J. polyanthum. It is commonly called as pink jasmine on account of its pale pink fragrant flowers which are produced in cluster during summer months. It is evergreen, rapid grower and produce rich green foliage.

Climate. For successful growing of jasmines, mild climate is required. Warm humid conditions favour for luxuriant growth, profuse flowering and high productivity of plants. The desirable temperature range is 15-30°C and beyond these, growth and flower production hampers. Ground forsts kill the shoots. Ample of sunny days are better for growth and flowering than cloudy day. Hence, the climatic conditions of South India is better than north conditions of India.

Soil. Jasmines prefer well drained rich sandy loam to clay soil. Soil should be deep friable and free from calcareous soil. Jasmines are very well adapted to different kind of soils. It can tolerate wide range of pH.

Propagation. Jasmines are commonly propagated by cutting and layering but seeds are also used to create variations and new hybrids. There is large variation in seed germination due to species and clones. Diploid species exhibit better germination 75-84% (*J. grandiflorum* and *J. auricuatcum*) whereas low seed germination of about 55% has been found in tetraploid species (*J. flexile*).

Cuttings. Jasmines are easily propagated through different kind of shoot cuttings and percentage of rooting depends upon maturity of shoots in different species. In *J. sambac* both semi hardwood and terminal cuttings root better exhibiting 92 and 94 per cent, respectively whereas terminal cuttings in *J. grandiflorum* exhibits as high as 98% rooting. In case of *J. auriculatum*, the semi hard wood cuttings and having terminal portion root better *i.e.* about 70%. Cutting made during summer months *i.e.*, May to

July roots better than January, April or October. Intermittent misting improves percentages of rooting as well as number and length of roots. Amongst rooting media coarse sand is better than other media. The number of days taken for rooting varies between 45 to 60 days.

Layering. Layering is successful in Jasmines but being cumbersome it is not generally practiced. Grafting and budding has also been tried experimentally. Micro propagation by taking apical and axillary buds as explants have been standardized and can be used for multiplying large number of plants.

Planting in the field. There are two main seasons of planting *i.e.*, rainy and spring season. If water facilities are available and plants are prepared through layering, spring season should be preferred for planting on account of its long growing period. When plants are prepared through soft or hard wood cuttings planting can be done during rainy season *i.e.* after 15 August onward through September.

Plant spacing differs from species to species. To obtain the higher yield proper spacing should be followed. It has been observed that for *J. grandiflorum* proper spacing is 2 × 1.5 m accommodating 3333 plants/ha. Similarly, the proper distance for *J. auriculatum* was found to be 1.8 × 1.8 m and for *J. sambac* 1.2 × 1.2 m which is followed for commercial plantings.

At the time of planting proper care should be followed so that there is low rate of mortality. The size of pit varies from 45-90 cm^3 depending on type of soil *e.g.* for sandy loam soil 45 cm^3 size of pit is sufficient whereas for clayey soils the size of pit should be 90 cm^3. After digging, the pit should be exposed to sun for atleast 7-10 days. In each pit 10-20 kg of well rotten farm yard manure should be mixed and pit should be filled. It should be followed by soaking irrigation containing diluted solution of chloropyro phosphate to keep the termites under control and soil settles down. After a day or two, planting in the center of pit should be done in the evening and should be followed by light watering.

Nutrition. For proper plant growth and flower production, jasmine plants need balanced nutrients. Since flowering continues for longer duration and, hence, regular supply of specially three major elements *i.e.* N, P and K is required. Any limitation of these nutrients will affect the flower production as well as quality of essential oil. It has been reported from T.N.A.U., Coimbatore that in case of *J. grandiflorum*, the application of N, P and K @ 60, 120 and 120 g/plant, respectively along with 10 kg well rotten farm yard manure found to be optimum. It should be applied in two splits *i.e.* once in January and again in July every year. For *J. auriculatum*, application of N at 120 g per plant and P and K at the 240 g/plant produced better flower yield. On the other hand, at Kalyani for *J. sambac* and *J. auriculatum* the response at N_{50} P_{25} K_{25} g/plant was better. Thus, the response of N, P and K varies due to species and locality, soil types and soil fertility level.

Irrigation. Jasmines require a regular supply of water during spring and summer season to maintain the luxuriant vegetative growth and for flower production. The scarcity of water during these months, will hamper the growth and flowering drastically. Except the rains during these months normally deep irrigation is required once in a week, however, depending upon soil texture the frequency may be adjusted. In sandy and sandy loam soils irrigation should be done at the interval of 3-4 days whereas for heavy textured soil, frequency of irrigation can be delayed to 10-15 days. During rainy season depending upon the intensity and frequency of rain fall, irrigation can be adjusted. During winter months *i.e.*, from November to February irrigation can be done at the interval of 10-15 days.

Pruning. In jasmines, flowering occurs on new shoots and hence, to get highest yield, the new growth should be encouraged by annual pruning or pinching of shoots. Pruning time and intensity ultimately affects the new growth of plants and flowering. It has been observed the pruning must be of much ahead of flowering. Flowering can be regulated by manipulating the pruning time. In general, pruning is done in late winter months and it consists of removing all dead, dry branches and cutting back past season branches upto 45 cm from the ground level. Thus plants remain in easy reach so that plucking of flowers can be done easily. In *Jasminum sambac,* it has been observed that pruning during December and January and upto 30 cm of ground level produced highest flower yield under South Indian condition whereas in North Indian conditions end of January is considered most ideal time of pruning.

On the contrary *J. auriculatum* and *J. grandiflorum* should be pruned in the end of January under both South and North Indian conditions.

Chemical defoliants like para quatdichloride, penta chlorophenol, potassium chloride, sodium chloride or ethrel at higher concentrations has been reported useful in alternate to manual pruning but, however, they have not come in use on large scale. Still manual pruning is being practiced.

Flowering and yield. Flowering occurs during summer months through rainy season and till autumn. It has been observed that peak season of flowering occurs during long days associated with comparatively higher temperature during day and night both. Thus peak flowering occurs during June and July. The yield varies due to species and their cultivars. In general yield is as follows :

Name of species	***kg/ha***
J. sambac	739-8129
J. grandiflorum	4329 to 10,144 (Pitchi)
J. auriculatum	4733-9152

JASMINE OIL AND CONCRETE

Fragrance in jasmine flower is due to the presence of volatile oil present in inner and upper surface of both sepals and petals and oil glands are deep seated in the cells and hence, steam distillation is not able to extract oil. Thus, solvent method is used to extract the oil. The fragrance begins to develop in the flowers soon after sun set and ceases within a few hours of sun shine. Therefore plucking of flowers is recommend early in the morning say upto 9.00 a.m. After this quantity and quality gets decreased. During transportation the factory care should be taken not to damage the flowers. The damaged flowers yield concrete with more percentage of wax.

The jasmine concrete is a wax like substance containing natural flower perfume as a volatile oil, plant waxes and colouring matter. Amongst various oil solvents like hexane, petroleum ether and carbon tetrachloride, out of these, food grade hexane was found better. The boiling point of food grade hexane has been observed to be 60-80°C but even 70°C is highly satisfactory. This hexane has to be purified before using for extractions. For rotary type extractor 2 litres of hexane is found to be optimum whereas for static type it can be 4-5 litres.

The flowers are treated with purified hexane at 2 liter per kg of flowers for about 30 minutes. The complete extraction of the perfume from the flowers can also be ensured by slowly rotating the container having flowers and solvent for about 20 minutes in the rotary type of extractor.

After complete extraction of perfume, solvent is filtered and distilled at bout 70-75°C to recover the hexane. At this temperature, hexane evaporates leaving concentrated liquid which contain perfume and waxes and it is further purified by vacuum distillation for removing total amount of hexane. The final form of floral concrete is in the form of molten wax.

Floral concrete to recovery % in different species :

J. sambac	0.14-0.19
J. auriculatum	0.29-0.36
J. grandiflorum	0.24-0.30

Insects, Pests and Diseases

Mites. Red spidermites are very common in jasmines during warm and dry summer months. Severe puckering and discoloration of leaves are found due to crown gall mites in *J. auriculatum*. The cv Parimullai of *J. auriculatum* resistant to crown gall mites has been released from T.N.A.U. Coimbatore. To control mites, application of Ethion 50 EC @ 2 ml/l ot Triazophos 40 EC @ 0.5 ml/l of water has been found useful.

Nematodes. *Jasminum sambac, J. flexile* and *J. nudiflorum* are commonly attacked by root knot nematode (*Meoidogyne incognita*) which cause knot formation in roots, yellowing of leaves and die-back of shoots. Application of Neem cake @ 1 ton/ha at planting can control root knot nematode.

Diseases

Wilt. It is caused by *Fusarium solani*, *J. sambac* is very susceptible to the disease. In the beginning, yellowing of lower leaves start and slowly spread up and finally result in death of plants. In the field disease spreads in patches and roots show black coloration. The disease can be controlled by drenching the soil will Bavistan @ 2 ml/l of water.

Rust. It is caused by *Uromyces* hobsoni and appears during rainy months *i.e.* July, August. It attacks severely *J. auriculatum* and other species like *J. sambac, J. angustifolium*, are also susceptible. This disease can be controlled by spraying Propiconazole 250 EC @ 1 ml/l of water.

Leaf Blight. It is caused by *Alternaria jasmini* and *Cercospora jasminicola* and results in formation of reddish brown circular spots measuring 2-8 mm which spread readily during rainy season. The disease appear in May-June and peak incidence occurs during rainy season. Flower production is adversely affected and may cause 50% loss yield. *J. grandiflorum* is severely attacked by this disease and others show stray incidence. It can be controlled by spraying Propineb 70 WP @ 3 ml/l or Propiconazole 250 EC 1 ml/l of water. Mosaic viruses and phyllody caused by mycoplasma are also observed some time.

12

Tuberose

It is commonly called as Rajni Gandha (Hindi, Bengali), Gul-e-chari, Gul-e-Shabab (Urdu), Nishigandha (Marathi), Nishi gandhi (Malayalam) and Sem pangi (Tamil) and scientifically—Polianthes tuberosa and belongs to family Amrylladaceae. Tuberose has long, narrow, light green leaves. Plant grows about 25-30 cm tall. The flowering stalk is 75-100 cm long having 25-50 florets per spike. It produces, waxy, white, highly fragrant single semi double or double flowers on a long spike which open from base to the top. Single variety has been named after the place of growing as Mexican single, Calcutta single, Hyderabad single, Coimbatore single, Bangalore single, etc. Tuberose blooms during summer, rainy season and autumn. Thus, it becomes very important cut flower of summer and rainy season of India. Its flowers are also used as cut, loose flower for making garlands, bracelets or for hair do.

Single white flowers are highly fragrant and hence, are used to extract its valuable essential oil which is sold on premium. It's essential oil is rich in geraniol, nerol, benzyl achohol, eugenol, benzyl benzoate and methyl anthranilate. Due to these compound, flowers emit a delightful fragrance and are source of tuberose essential oil. These compounds are present in oil glands which are deep seated in petal and hence, solvents like hexane is used to extract the oil.

White flowers of tuberose can be coloured by dissolving coloured dye in vase water like Bromocresol green and Bromophenol blue which will impart different shades of blue colour. Phenol red imparts yellow colour and erythrosine red will impart rose colour to flowers. Different shades and intensity of colour will depend upon the concentration and quantity of chemical used.

Origin. It is native of Mexico from where it reached to different parts of world during 16th century when Spanish conquered it. At first it reached Europe than to Africa and Asia. It's arrival to India is not well documented but Britishers might have introduced it. Now, it is a commercial crop of France, Spain, Morocco, South Africa, Israel, New Zealand, U.S.A., tropical and sub-tropical areas of India and Cylone.

Area and production. Tuberose is well adopted and grown commercially in various parts of India. It is prominently grown in **West Bengal**—Bargan, Golaghat, Midnapur, Panskura, Ranaghat, Thakur Nagar; **Karnataka**—Mysore, Bangalore; **Andhra Pradesh** —East Godavari, Guntur, Chitoor, Nellore, Visakhapatnam, Krishna districts; **Tamil Nadu**—Coimbatore, Madurai; **Uttar Pradesh**—Meerut, Saharanpur, Muzaffarnagar; and in limited areas of Haryana and Punjab.

In all total area under tuberose in India is estimated to be about 12,000 ha.

Species and Cultivars. Polianthes is a monotypic genus and is closely related to Bravoa. There are about a dozen of species of Polyanthes but they are not clearly distinct. The only close relative of *P. tuberosa* is considered to be *Polyanthes gracilis*. Other species are *P. palustris, P. durangensis, P. longiflora, P. platyphyla, P. graminofolia, P. geminiflora, P. blessii, P. pringlei, P. sessiliflora. P. nelsonie,* etc.

The haploid number of chromosome of tuberose is 30 of which 5 are large and remaining 25 are small. Seed setting in tuberose is quite erratic in single flowered cultivar whereas it is not observed in the double type. The variegated cultivars, however, exhibit high degree of seed setting. The exact reasons of poor seed setting is not known. It is not due to any defects in the formation of the pollen grains or development of the embryo sac. It might be due to lack of pollinisers because flower are not colourful and attractive.

Cultivars. There are four type of varieties *viz.* Single—it contains one row of petal; semi double—it contains two to three rows of petals ; double having more than three row of petals; and variegated—in these leaf margins are variegated which there highly suitable for landscape planting.

1. **Single Flowered.** The cultivars have single row of petals. Floral buds are greenish white whereas flowers are waxy white. Loose flowers are used for decorations and floral arrangements. Flowers are highly fragrant and, hence, are used for extraction of essential oil which is present at about 0.08-0.11%. Seed setting is observed. Single types are identified with the name of locality like Mexican, Calcutta, Pune, Hyderabad, Coimbatore, Bengaluru.
2. **Rajat Rekha.** It has been developed through mutation breeding at NBRI, Lucknow. Leaves are variegated with slivery white streak along the middle of leaf blade. Flowers are single and oil content is about 0.089%. In garden, it is planted for its variegated leaves and flowers.
3. **Shingar.** This variety has been developed from a cross between single × double. The flower bud is slightly pinkish tinged. Florets are bigger than Calcutta single. The average yield of loose flowers is about 15 tonnes/ha which is 40% higher than Calcutta single. It is considered resistant to *Meloidogyne incognita* nematode. Therefore, it is being preferred by growers.

4. **Prajwal.** It is also a hybrid between single × double. It's spikes are tall and stiff. Flower buds are slightly pinkish in colour but flower are waxy white. The individual flowers are large and yield is about 12 tonnes/ha. It is grown for loose flower as well as for cut flower purpose.

5. **Semidouble flowered tuberose.** In these type flower petals are in 2-3 rows. Flowers are white.

6. **Vaibhav.** This is a cross between Mexican single × IIHR-2. The flower buds are greenish in colour. Flowers are white. Spike yield is comparatively higher. It is recommended for cut flower production.

7. **Double flowered tuberose.** As the name indicates, flowers have more than three rows of petals on long and sturdy spikes. Flowers open slowly. Therefore, it is highly suitable as cut flower. It has pinkish tinge at the top of flower buds, flower colour in white. The 'Pearl' variety does not open well.

8. **Swarna Rekha.** It is a mutant of double type developed at NBRI, Lucknow through gamma radiation. Leaf margins are golden colour. Flower buds are pink but flowers are white and double.

9. **Suvasini.** It has been developed by crossing between single × double. Flowers are double, white, bold and spikes are longer. Highly suitable for cut flower production. It yields about 25% more than the Pearl.

Climate. Tuberose is a half hardy bulbous crop and happily grows in mild climate. It can tolerate wide range of climatic conditions and areas having a range of 20-35°C temperature are considered good for its successful growing. For its luxuriant growth and high productivity the mild temperature of about 30°C with high humidity is most ideal. The temperature above than 40°C reduces the productively and quality of spikes. Much lower temperature *i.e.* 1-2°C with ground frosts kill the vegetative parts of the crop completely. However, sprouting and flowering is normal in ratoon crop with the onset of spring.

Soil. Tuberose is a bulbous plant, hence, loamy and sandy loam soils having pH of 6.5 to 7.5 with good aeration and drainage are highly pertinent for its successful cultivation. It has been observed that it can tolerate higher degree of pH *i.e.* upto 8.2 to 8.3. It's cultivation can be extended in marginal and unproductive soils affected by salinity and alkalinity. The soil should be rich in organic matter and good water holding capacity for proper growth of tuberose.

Preparation of field. The field should be well prepared by giving 2-3 ploughings and incorporating 40-50 tonnes/ha of well rotten farm yard manure before planting. The application of 100 kg N and 400 kg P_2O_5 and 200 kg K_2O/ha is found to be optimum which should be applied before planting the crop alongwith 8 kg of thimet or 2.5 kg carbofuron/ha should be mixed to protect the crop of tuberose against nematode.

Planting time. Tuberose is generally planted in February-March (East India) and March-April (North India). In the hills it is planted in April-May. In South India, the crop is planted in July-August. Under Bengaluru conditions, the planting can be done round the year but April-May planting is good for better crop.

Bulb selection and planting. The size of bulb, its diameter and weight reflects the storage of food material which will be utilized for the growth, flower production and subsequent production of new bulbs. In tuberose, the production of daughter bulbs and flower production go simultaneously and hence, sufficiently good sized bulbs should be used for planting the crop. On an average the bulb size of 2-3 cm is good for raising healthy crop. But relatively larger bulb *i.e.* 3.1-3.5 cm weighing about 40-50 g exhibit better result and produced good quality spikes.

After field preparation, the field is divided into plots leaving the path and water channels. Bulbs are planted in the field at the distance of 20 × 20 cm and at 4-5 cm deep. Little more wider distance of 30 × 30 cm is followed in Maharashtra. Close planting at the distance of 15 × 15 cm has been found satisfactory at many locations including Assam. After planting field is irrigated lightly.

Manuring. Tuberose is heavy feeder and very exhaustive crop and thus responds well to the applied organic and inorganic nutrients. Fertilizer application of 200, 400 and 200 kg N, P_2O_5 and K_2O per ha, respectively is recommended for quality flower production and bulb production. Half of N and full dose P and K to be applied before planting and remaining half dose of N to applied one month after planting. This can be increased or decreased depending upon fertility of soil.

From Assam, it has been reported that the application of 80-40-60 g/m^2 of N, P_2O_5 and K_2O in plant crop and ratoon crop was found highly beneficial in terms of flower production and bulb production of single variety of tuberose.

Irrigation. Tuberose requires sufficient moisture for its growth and quality flower production. It is advised that a light irrigation should be given before planting which helps in early sprouting and growth of plant. Too much moisture at the time of sprouting of bulb result into rotting of bulbs. Similar conditions during flowering affect the quality of spikes adversely. The frequency of irrigation depends upon soil types, weather conditions and stage of growth. In general weekly irrigation is required during April-June. Irrigation interval during July-September depends upon rains and their intensities whereas during October to March irrigation at the interval of 10 days keeping the rains in view is deficient to meet its requirement.

Weeding. It is advised to keep the tuberose field free from weed otherwise they compete for nutrient and moisture with the main crop. Since the tuberose is in the field for full 1-3 years and different kind of weeds appear season wise. Hand weeding is laborious, time consuming and expensive. Generally after each irrigation, weeding is done. Hoeing is also done to loose soil and destroy the weeds. Preplanting application of atrazine @ 3.0 kg/ha has been found effective in controlling weeds in tuberose.

Harvesting, Packing and Marketing

Depending upon the purpose and the use of tuberose flowers, harvesting is done accordingly. For loose flowers, individual flower is plucked regularly which are used for various purpose whereas for cut flower the spike are cut from the base so that longer spike are available. These operations are done in cool hours either early in the morning or in the evening keeping the harvest of spike taken for the travel of these spikes. Similarly the stage of harvest of spike is decided upon distance of travel. For local market and immediate purpose spikes are harvested when two-three rows of basal flowers are open. For distant market, spikes are harvested when basal flower just going to open.

Loose flowers are generally packed in bamboo baskets, holding about 10-15 kg flowers and are sent to the market. These are sold by weight.

The quality of cut spikes are judge according to the length of spikes, length of flowering zone, number of florets, quality of florets. These cut spike are bunched in round bundles of 100 spikes. Basal portion of the spikes wrapped with newspaper. To avoid any injury to flowers and buds, the whole bundles is wrapped in soft, white tissue paper or perforate polythene of 300 gauge. A holding solution of 2% sucrose + 300 ppm aluminum sulphate was found best for the post harvest life and quality of cut spikes of tuberose cv Shringar. Another useful combination is 1% sucrose + 300 ppm citric acid + 50 ppm calcium nitrate at 4°C for 48-72 hours to improve the vase life of spikes. These spikes can be stored at 4°-7° for 5-7 days.

Lifting and curing of bulbs. At the time of complete maturity growth ceases completely, leaves become yellow and dry out. At this stage irrigation is with held and after two weeks, clumps of bulbs are dug out from the ground. The bulbs are separated out and grouped according to their size. Cleaned and graded bulbs are stored on shelves for curing. The position of these bulbs are changed after two weeks to prevent fungal attack and rotting. Then bulbs are stored in cool and shady place at about 18°C for four weeks before planting which result in production of good quality of bulb and flower spikes of subsequent crop. Longer storage at 30°C or above before planting advances flowering but quality of spikes and bulbs deteriorates.

Flower and concrete yield. Flower production depend upon climatic conditions, soil type, its fertility level and agronomic practices followed, etc. and hence, different level of production has been reported from various parts of India. Under optimum conditions single tuberose yields about 5,00,000 flower spikes/ha or 10.5 tonnes/ha of loose flower. In a cycle of three years, first two years flower yield is high and in third year it is comparatively lower. In one year about 9.2 to 10 kg concrete/ha can be obtained and in turn 1.75 kg of tuberose absolute can be obtained.

Method of extraction

The simple method of steam distillation for extracting the volatile oil of tuberose can not be adopted as tuberose do not yield the essential oil on steam distillation. Hence, the solvent extraction method is employed. Freshly plucked fully opened flowers in cool hours of the day (before 8.00 a.m.) are used. The purified hexane is a colourless liquid with a boiling point of 60-80°C. The flowers are immersed in purified hexane for 30 minutes. The complete extraction of the perfume from the flower can be ensured by slowly rotating the extractor having flowers and solvent for about 20 minutes. After the complete extraction of the perfume from the flowers, the solvent is filtered and concentrated at a constant temperature at 70°C. At this temperature solvent evaporates and condensed for reuse, leaving the perfume and other plant waxes. To purify further the remaining vapours of solvent containing perfume, wax and pigments is distilled in a vacuum distillation unit where the complete removal of solvent takes place, leaving floral concrete. In any unit, four components are there : (1) extractor, (2) evaporator, (3) condenser and (4) vacuum distillation.

Important Insects, Pests and Diseases

Aphids. These are very small size insects feed on flower buds and on growing tips. These can be controlled by spraying Imidacloprid 200 SL 0.4 @ ml/l or Thiamethoxam 25 WG @ 0.54 g/l of water.

Thrips. These feed on leaves, flowers stalks and flowers and suck the sap and ultimately damage the crop. These can be controlled by spraying Ethion 50 EC @ 2 ml/ l or Triazophos 40 EC @ 0.5 ml/l of water when the attack starts and repeat after 15 days if necessary.

Red spider mites. Yellow stripes and streaks appear on the foliage due to attack of mites. Spraying with Ethion 50 EC @ 2 m/l or Triazophos 40 EC @ 0.5 ml/l of water has also been effective on controlling mitles.

Root knot nematode. It is caused by *Meloidygyne incognita*, and has been observed that root knot nematodes reduce crop yield considerably and make the plants highly susceptible to the attack of *Fusarium oxysporium* sp. *dianthi*. Various methods like crop rotation, treatment of bulbs before planting (dip bulbs in 0.55 Nuvacron 40 EC for half an hour) or application of Carbofuran 3 G @ 1 g/plant near root zone followed by irrigation or treatment of soil with Basamid @ 40 g/m^2 or addition of neem cake @ 1 ton/ha at planting can control root knot nematode. Recently it has been reported from IIHR, Bangalore that neem based formulation of Pochonia chlamydosporia, (verticillium chalamydosporium) (40 g @ 10^6 CFU/g) and Trichoderma harzianum (40 g @ 10^6 CFU/ g) per 2 × 2 m sized plots before planting and subsequently once in 4 months decreased the incidence of root knot nematode and wilt disease complex significantly. These treatments increased the number of flotes per spike and spikes/plot, significantly.

Besides, the use of nematode free planting material and hot water treatment of bulbs at 57-58°C for 30 minutes is recommended. Further, the population of nematode in the field can be kept under check when intercropping with marigold is followed.

Diseases

Botrytis spots or leaf blight. This disease is caused by *Botrytis elliptica* and causes serious damage to the leaves. It can be controlled by spraying Chlorothalonil 75 WP @ 2.5 ml/l of Iprodione 50 WP 2 ml/l of water.

Alternaria leaf spot. It is caused by *Alternaria polyanthii*. Faint concentric rings appear on mid rib of leaves but rarely near the margin of leaves. The leaves become necrotic and dry up with the infection. It can be controlled by spraying plants with 0.2% of Dithane M-45.

Stem rot Sclerotial wilt (*Sclerotium rolfsii*). Main symtomps are the appearance of spots of green and brown round spots on the leave and leaves become yellow and ultimately dry up. The fungus ultimately attacks the roots and infection gradually spreads to bulb and collar portion of the stem which cause rotting. Drenching soil with Bavistin (0.5-1%) is effective in controlling the disease. If necessary, repeat at the interval of 20 days till it controls.

13

Lilium

The name Lilium has been derived from Greek work 'Leiron' which means as true. Lilium belongs to bigger group of Lilies. In general, Lilium symbolizes humility and devotion. Lilium are commonly known as 30th anniversary flower. They are also associated with funerals and thus also symbolize departed souls that have been restored to innocence after death.

Lilium is a herbaceous plant and produces often fragrant, very big flowers measuring about 10-15 cm on long stalks in different colours which symbolize differently. White colour of Lilium represents purity and modesty whereas pink colour symbolizes admiration and infatuation. The beautiful red Lilium conveys the message of love whereas yellow colour stands for happiness and cheerfulness. A harmonious blend of white and orange conveys the feelings of respect and honour.

Origin

Most of species of Lilium are native to temperate northern hemisphere *i.e.* extending to northern sub-tropics *i.e.* Europe, across Asia to Japan, India and East of Indochina, Philippines, Southern Canada through much of United States. Overall, the species are distributed in North America (24) and Europe (12).

Classification of Lilies

Lilies are classified into 9 groups according to their species.

Division I. Asiatic hybrids—This group is derived from central and east Asian species. The plants of this group are of medium size, upright and flowers outward and unscented. This group has been evolved by hybridizing different species like *Lilium amabile, L. davidii, L. maculatum, L. dauricum, L. hollandicum, L. concolour, L. leichiteni, L. putmilum* and *L. tigrinum*. These have extended range of flowering period. Leaves are narrow and upward growing. Flower colour is orange, red, yellow, etc. Bulbs are generally small, of 10-16 cm size. Important export cultivars of this group are : Connecticut King, Paradiso, Elite, Polyana, Prato and Solemio.

Division II. Martagon hybrids—These hybrids are derived from *L. martagon* and *L. hansonii*. The flowers are nodding type with petals strongly recurved.

Division III. Candidum hybrids—In this group, hybrids are of *L. candidum* and several European species.

Division IV. American hybrids—These are mostly taller growing types derived from *L. paradpinum*. Many are clump forming and rhizomatous rootstocks.

Division V. Longiflorum hybrids—The species and sub species of this group are cultivated. This group is preferred mostly for cut flower growing.

Division VI. Trumpet lilies—In this group, mostly hybrids of *L. regale* and *L. aurelians* are found. The flowers are trumpet shaped and facing outward or downward. The flowers are strongly fragrant especially in the night.

Division VII. Oriental hybrids—This group has been derived from *Lilium aurantium, L. speciosum* and *L. rubellum*. These are late flowering. Leaves are broader and almost parallel to ground and flowers are mostly white and pink. Bulbs are large 16-22 cm. Important Oriental varieties are Star Gazer, Marcopolo and Casa Blanca.

Division VIII. Other hybrids—This group is comprised of other garden hybrids.

Division IX. All natural species and all naturally occurring forms are included in this group.

Breeding of new types

The necessity was being felt for the last twenty years to breed new types suitable for cut purposes, for pot plants purposes, for park, garden and landscape decoration. Varieties developed from *L. longiflorum* and Asiatic hybrid like LA and *L. longiflorum*, oriental type LO and between oriental and trumpet species have been developed. This has increased the availability of cut and pot cultivars.

Soil

The good soil for lilium cultivation is porous, sandy to loamy and having good organic matter with good water drainage system. Lilium is also pH specific. Asiatic lilies prefer pH between 6-7 whereas oriental hybrids like little acidic soil pH *i.e.* 5.5-6.5.

For growing lilium in pots, adequate growing medium containing soil : peat moss 1 : 1 or soil : peatmoss : perlite 1 : 1 : 1 is considered good.

Climate

For growing lilium congenial temperature conditions and partial shade is required. Most suitable temp for growing lilium, is 15-25°C in day, and 12-18°C in night. Partial shade (40-50%) is ideal. If it is grown in plastic houses, good aeration/ventilation is required for proper growth of lilium plants.

Area and production

The cultivation of lilium has spread in hilly regions of India. These areas are highly suitable for flower as well as bulb production. However, bulb-production has not yet started commercially.

In India, due to limited procurement of bulbs from Holland, very limited cultivation is being done in different states. Normally, it is being grown in Himachal Pradesh, Haryana, Maharashtra, Karnataka, A.P. and North East hill states like Mizoram, Arunachal Pradesh, Nagaland, Manipur and Sikkim.

Data is available only for two states *i.e.* Himachal Pradesh, where it is being grown in around 20 ha and Haryana, where it is being grown in 5-6 ha only. The total value of the crop is ₹ 4.5 crores.

Varieties

There are number of varieties grown and every year these varieties get changed due to their availability in Holland. Most common varieties are as follows.

Asiatic varieties. Brunello, Tresor Bach, Minorea, Flestone, Literawen, Party Diamond, Courier, 8. Hyde Park, Nashville, Prato, Pollyana, Mercedes, Noracento, etc.

Oriental varieties. Mothers choice, Rubeto, Rialto, Sibenia, All Choice, Snow Petrol, Ravena, Purple Palace, Asterium, Star Gazer, Marcopolo, Casa Blanca etc.

Oriental Trumpet (O.T.) hybrids : Ice Crystal, Justina, Martegobay, Santender, Sappora, Torvengo, Viviana, Zambasie, etc.

Propagation. Lilium is easy to propagate. It is propagated through seeds, bulbills, bulblets, and bulbscale. For breeding new varieties, after hybridization, mature seeds are collected and raised into plants. It is very lengthy and cumbersome method but is very rewarding. Propagating lilium in vitro has many advantages but huge investment and skill is required. The two methods, bulb scales and stem bulblets are chiefly used in commercial propagation.

Preparation of field

The field should be well prepared by deep ploughing about 60 cm. So that soil become porous and well drained. A good amount of organic manure should be mixed to make soil rich.

Planting of bulbs

In plains and low hills, bulbs are planted from October to November whereas in mid hills including Bengaluru, two plantings are done first in February-March and second planting in July-September. In high hills, planting is done in March-April. The basal roots of bulblet penetrates in the soil upto a depth of 10-15 cm. The planting distance or density per m^2 depends upon size of bulb and group it belongs which is generally as follows :

Oriental bulb (cm)	16-18	18-20	20-22	<22
Bulbs/m^2	40-50	35-45	30-40	25-35
Asian bulb (cm)	10-12	12-14	14-16	16
Bulbs/m^2	65-90	55-80	45-70	40-65

Weeding and Hoeing

Normally, 3-4 weedings are required to remove the weeds by hand and it does hoeing also. Chemical weedicides also can be used depending upon season and location.

Fertilization

Lilies do not require feeding during first three weeks but after that they need to be fed properly to obtain good quality flowers and bulbs. The application of 20 g NPK per sq. m. in the ratio of 30 : 20 : 20 is required to obtain good quality flowers and healthy bulbs. Liquid feeding can be carried out by dissolving NPK in the ratio 10 : 20 : 20 or 10 : 25 : 25 @ 2 g/L at weekly intervals.

Irrigation

Lilium needs uniform moisture level especially during first three weeks after planting when the shoots are 7.5 to 15 cm tall. Regular light irrigation is required to produce healthy crop. Avoid the wetting of foliage.

Harvesting and Yield of Flowers

Stems are normally harvested when first or at least five buds are just opening and showing colour. Spikes are cut about 15-20 cm above the ground. The remaining lower portion of the plant is allowed for development of bulbs. Lilies spikes are harvested in the morning to avoid desiccation. Average yield of lily flower is one stem per plant.

Insects, Pests and Disease

Lilium is attacked by various insects, pests and diseases. So, to raise healthy crops, these should be controlled by spraying suitable pesticides and fungicides.

Insects and pests. Aphids and thrips suck the cell sap and disfigure the plant. These can be controlled by spraying 0.2% Dimethoate 30 EC or 0.04% Imidachloprid 200 SC. Mites can be controlled by dusting wet table sulphur.

Important diseases. Lilium is affected by various types of moulds which results in rotting of bulb and production of quality crop gets reduced. Moulds like grey (Botrytis cinereae, Botrytis elliptica), soft bulb rot, Fusarium bulb rot, root rot (Pythium splendens) and bacterial soft rot, etc. can be controlled by dipping bulbs in 0.2-0.3% Bavistan solution for 30 minutes.

Viral mosaic disease. To check the spread of viral disease, vectors like aphids, thrips etc. should be controlled in time as suggested earlier.

***

14

Dahlia

Dahlia is a beautiful perennial plant and relatively easy to grow. This is one of the most important garden flower owing to a large variation in plant height (20-150 cm), flower size (5-30 cm) and wide range of colour of flowers varying from white, cream, yellow, orange, scarlet, pink, mauve, purple, red, deep crimson to chocolate and various intermediate shades.

Origin and Etymology

Dahlia is native of Mexico and was discovered in 1525 by Spaniards. It was introduced into Europe in 1789 and was flowered successfully. The name dahlia was given by Carl Linnaeus in the honour of his late student Ander Dahl—the author of "Observations Botanicae". Some species are native to other South American countries like *D. australis* to Guatemala whereas *D. coccinea* and *D. imperales* to central America. Britishers introduced Dahlia in India.

Botany. It belongs to the family Asteraceae (compositae). There are about 42 species of dahlia and several hybrids grown in gardens. Modern varieties of Dahlia are Octoploid. Seed setting is not a problem. Honeybees are good pollinizers.

Classification of Dahlia

According to flower size and shape of petals, it has been divided into 14 groups by Royal Horticulture Society, England.

Group 1. Single flowered dahlia (Sin)—Flower has a prominent central disc with a single outer ring of florets which may be over lapping, encircling it, rounded or pointed tip of petals. The very famous variety of this group is Twgning's Eight.

Group 2. Anemone flowered dahlia (Anem)—Disc is absent because central petals of flower consist of densely elongated tubular florets, longer than the disc florets while the outer petals have one or more rings of flat ray florets. Boogie Woogie is awell-known variety of this group.

Group 3. Collertte dahlias (Col)—There are large flat florets forming a single outer ring around a central disc. There is an overlapping of smaller circle of florets close to the center which look like a collar. The well-known varieties are Star Sisters, Lilian Alice and Apple Blossom.

Group 4. Waterlily dahlia (WL)—Flowers are fully double with slightly curved or flat florets and very shallow in depth compared with other dahlias. Cameo and Delhi Sun Set Pink are good varieties of this group.

Group 5. Decorative dahlias (D)—Flowers are double, ray florets broad, flat, involute, slightly twisted and usually bluntly pointed. Central disc is not visible. Berliner Kleene, Woodland Merinda are good varieties of this group. In India, this group is very popular, and it is grown in pots for exhibition purposes. It is further divided into three subgroups according to their size of bloom. Important subgroups are as follows :

Giant decorative. The size of bloom is 25-30 cm. Important varieties are : Kenya (different colours are available), Powder Puff (Pink), Kelvin Rossey (Pink), Prime Minister (Purple and White), Yellow Monarch (Yellow), Kennedy (Magenta).

Large decorative. The Size of bloom is 15-20 cm. Important varieties are : Amrita (Pink), Croydon Apricot (Coppery Orange), Oxford (Gold), Cabaret (White and Red), Krishna (Red), Challenger (Red).

Medium decorative group. The size of bloom is 10-15 cm. Important varieties are: Delicate (Lavender), Barbara Marshal (Red), Vineyanand (Yellow and White), Eternity Sport (Orange, Red and White), Trio (Bicolour), Dorothy Stanton (Red), Silver Wedding (White).

Group 6. Ball dahlias (BD)—Flowers are fully double, ball shaped or slightly flattened. Ray florets are blunt or rounded at tips. Margins arranged specially, involute for at least seventy-five percent of the length of florets. Flowers are bigger than pompoms. Very famous variety of this group is Barbarry Ball.

Group 7. Pompom dahlias (Pom.)—Double spherical miniature flower made up entirely from florets that are curved inwards (involute) for their entire length resembling a pompom. 'Small World' is a famous variety of this group. Other varieties are Golden Star (Yellow), Kasturi (Bicolour), American White (White).

Group 8. Cactus dahlia (C)—Flowers are double, ray florets are pointed with majority of them revolute (rolled) over more than fifty per cent of their longitudinal axis and straight or in curved florets are narrower than semi cactus. Jaldee Joker is a good variety.

Group 9. Semi cactus dahlia (S-C)—Flowers are double and ray florets are very pointed. These ray florets revolute for greater than 25 percent and less than 50 percent of their longitudinal axis, moreover they are broader at the base and straight or incurved and also spiky in appearance. Very famous variety is Micks peppermint.

Group 10. Miscellaneous Dahlias (Misc.)—These types do not fall in any above category. Flowers are single and small, and the arrangement of petals is different. Moon fire and Carolina Burgundy are good varieties of this group.

Group 11. Fimbriated dahlia—Ray florets are evenly split or notched in two or more divisions uniformly throughout, the bloom, appearing like fringed effect. The petals may be flat, involute revolute, straight, incurving or twisted. Known variety of this group is "Marlene Joy'.

Group 12. Single orchid (Star) dahlia (Sin O)—In this type there is a single outer ring of florets surrounding central disc. The ray florets are either involute or revolute. The important variety of this group is Alloway Candy.

Group 13. Double orchid dahlia (Db/O)—Central disc is absent, however, flowers are double with triangular centers. The ray florets are narrowly lanceolate and are either involute or revolute. The famous variety of this group is 'Pink Giraffe'.

Group 14. Peony-flowered dahlias (P)—Flowers are large with three or four rows of ray florets that are flattened and are arranged irregularly. The ray florets are surrounded by a golden disc similar to that of single dahlias. Bishop of Llandaff is an important variety of this group.

Climate

Dahlia being native to Mexico and South American countries which are tropical in climate, therefore, it prefers mild warm weather for growth and flowering. However, it can tolerate low temperatures, but frosty conditions should be avoided.

Selection of site

The site for dahlia growing should be sunny at least for 6-8 hours a day. They also need protection from strong wind, therefore, plant them in protected place. The large flowered dahlias become heavy when mature and can snap off by strong wind.

Soil

Dahlia thrives well in rich, deep, well-drained soil. The pH level of soil should be around 6.5-7.0 or towards slightly acidic side. In heavy or clayey soil, bulb development gets hampered. In such soil, mix sufficient quantity of compost, vermi compost, sand, leaf mould, barks, etc., to make the soil loose and friable.

Propagation

Dahlia is easily propagated through seeds, division of tubers and terminal cuttings.

Seeds

Abundant seed set is there in dahlia and seeds easily germinate and healthy plants are developed very easily.

Division of tubers

After flowering and maturity of plants, tubers are dug out from soil and after giving fungicidal treatment, tubers are stored in sandy soil, and water is sprinkled regularly to keep soil moist till next planting season. In plains of India, tubers are dug out in the month of March-April and are stored till September. Whereas, in hills, tubers are dug out in the month of November and re-planting is done in March-April.

Cuttings

Terminal cuttings strike roots well in sand and it takes 2-3 weeks for rooting and root development. Care should be taken that terminal cuttings are solid and 7-10 cm long and having at least four nodes. Two basal nodes should be in soil and regular watering will help in profuse development of roots. Cuttings can be planted in beds or pots.

Planting

Dahlia is a versatile plant and it grows very well in beds and in pots. Plants raised through seeds or cuttings are very successfully grown in beds and pots. Beds of appropriate size *i.e.* 5 × 3 or 6 × 3 m can be made depending upon area available. Distance for bed planting can be about 60 cm whereas for large Dahlias. Row to row and plant to plant distance can be 60-90 cm. Pots of 30 cm size of earth, cement or plastic can be used for small or medium dahlia.

Fertilization

For better growth and flowering, low level of nitrogen is required. For balanced feeding, liquid fertilizers should be used. Beneficial effects have been observed by applying 50-100 gram of fertilizers in 5 : 10 : 10 or 10 : 20 : 20 strength fortnightly or monthly depending on general fertility level of soil. In case of specific nutrient deficiency, particular fertilizer can be used. Once or twice, micro-nutrient mixture can be sprayed for better colour development.

Weeding and hoeing

Regular removal of weeds and timely hoeing will help in growing robust and healthy dahlia plants. Manual labour or weedicides can be used if area is large.

Staking

Tall dahlias need sufficiently strong stake to provide support when the plant is a month old. It should be tied properly at two/three places. Care should be taken that support is long enough, so plants get full support if there are more number of branches. Each branch should be provided with separate support.

Pinching and disbudding

These two steps are very necessary to get desired effect. If flowers of somewhat small sizes are required, pinching is done before central bud appearance. On the contrary, to get a bigger size single bloom, side buds are to be removed regularly.

Collection of seeds

Dahlias produce seeds abundantly. They should be collected from fully dry flowers on the plant itself. These plucked flowers are further dried in shade. Seeds are separated from flowers by pressing with small roller. Seeds are stored in clean bottle till next use.

Storage of tubers

Tubers are lifted from the ground when plants have fully dried. After drying in shade and removal of dust, tubers are dusted with fungicide and stored in dry and cool place.

Insects, Pests and Disease

1. **Aphids** (black and green fly) attack the young foliage and cause twisting of foliage. They suck cell sap and spread viral disease. It can be easily controlled by spraying Dimethoate 30 EC @ 0.2% or Imidachloprid 200 SC @0.04%.
2. **Capsid bugs** are also sucking insects and feed on foliage and flowers buds. Regular spraying of any insecticide from above will also control it.
3. **Slugs and snails.** Dahlia Plants are attacked by slugs and snails. These can be controlled by applying 'Slug go' which is an organic compound. Because, Slug Go after use, breaks down into harm less fertilizer (iron phosphate), it attracts slugs and snails and lures them from the hiding places.

Diseases

1. **Spotted wilt.** It is the most serious disease. It causes mottling of leaf blade and black spotting. Weak stem and poor flowers are observed in advanced stage. Remove and burn any plant showing such symptom immediately as the thrips and aphids may carry the infection to healthy plants.
2. **Dahlia smut.** It generally attacks the plants in cool, damp, weather, causes light coloured spots on the leaves which later on turn brown. Spraying with any copper fungicide helps in controlling the disease.

15

Bird of Paradise

It is considered one of the best ornamental plant of tropics. It is a cousin of banana. It is an evergreen perennial plant and cultivated for its most beautiful flowers. The generic name—***Strelitzia*** is given in the memory of British Queen, Charlotte of Brig-Strelitz, wife of Gerorgia III. The specific name ***reginae*** is of Latin and means of the queen.

It is commonly called as 'Crane flower' or 'Bird of Paradise' due to the resemblance of opened flower to head and beak of bird. They are big size plants and grow up to 120-150 cm tall, with large, strong leaves and arranged in two ranks making a fan shape crown. After emerging flower from spathe, it looks like head of the bird and beak. It is strong enough for holding sun birds which pollinate the flower.

The flowers consist of three brilliant orange colored sepals and three purplish to white petals. Two of blue or white petals are joined together and form an arrow like nectary. Sun birds when sit on them to drink the nectar, the petals open the cover and their feet get covered with pollen.

Origin and species

It is native to South Africa. There are five species but the main species is *Strelitzia reginae* (The orange bird of paradise). Another important species is *Strelitzia nicolai* (White bird of paradise). There is a yellow flowered cultivar which is known as Manadela's Gold. The other species are : *S. juncea*, *S.augusta*, *S.caudata*

Area in India

In India it is being grown on very limited scale in hilly areas of Karnataka, Kerala, Tamil Nadu, H.P. and other places. Main growing areas are Hosur, Ooty, Kodai Kenal, Coonoor, Trichur (Tamil Nadu) and Mudigree (Karnataka). It is growing in about 20 ha. However, the area is highly scattered.

Climate

Bird of paradise being tropical in origin, needs ample of sun light, however, bright sunshine of afternoon may cause burning of foliage. Low winter temperature may hamper its growth. Therefore, suitable lowest temperature is around 10°C and above.

Soil

It is grown in variety of soil but most ideal soil is rich, loamy, porous, well fertilized and needs plenty of water throughout the growing season. Strelitzia needs regular supply of nutrients. For this, slow release fertilizer and compost is applied.

Propagation

It is propagated by seeds as well as division of root suckers.

Flowering and flower production

It is a slow growing plant and takes about 4-5 years to bloom. But once flowering gets started, it keeps on producing flowers for many years. Peak flowering occurs in winter and early spring. The flowers have a long vase life and remain fresh for more than 10 days.

Insects, pests and diseases

Bird of paradise is comparatively free from any major insects and pests' attack and has a low incidence of disease. However, occasionally, following insects and pests may attack, so care should be taken and before severe attack, suitable control measures should be applied;

1. **Mealy bug.** These are dull white coloured sucking pests and can be controlled by spraying polo.
2. **Scales.** These tiny insects can be controlled by spraying Rogor @ 2-3 ml/L of water.
3. **Spider mites.** Mites can be controlled by spraying Omite.

16

Protected Cultivation of Ornamentals

Different type of structures have been used time to time for modifying the environmental conditions keeping the requirement of the crops, flower forcing, raising of seedlings or propagation of plants. Though India enjoys wide range of climate conditions and various agro-climatic zones which favour the production of various kinds of flowers, seeds and ornamental plants in one or other zone round the year. However, there is a wide range of daily, weekly, monthly in seasonal fluctuations in temperature, light conditions, humidity which not only affect the quality of crops but also limit their availability for a particular period of the year. Wide climatic fluctuations are constant features of 23½ latitude towards north or south of equator line. Such conditions are not favourable for large scale productions of high value crops especially cut flowers. This necessitates the modification of the environment to the congenial limits by artificial management of environment condition. Due to increased internal and external demand of high value cut flowers, the protected cultivation is gaining popularity and Area is increasing every year. Different modified structures have been widely adopted by flower growers, nurserymen and seed growers for producing cut flowers, ornamental plants and seedlings. The area under green house in different states is as given in Table 16.1.

CLIMATIC ZONES OF INDIA IN RELATION TO PROTECTED CULTIVATION

Broadly, India can be divided into three main climatic zones from protected cultivation view point and structures have been modified in three zones, accordingly. These zones are (*i*) Wet humid zone – It is found in south and coastal areas which is characterized by frequent tropical rains and warm, temperatures; (*ii*) The central plains and desert area – characterized by moderate rains and wide fluctuating temperature *i.e.* low temperature

TABLE 16.1
Area and crops grown under Poly houses and plastic tunnels in different states of India*

States	Area ha	Crops
Maharashtra	450	Rose, Gerbera
Tamil Nadu	380	Rose, Carnation, Gerbera, Lilium
Karantaka	370	Rose, Gerbera, Carantion, Orchid, Anthurium, Calla Lily.
Andhra Pradesh	160	Rose, Gerbera
West Bengal	150	Gerbera, Carantion
Uttarakhand	100	Gerbera, Carantion, Lilium, Rose
H.P.	70	Carantion, Rose, Gerbera, Lilium
U.P.	20	Gerbera, Rose

In other states like, Punjab, Haryana, J&K, North East States, area is very limited.
*This information is based on personal enquiries.

(0°C) during winter and very high temperature (42-45°C) in summer with very low humidity; (3) Hilly area of north and northeastern India, enjoys moderate to high rainfall with freezing or below freezing temperature in winter to moderate temperature in summer.

Therefore, different geographical location, crops and purpose of growing will decide the selection of types of structures, covering material, irrigation system, heating, lighting and other basic facilities. Briefly for warm humid areas low plastic tunnels and Polyhouse with fan – pad cooling system will be highly useful, for raising crops, transplants and bedding plants. Hence, ground to ground Quonsets type polyethylene structures of 5 × 20 m are highly suitable with sides folding. Under central plains and desert areas, structures like polyhouses having fan pad cooling system will be highly beneficial for raising crops and bedding plants. In hill areas, due to prevailing extremely low temperature in winter green house made up of polycarbonate of single layer, having heating system and CO_2 enrichment facility will be useful for raising crops and bedding plants.

GREEN HOUSE

In strict sense, green houses are the structures in which optimum environmental conditions are maintained round the year so that regular supply of flowers is maintained. The definition of green house in Europe and America is different. According to European definitions green house is a structure in which little or no artificial heating is done while glass house refers to an artificially heated structure. However, in United States, green house refers to a structure covered with transparent material allowing natural light to enter in it which is necessary for plant growth. Such structures are usually heated artificially. India, being, tropical country, green houses are the structures covered with transparent material and are cooled artificially mostly with fan pad cooling system. At higher elevation heating is required in winter and only fan cooling during summer is required to raise successfully crop.

Commonly used Different Types of Structures

Before commencement of the advanced green houses, simple structures were commonly used by nurserymen and growers to modify the environment of limited area. Activity was limited due to simplicity of the nature. Amongst such structures are cold or storage pits, cold frames, hot beds, sash bed, etc. These structures are mostly used to save the crop for limited period only *i.e.,* few weeks to few months only. When conditions become favourable, these structures are removed. Normal cultivation practices are followed to raise the crops. These structures help to overcome the unfavourable weather conditions.

I. Temporary Structures

Cold or Storage pits. In early times, pits used to dig about 90 cm deep to store the pot plants or germinating seedlings, the top was covered with transparent material or glass in frame in slanting position keeping higher side in north side and slope on south side. Thus the plants were remained to much below of frost line so they were not damaged due to prevalence of low temperature.

Hot Beds. In such beds, farmyard manure is spread on the ground ranging from 30-45 cm thick for mild climate and 60-70 cm thick for severe climatic conditions. On fermentation, manure provide heat for three to six weeks.

Sash beds. Under severe adverse climatic conditions to save the crop from top, crop is covered with transparent glass fixed in frames. The frames are made up of wood. The length of frame is about 180 cm and 150 cm wide. These frames are made in such a way that one or more layers of glass can be accommodated. The usual height is 30 cm on one side and 15 cm on other side. Higher side is placed on north side so that sloppy side faces south side to get the maximum benefit of sun light. The thickness of glass is 4 or 6 mm. The thicker will be more heavier to handle but will be more durable. Under severe winter conditions double glass layer is used. Generally these frames are heated with farmyard manure, flues or electric cables.

Cold Frames. When sash beds are not heated, these are called as cold frames. These are successful in mild climatic conditions.

Forcing boxes. These are small cold frame with a single glass or glass substitute like polythene, polyvinylchloride or Rein forced plastic fiber which is over a single plant in the field during early spring. They provide sufficient heat to help in establishing them and maintain them in healthy conditions.

II. Permanent Structures

(*a*) Lean to Type. It is simple to construct and maintain and is commonly called as shed roof house. It is constructed on the south side of the building so that it get maximum sunlight during winter when it is required most. Since it is constructed against the southern wall of the house, its three sides are made with glass structure. Slope is made towards south side. Lean to types are mostly used to protect the potted plants, growing plants which require less light like ferns, foliage plants or for starting early plants or raising nursery seedlings. Being simple to construct, it has several advantages as follows.

Advantages

(*i*) Less expensive

(*ii*) Less cost of maintenance

(*iii*) Glass area is less so less breakage

(*iv*) Radiates less heat

(*v*) Highly suitable for amateur use as they are attached to the house.

Disadvantages

(*i*) Less productive

(*ii*) Receive sunlight mainly from one side.

(*b*) Even Span Type. As the name indicates such type of green houses have equal length on both the sides from the centre. These type of houses are very popular and more than 80% of this type are being constructed. The width of such green houses varies from 4-20 m and length is 50 m. More length then this result in too much walking of workers.

Larger green houses should have a larger garage type door in the centre so tilling equipment and small trucks can move in and out readily.

Advantages

(*i*) Heating or cooling is achieved uniformly and efficiently.

(*ii*) Full advantage of sunlight is achieved.

Uneven Span House. In these type of houses one side of roof is longer than the other. It makes a small angle with the horizontal and thus has a lower pitch. Such type of green houses are very popular on sloppy terrain and to utilize the maximum sun light, smaller arm kept on north side whereas longer arm is facing south side. The shorter arm is normally one third of longer arm.

Curvilinear and curved eave type. These types are highly suitable for conservatories and are features of mainly city parks, conservatories, botanical gardens and of rich individuals. Such type became part of landscape plan and improve the aesthetic of surroundings. Many of these conservatories are used for seasonal displays and exhibition of floral crops. These types are rarely used for commercial flower production but add much landscape value.

Ground to ground types. These types are invented during World War II. The frame work of such structure are galvanized pipes of various diameter varying from 1.25 cm to 4 cm and central height is kept 2.7 to 3.0 m. The length is kept 16 m to 25 m and width is 6.5 m. These pipes are fixed in the ground or fixed grouted pipe of larger diameters at the gap of 2 m from pipe to pipe. In the centre pipes are connected with main pipe so structure become rigid and stable. This structure is covered with polyethylene or rigid fibre glass. On one side gate is kept at entrance. In these type, some land of sides used to remain unutilized so after improvement new type have been made which is called as Quonset type.

Quonset Type. In these type pipes are straight from the ground about 2 m so that upto the last end of ground is utilized. Now a-days these are most popular and being fabricated for growing cut flowers. Length, width and height are the same as above.

Above types can be made separately (single) or detached type or many are attached with each other. There are certain advantages of each system.

Advantages of detached green house or single units

(*i*) It is easier to programme and maintain the temperature to meet specific requirement of crop.

(*ii*) It is easier to ventilate without exposing the plants to blast of cold air.

(*iii*) Fewer ventilators and less ventilating machinery is required.

(*iv*) Light distribution is uniforms.

(*v*) Care of the house is easier.

(*vi*) Appropriately spaced detached green houses are easily cleared of snow without damaging structures.

Disadvantages of single houses

(*i*) The greater height of a single detached house requires more energy of heat the large volume of contained air.

(*ii*) More height is also exposed to winter winds and hence, greater loss of heat than low houses.

(*iii*) They have higher initial cost.

Now-a-days for commercial production larger area is covered under green house both detached or many are attached at single location. Such units should be economically viable. By setting up such units, over head expenses are reduced drastically. There are certain considerations which should be kept in mind before starting up such commercial activities.

Consideration for setting up green houses

After making decisions of geographical location, proximity of market, fuel supply and investment in land for setting up green house, following points should also consider about the location of green house.

(*i*) It should be sunny and sunlight should not be obstructed at any time during the day.

(*ii*) Soil should be well drained and area should be free from flooding.

(*iii*) It should not be exposed to cold, bleak, winds otherwise fuel bill may escalate for this a wind break will be very effective in protecting from winds.

(*iv*) The land should be well leveled or gently sloping toward south or south east side.

(*v*) There should be regular water supply.

(*vi*) There should be sufficient land for future expansion.

Other considerations

Arrangement of houses. It depends upon the number of houses, size of range, and purpose for which it is to be used, the arrangement should be such that it saves the movement of workers and all the houses get sunlight whole of the day without interruption. For this direction of the houses to be constructed, pitch of the houses and distances between two houses are important factors.

Direction. The single houses are constructed in N-S-direction enjoying the sunlight whole of the day on both sides equally. However, during afternoon when sun light strikes obliquely, much heat and light is reflected back in the atmosphere. Frame works will also caste lot of shade, on the contrary, when houses run east and west side, the sunlight enters in the green house from the sides. In the morning and afternoon, it strikes roof obliquely. During the middle of the day, light strikes at right angle so green house get maximum light and heat, however, southern part of green house will be comparatively more warmer then north side.

Normally attached houses run north and south so all the houses get full advantage of light and heat whereas single houses run east west side.

Connected Houses. Many single even span style houses are joined together so they make ridge and furrow house. In such type houses there is no side walls between individual units. The number of houses is based on requirement of crop or crops to be grown. According to the construction, connected houses are as follows.

Saw-tooth houses. As the term indicates, these type of houses consists of a series of lean to green houses connected together. The elevated portion of saw tooth opening is directed away from prevailing winds to aid its ventilation generally, these are covered with plastic sheet.

Multi chamber. These types are modification of Quonset type detached green house and called as barrel vent green house consists of Quonset type structure set on side wall. The height of side wall is about 2.10 to 3.0 m high.

Shade Net Houses. In tropical climatic countries where sunlight is very bright and to reduce it shading nets of different colour and density used according to the need.

Normally green coloured shade nets are very popular followed by black or white. Saran cloth made up of white plastic is also used. These are not a true green houses but are very useful where frost is not a problem. Green foliage crops are commonly green.

Lath House. These are generally made up of wood or wood with a metal frame work. The degree of shade is determined by the distance between two laths which is generally kept about 5 cm. The width of individual lath is 5-7.5 cm and length is about 3.0 to 3.50 m. In such structures green foliage crops, ground covers or other crops are grown which require partial shade or broken sun light.

The advantages of detached type are disadvantages of attached type and *vice versa*.

Low Tunnels. These tunnels are 1-1.6 m in height covered with clear polyethylene. Some time very low height tunnels of 30-60 cm in height only are made keeping the height of crop and duration into consideration. These have several advantages over polyhouses and green houses and have been found very useful being simple and of low cost in nature. These are mostly used when adverse weather conditions are limited to short duration. Such conditions are found in north-west plains from 15th Dec. to 15th March or during May to August. During winter months polyethylene of 100 mil is used whereas during summer months, black, white or green netlon is used to cut the sunlight and to provide partial sun light. The selection of covering material depends upon months and duration. Structure is made of steel rods of 4 mm thickness which can be easily bended to give a desired shape. These are fixed in the ground at distance of 1-2 m, after use, the structure and covering material is stored for the next time.

Construction of green house

After making decision of types of green houses to be constructed various considerations are to be made regarding selection of material for frame work, covering, heating, cooling, lighting, ventilation, fertilization, methods of raising crops, etc.

Foundation. It provides support to the structure. Small green houses may have shallow foundation continuously around perimeter of the building. A typical foundation consists of concrete footings located below the frost line. The spacing between footings is 2-3 m. The posts are embedded in concrete piers which rests on footing. This system will provide the main support for the green house. The curtain wall of desirable height is made up of bricks.

Frame work. The frame of green house is to provide support for covering material and to other accessories like water pipes, electrical fittings ventilators and other. Thus the frame work of green house should be light, strong but casts less shade, and has long life. Earlier wood was used for framing the green house. It has several advantages like bad conductor of heat, cheap and easy to use but frame work of wood will cast lot of shade and its moving shadow affect the performance of crop.

On the contrary, cold roll steel or aluminium is used now-a-days. It has all desirable characteristics like it is light, strong, casts less shade and has long life, but being bad conductor of heat, it became hot or cool quickly.

Pitch of Roof. The angle of divergence from the horizontal will decide the degree of slant and is called as pitch of roof. The glass or other transparent material of green house will allow light to penetrate through it but some light will be reflected back in the atmosphere. This will depend upon the angle of incidence which makes the light over the transparent material.

The amount of light lost by reflection is inversely proportional to the angle of incidence. Thus, if the sunrays fall upon the roof at right angle, little or no light is reflected back. Similarly, if light falls at low angle, loss of light by reflection will be more.

***TABLE 16.2**

Loss of sunrays when light strikes the covering material at different angles

Angle	% loss by reflection
60	2.7
50	3.4
40	5.7
30	11.2
20	22.2
15	30.0
10	41.2

*Nelson, P.V. 1991 Green House operation and management. Prentice Hall Inc. New York.

Thus, it is vivid that maximum amount of sunrays may be conserved by a roof when it makes a 90° angle by the falling light. During winter it becomes much more important. To make a such roof, it would need a pitch of 65 degrees. The roof with such a high pitch of angle will have certain disadvantages like (*i*) It will be too high for house having width of 3 or 3.5 m. (*ii*) It well be very expensive to build and maintain (*iii*) It will expose too large surface area for radiation of the space covered.

To overcome above disadvantages if pitch is reduced to 35 degree, the sun rays will strike the roof at an angle of 55 degree then loss of light through reflection will be only 2-3%. Such roof has not many disadvantage. Roof having a pitch of 26° or less, they are not satisfactory on accounts (*i*) loss of light through reflection is more (*ii*) snow does not clear off efficiently (*iii*) condensed drops of water will fall on the plant like bullet which will impair the quality of product.

Determination of the pitch. The pitch of the house can be calculated trigonometrically when the width of the house and the height of the ridge is known. Half width and height will make a triangle so the pitch can be determined with the formula

$$A = \frac{b}{a}$$

where A is pitch, *a* is half width and *b* is height. However, it is more desirable to determine the length of rafter to make a roof of a house on given pitch. This may also be trigonometrically solved be applying a formula

$$A = \frac{\text{Half width } (b)}{\text{cosine of pitch angle } (c)}$$

whereas A is the length of rafter, (*b*) is half width and (*c*) is cosine of pitch angle.

ROOT MEDIA AND ITS PASTEURIZATION

Root Media

Healthy root media is very important for the successful raising of green house crops. It is a reservoir of nutrients, hold water, gases and provides anchorage or support to plants. In nature perfect root media is not available. One of root media may be good for aeration or in water holding capacity or nutrients. Therefore to make complete media two or three media are mixed in certain proportion so that desirable media can be obtained. Apart these, four main functions of media, there are other characteristics which should be kept in mind while selecting or making media specially for pot culture. These are as follows :

Stability of organic matter. Pot plants raised in green house are held for two to four months (flowering and foliage plants) or one to two years. (shrubs, trees, climbers, etc.) and then they are transported for market. If organic matter of media decomposes at faster speed, there will be reduction in significant volume which is highly harmful to the plant specially small plants. Moreover, decomposition of organic matter will lead to the formation of finer texture which will reduce aeration. Addition of straw or saw dust is not desirable because they decompose rapidly, however, where volume of medium is large enough, it permits shrinkage. To avoid such problem addition of sphagnum peat moss, or coarse powdered bark is desirable.

Bulk Density. It provides support to the plant. Medium should be sufficiently heavy to prevent a potted plant from falling over due to weight of the plant. A mixture of sphagnum, peatmoss and perlite is sufficiently heavy. But after some time when larger plants have taken up water from such media they topple very easily when are handled during transportation. An acceptable range for bulk density of potting media is 640-1200 g/d m^3 just after watering at container capacity.

Carbon : N ratio. The amount of N in root medium in relation to organic matter decides the availability of N to crop for its growth and ultimately productivity. Normally when ratios is 30 C : 1 N, decomposition of organic matter will be ideal whenever this ratio exceeds, N already present in root medium will be utilized by micro organisms rather than by plant. Crop will be deficient in N thus N should be added. The C : N ratio for straw is 1000 : 1, whereas for tree bark is 300 : 1.

Moisture retention and aeration. The primary function of media to make balance between moisture and aeration. This is also necessary to ensure suitable long interval between watering and to provide adequate aeration. A heavy clay soil results in a high available water content at the expense of aeration. A larger portion of sand is required to improve aeration. Addition of vermiculite, peat or sphagnum is also good. In pot culture depth of pot also govern the retention of moisture. In shallow or deeper pots capillary force holding water may be equal but in large pots, gravitational force pulling water out of pot becomes greater as the depth of pot is increased. Thus well formulated root media with high aeration and water retention values are suitable for wide range of pot sizes.

pH. The pH level of root medium controls the availability of nutrients to the plants. Most of green house crops are grown in slightly acid pH range of 6.2 to 6.8 in soil based media and at 5.4 to 5.6 in soil less media. Some crops are acid loving and they grown happily in a strongly acid pH range of 4.5-5.8. Acidic media are moss and pine bark. Sand, rock wool and perlite are neutral. Vermiculite and hard wood bark are alkaline. Field soil can range from acidic (3.5) to alkaline (8.5). So depending upon the crop requirement mixture should be made.

Cation exchange capacity (CEC). It is a measurement of the magnitude of fixed negative electrical charges and is measured as multiequivalent per 100 cubic centimeter (me/100 cc) of dry root medium. A level of 6-15 me/100 cc is considered most desirable for green house medium. High levels are desirable because it will act as good reservoir of nutrient. Thus in preparing root medium it is desirable to include a component with a high C.E.C. Root media components and fertilizers have negative electrical charged which will attract and hold positive electrical charges.

Components of root media. There are many natural and artificial materials available by mixing them in pertinent proportions, a desired root media with desirable characteristics can be prepared. Selection of components is based on the required functions, cost and availability.

Field soil. Most desirable soil for green house crops is having high degree of structure and loamy texture. After deciding the location for setting up green house range, availability of good soil texture is an important factor. Good structure is further improved by growing mixed corps desirable leguminous crops for 2-3 years which improves the soil fertility and improve the structure as well. Decomposition of root residues builds up humus. These soils when used for growing the green house crops result in better production.

Normally flower producers do not change the soil every year so it is important to make a good soil in the beginning, on the contrary, pot growers need good soil regularly. However, mostly pot growers use artificial media for raising the flower or pots plants.

Sand. It is mixed in soil to improve drainage and aeration of media. Mostly river sand or coarse sand is used. It should be washed before use. Sand is mostly free of clay, silt and organic matter.

Sphagnum Peat moss. There are different types of peat. It is light tan to brown colour and least decomposes. It is formed from **Sphagnum** or **hyphnum moss**. It decomposes slowly and contain about 0.1-1.4% N. It is good water holding capacity (holding upto 60%) and acidic in nature (3.0-4.0 pH) and needs the application of lime to bring pH up. But when it is used with vermiculite no liming is needed. In India, now-a–days coco peat is being used on large scale.

Vermiculite. It is mica like silica mineral and is very light in weight and thus is highly desirable to use for pot plants. It contains good water holding capacity. Mostly expanded form is used and it contains about 16 times water holding capacity. Drainage and aeration qualities are also good because of large pores between particles. It is charged negatively and contains potassium, magnesium and calcium. The reaction of various vermiculite varies from slightly alkaline to alkaline but there is no problems in mixing with acidic components like moss or pine bark but generally not mixed with soil.

Perlite. It is siliceous volcanic rock when crushed and heated to 980°C expands to form white particles with numerous closed air filled cells. Water will adhere to the surface of perlite but it is not absorbed into perlite aggregates. Perlite is sterile, chemically inert and has a negligible CEC (0.15 m e/100 cc) and is nearly neutral with a pH value 7.5. Thus, it does not appreciably affect the pH level of root media. It is good substitute for sand for providing aeration in root media but may be little more costlier than sand. It is better than sand in root media because of its light weight.

Synthetic root media

***(i)* Rock wool.** Horticultural rock wool is formulated to a prescribed density to provide air and water holding requirements of plants. Insulation rock wool is not good for plant growth. It is produced by burning coke, basalt lime stone and possibly slag from iron ore. The mixture is heated at 1600°C temperature and it melts in furnace. A steam flows onto a high speed rotor. Droplets from rotor when it falls on heated mixture it lengthen into long fiber. The fibers are sprayed with binding agent in an air stream which cools down and carry them into conveyer belt and are deposited. The cubes are made of this wool of different size with holes or without holes in different sizes. Keeping the duration of crop in view, the lengths can be 75-100 cm whereas the width may vary from 15, 20, 30 to 45 cm while the height is 7.5 cm.

Polystyrene Foam. Commercially it is known as Styrofoam , Styropor and styromull. It is very good for improving aeration, and light weight to root media. Further it does not absorb water and has no appreciable CEC. Being neutral it does not affect pH. It is a white synthetic product containing numerous enclosed cells which are filled with air and available in beads or in flakes about 1-13 cm size. Larger particles are used in benches or for epiphytic orchids.

Pasteurization

Prevalence of mild temperature conditions in green house and raising of a crop for longer duration or raising of few crops repeatedly accentuate the disease problem by providing a continuous host on which disease organism can build up. Thus, necessitates the pasteurization of root media. It has become a standard practice virtually for all green house and is done on a yearly basis. For shorter duration crop it can be frequent. Generally summer time is good for pasteurization as there is relatively low production. Apart disease (fungus) control, pasteurization also controls nematodes, insects and weeds.

Pasteurization is done by two methods *viz.* steam pasteurization and chemical pasteurization.

Steam Pasteurization. In green house where heating during winter is done through steam, same facility can also be used for root media pasteurization. However, during this process useful bacteria also get killed and develop later on after few weeks. Now-a-days, equipment is available that will mix air with steam and adjust the temperature to a desired level below 100°C. Normally, temperature of 71°C is kept for pasteurization.

Before pasteurization rooting media should be loosened and should be sufficiently wet. This means that moisture content should be similar to the planting time. In wet and loosened soil steam passes rapidly than dry soil. Too wet soil also act as a bad conductor of heat. Steam is distributed in fresh flower ground beds through buried perforated pipes. For beds measuring 0.9 m wide one row is buried and whereas for 1.2 m wide beds two rows are buried.

Raised benches filled with a root medium may be pasteurized with or without buried steam conduction pipes. Pipes are buried at half the depth of soil. Some growers instead of buried pipes, inject steam between cover and medium. Once the cover is inflated, steam readily penetrates the loosened root media. This system is easier to set up but require long time for steam to penetrate the root media.

While during pasteurization, the desired temperature should reach of cold spot which is at the end of bench. Temperature should be measured at the point. This cold spot can be corrected by applying an extra quantity of steam at that point.

After steaming, it should be covered so that it does not escape quickly and lost in air and thus, efficiency will be reduced. Cover helps in holding the steam in close contact with the root medium and pasteurization is done effectively. There are three types of covers namely polyethylene, vinyl and neoprene coated nylon fabric. Out of these nylon has the shortest life but it is cheapest.

After Steaming Problem. There are two types of toxicity which appear after steaming *i.e.* ammonium and manganese toxicity. The organic matter of root media releases ammonium after steaming and may continue to release up to few weeks more. Ammonifying bacteria convert nitrogen of organic matter to ammonia and nitrifying to nitrate nitrogen. During pasteurization the populations of both ammonifying and nitrifying bacteria are reduced to low level and develop again after 3-6 weeks. During this period of conversion ammonia to nitrate, ammonia may result toxic to certain crop *e.g.* roses and poinsettias.

In certain soils where parental manganese is more but is in unavailable form which becomes in available form after steaming and becomes toxic to crop. The longer the soil is steamed the greater is build up of manganese, and hence, the greater risk of manganese toxicity. To avoid this soil should be pasteurized for only the length of time necessary *i.e.* 30 minutes. Toxicity results in tip burn of older leaves. Iron deficiency is also commonly caused by high levels of available manganese.

Chemical Pasteurization. Where there are no arrangements of steam pasteurization, chemicals offer a good alternative. However, these chemicals being hazardous, injurious to human health and environment but still followed. There are number of chemicals like Methyl bromide, Chloropicrin, Basamid, Formaldehyde, etc. Out of these, most commonly preferred are Methyl bromide and Chloropicrin.

Methyl bromide. It is liquid under pressure and turn to gas when released. It is available in cans of 454-680 g or in cylinder for tractor mounting. At the time of application root media should be loose for rapid penetration of gas and should have moisture content as desired at the time of planting. For benches or rooting media it should be applied at the rate 6 kg/m^3. It is applied through an applicator. A layer of polythene is placed over bench or pile of media. Methyl bromide is injurious to carnation plants so it should not be used.

It can also be applied on larger area with the help of tractor over which cylinder and polyethylene is mounted. Methyl bromide is released by plastic tubes at the bottom of chiesel (10-15 cm deep) and immediately covered by polythene. The leading edge is buried by soil and subsequent sheets are glued.

Chloropicrin. It is also called as tear gas and is popular choice for carnation plantation. It is mostly applied at the rate of 32 cc/m^2 of bench or field surface. Chloropicrin is injected into media by a hand applicator. It is covered by polythene. On large area, chloropicrin can be applied by tractor as in case of Methyl bromide is done and field is covered with polythene.

Planting of the crop should be done when injurious gas is below injury levels which can be checked by sowing presoaked seed of lettuce on limited quantity of media.

Basamid. It is available in micro granular form having active ingredient dazomet (DMTT) 98-100%. It is considered good soil fumigant for the control of nematodes, root diseases soil insects and weeds. It is suitable for fumigation of open land, media for pot plants and all kind of seed beds, nurseries and green houses. It is applied at the rate of 30-50 g/sq. m. After application irrigate the field or cover with polythene.

Under Indian condition where summer is very hot and temperature is going as high as 40-44°C, by leaving the field fellow during summer limited pasteurization can be achieved. Higher temperature can also be obtained by covering with black polythene and solarization will pasteurize the root media to certain degree.

Fertilization. Prior to invention of chemical fertilizers, nutrients were made available to the crops from the organic matter present in the soil. Annual application of organic manures provided the additional nutrients. In green houses, flower production requires more frequent application of inorganic fertilizers. Instead of dry forms their application, along with water or water containing soluble fertilizers enhances the productivity and thus found highly profitable. Further automation in fertilizers solution mixing in irrigation and automated water application reduces much expenses and increases productivity.

Now-a-days it is a standard practice to dissolve high analysis fertilizers carriers into concentrated solution. The concentrate is then proportionated by means of fertilizer injector into the water line of the green house at the final concentration desisted for crop application. Fertilizers most commonly are applied on a seven day basis or with each watering which is called fertigation. This fertigation system is getting popular in the green houses.

Preplant Fertilization. Soil analysis becomes the basis to decide about type and quantity of each nutrient to be applied keeping the pH and amount of nutrients present in the root medium. Normally, N, P, K, S, Ca, Mg and micro nutrient in desired quantity are applied. Essential micro nutrients like iron, manganese, zinc, copper, boron, molybdenum, etc. can be applied in a single application. Several commercial preplant mixtures are available for incorporation in root medium. Liquid application of micro nutrients after planting provides an alternative to dry preplant application. Most of complete N-P-K green house fertilizers contain a full complement of micro nutrients.

Fertilizer concentrations are expressed in terms of either ppm or g/liter. It is easy to convert from one expression to another. The first method is to determine the amount of fertilizer carrier required by following equation.

OZ of fertilizer carrier per 100 gal

$$= \frac{\text{Desired ppm} \div 75}{\text{Decimal fraction of desired nutrient in fertilizer}}$$

Let us assume we need 200 ppm N from a 20-10-20 fertilizer grade thus with above equation it will $\frac{200 \div 75}{0.20}$.

Further similarly it can be worked out the amount

= 13.33 Oz/gal of water which can be converted in g/L equation.

ppm = Oz of fertilizer carrier per 100 gal × 75 × decimal fraction of desired nutrient in fertilizer carrier

Concentration = 13.33 × 75 × 20 = 199.95 ppm

Continual fertilization. All essential elements are retained by root media except N&K which are not retained long in root media and need their application regularly. These are most commonly applied as a solution continually through crop production. Thus, it should be applied weekly or with each irrigation. It has been found that for most of crops, concentration of 200 ppm works well with each irrigation whereas 240 ppm N&K to 720 ppm observed better for sensitive crops such as bedding plants and elatior begonia, respectively.

TABLE 16.3
Conversion table for ppm of desired nutrient to gms/ litre or *vice versa*

Gms of fertilizer carrier in 1 litre	*ppm*													
0.1	12	13	14	16	16	20	20.5	21	33	44	45	53	60	62
0.2	24	26	28	31	32	40	41.0	42	66	88	90	106	120	124
0.3	36	39	42	47	48	60	61.5	63	99	132	135	159	180	186
0.4	48	52	56	62	64	80	82.0	84	132	176	180	212	240	248
0.6	72	78	84	93	96	120	123.0	126	198	264	270	318	360	372
0.8	96	104	112	124	128	160	164.0	168	264	352	360	424	480	496
1.0	120	130	140	155	160	200	205.0	210	330	440	450	530	600	620
1.5	180	195	210	233	240	300	307.5	315	495	660	675	795	900	930
2.0	240	260	280	310	320	400	410.0	420	660	880	900	1060	1200	1240
2.5	300	325	350	388	400	500	512.5	525	825	1100	1125	1325	1500	1550
3.0	360	390	420	465	480	600	615.0	630	990	1320	1350	1590	1800	1860
3.5	420	455	490	543	560	700	717.5	735	1155	1540	1575	1855	2100	2170
4.0	480	520	560	620	640	800	820.0	840	1320	1760	1800	2120	2400	2480

* Adapted from J.W. Love, Department of Horticultural Science, North Carolina State University, Raleigh, NC 27695-7609.

*TABLE 16.4
Standard concentration requirement of fertilizers containing 20% N for several green house crops

Crop	*Concentration Category*	*Concentration Weekly g/litre*	*Constant g/litre*
Rose	Moderate	2.4	08
Carnation	Moderate	2.4	1.0
Chrysanthemum	Heavy	3.0	1.0
Easter Lily	Moderate	2.4	1.0
Poinsettia	Heavy	3.0	1.0
Elatior Begonia	Very light	1.3	0.6
Bedding plants	Very light	1.2	1.0

* Nelson, P.V. 1991 Green House operation and management. Prentice Hall Inc. New York.

Specific Recommendation. Required concentration of fertilizer varies widely with the crop. However, most commonly used fertilizer in the green house is 20 : 20 : 20. Due to ammonical injures in above a 20 : 10 : 20 fertilizer is more commonly used. This is also commonly available. Other combinations are also available in the market. Other formulations like 15-15-15, or 15-0-15, 25-0-25 are also available. It has been observed that most crops grow well on equally balance N and K but there are few exceptions *e.g.* Elatior Begonia or Petunia grow faster and develop more side shoots when they are fertilized with a ratio of 2N : 1K. The requirement of azalea is 3 : 1 whereas for carnation it is 2 : 3.

Other forms of fertilizer application are through slow release fertilizers in the form of plastic encapsulated fertilizers, slow soluble fertilizers, urea formaldehyde, sulfur coated fertilizers, chelated micro nutrients, etc.

Nutritional Monitoring

Progressive growers make use of following three systems to monitor nutrient status and are able to forecast problems as they arise and their remedies.

I. Visual Diagnosis. By observing plants in terms of growth, foliage, colour, branching pattern, guess is done about deficiency or toxicity symptoms of particular nutrient. When these symptoms, develop, by that time damage has been already occurred to the crop. Each nutrient exhibits certain typical nutrient deficiency symptoms which are as follows.

1. ***Nitrogen :*** Entire leaf blades of lower leaves are chlorotic followed by necrosis and leaf drop. Growth remained stunted.
2. ***Phosphorus :*** At first, all basal leaves are dark green and growth is stunted. Purple pigmentation on leaves develop.
3. ***Potassium :*** Margins of older leaves become chlorotic and then burn or develop small chloritic spots progressing to necrosis appear scattered on old leaf blades.
4. ***Sulphur :*** Leaves on all parts of plant are affected and sometimes have a yellowing of plants.
5. ***Calcium :*** Margins of young leaves fail to form and look like strap leaves, growing point fail to develop leaving a blunt end. Young leaves exhibit light green colour or uneven chlorosis.
6. ***Magnesium :*** Recently mature or older leaves develop interveinal chlorosis.
7. ***Iron :*** Only young leaves show interveinal chloriosis.
8. ***Manganese :*** In addition to interveinal chlorosis, the tips and lobes of young leaves grey or tan necrotic spot develop in chlorotic areas.
9. ***Zinc :*** Young leaves are very small and typical symptom is called little leaf. This is associated with short internode.

10. *Copper* : Young leaves have interveinal chlorosis, the tips and lobes remain green and necrosis of leaf blade occurs.

11. *Boron* : Young leaves become leathery and chlorotic. Terminal buds die giving rise to a witch's broom.

II. Soil testing: Regular soil testing of green house crop for various parameters like pH, CEC, available quantity of N, P & K will help grower to follow proper productivity.

III. Foliar analysis. Like soil analysis, it is very useful to asses a problem before damage occurs. It is a analysis of representative leaves to determine the quantity of nutrients present in the plant. Then, these quantity are compared with available standards. On comparison, treatment is decided. Foliar analysis works well because of a strong relationship between leaf composition and plant response exist.

Leaf analysis should be done frequently depending upon the duration of crop.

Rose. Plants are sampled by picking two uppermost five leaf lets from a stem whose flower calyx is cracking and colour is just beginning to show. Thirty leaves with petioles attached should be collected.

Carnation. Leaf samples from fourth to fifth pair of leaves from unpinched plants are collected. After pinching, leaf sampling continued on secondary to tertiary branches as the case may be but before flower development starts.

Chrysanthemum. Youngest fully developed leave are sampled. Chrysanthemum are generally sampled five to six weeks after planting for a single stem crop or five to six weeks after a pinching for a pinched crop.

***TABLE 16.5**

Maximum Critical foliar levels of nutrients for florists crops

Nutrient	*General Crop*	*Rose*	*Carnation*	*Chrysanthemum*
N%	—	3.0	3.0	4.5
P%	0.3	0.2	0.45	0.3
K%	—	1.8	3.0	3.5
Copper	—	1.0	1.0	1.0
mg%	0.3	0.25	0.30	0.3
Fe ppm	50-60	—		
Mn ppm	30	—		
Zn ppm	20	—		
Cu ppm	5	—		
B ppm	25	—		7

* Nelson, P.V. 1991 Green House operation and management. Prentice Hall Inc. New York.

In green houses there is year round production of pot plants or flowers, and hence, there is a continuous demand of nutrients by the crop.

Cooling of Green House

India, being tropical and hot country, cooling of green houses is more important than heating. Cooling mostly is required during summer months specially when outside day temperature is above than 35-38°C which will result much higher temperature (47-50°C) inside the green house. The effects of high temperature on crops are loss of stem strength, delay in flowering, reduced flower size and even may be bud abortion. Hence, it is impossible to raise the successful crop without proper cooling system. Under such circumstances, evaporative cooling system has been found highly successful which lowers down 8-10°C than the temperature outside depending upon the presence of relative humidity in atmosphere.

Fan Pad Cooling System. This system is available since 1954 and still today it is most ideal for summer cooling. In this system on one wall of green house water is passed through a pad which is placed vertically in wall. On the opposite wall exhaust fans are fixed. Warm air from outside of the green house is passed through wet pad and pushed outside from the other side. During this process water absorbs heat from surroundings and evaporates thus reducing the temperature of green house. This reduction in temperature depends upon (1) the rate at which warm air is being removed from the green house, (2) the type of pads used for evaporation, (3) the size and placement of fans, (4) the path of air stream. The normal rate of exchange of air is 2.5 cubic meter per meter which has been found satisfactory.

Pad Size. The pads of different types and sizes *i.e.* thickness, length and width are used, keeping various factors in view. Earlier wood fiber (shreddings) type pads were used which were 2.5 to 4 cm thick and called is excelsior type. These are to be replaced annually. Now-a-days, pads used are made up of wax impregnated cellulose material available in the market of great thickness varying from 5, 10, 15 and 30 cm. Similarly these are available in different heights are : 30, 60, 90, 120, 180 cm etc. Holes of pads are in cross fluted design. The height of pad is determined by dividing total area of pad by length of the pad. The pad should be placed inside at the end wall of entire width. The pad wall should be equipped with ventilators exterior to the pad to permit the air entry during hot weather and for sealing off the outside air during colder period. Since the fan and pad are on opposite wall and should be at plant height to keep the cooling air on the plants.

To keep the pad wet for efficient cooling, water should be delivered at the top of 10 cm thick pad at the rate of 6.2 L/m/min of pad (it is 0.5 gpm). For pad length of 9 to15 m, 32 mm water pipe is required while for the length of 15 to 18 m, 38 mm pipe is used. For more longer pad length, water should be supplied from the mid point. Pipes should have holes at every 7-8 cm of 3 mm size for efficient wetting of pad.

Fog Cooling System. The rate of cooling of air increases proportionately as water drop- let size decrease. The droplet size of mist is 1000 micron whereas the size of droplet of fog is 40 micron. The droplets of this size remain suspended in air while they evaporate to cool the air which occurs without water condensation at surface of leaves.

The fog cooling system is used green houses to be cooled by ventilators only. Fog nozzles are spaced above plants throughout the green house. Fog emits intermittently through nozzles to cool the air which has entered green house through ventilators. The cooled air warms up and leave the green house through ventilators. More outside air is drawn in and is cooled by subsequent fog.

Green house fog cooling systems are available that can convert 95% of water to 40 micron or small with an average droplet size lever than 10 micron. These droplet evaporates 40,000 time faster than water evaporates from crop. Such a fast evaporation results cooling nearly to 100 per cent. For such system water should be of good quality. Generally clogging of foggers is a problem, and hence multiple filters capable of screening down to particle upto 5 micron are used. Different timers can help to regulate the fogging system.

Fogging system can also be used for seed germination and propagation of plants through cuttings by replacing misting. Foliar fertilization can also be done efficiently through fogging system.

Heating of Green House

During winter months when temperature in green house goes down below the optimal range of temperature required to produce quality blooms. Heat loss in green house is through conduction, infiltration and radiation. Metalic frame, polythene, sash bars, etc. are the sources of heat loss through conduction whereas leakages and holes are the sources of infiltration. In radiation warm objects emits radiant energy which passes through air to colder objects without warming air significantly and objects become warms. Glass, venyl plastic, and water fiber reinforced plastic (FRP) are relatively opaque to radiant every whereas clear polythene is not. Therefore, polyethylene used as covering material for polyhouses losses heat rapidly through radiation.

Heat Loss Through Various Green House Covering*

Covering material	*Heat loss Btu/hr/m²*
Glass 3.2 mm (single layer)	6.40
Glass 3.2 mm (Two Layer)	3.97
Polyethylene (Single layer) of 6 ml	6.80
Polyethylene (double layer) of 6 ml	3.97
Poly vinyl chloride (rigid)	5.21
Fiber glass reinforced plastic corrugated	6.80
Polycarbonate (Fluoride) 6 mm	4.08

*Nelson, P.V. 1991 Green House operation and management. Prentice Hall Inc. New York.

The requirement for heating the green house is adding of heat at the rate it is being lost.

Source of energy. There are three sources of energy *viz.* fossil fuel, solar, and electricity. Among these, fossil fuel source of energy is mostly used for heating the green houses. Among fossil fuel, coal or gasoline can be used to heat up the water or steam. The hot water or steam is then circulated through pipes inside the green house and desired effect is obtained. For this the central boiler system is installed. The capacity of boiler depends upon the size of range of green house. It is known that much heat is required to warm the boiler and its associated plumbing material before actual heat is received in the green house. It is therefore, very necessary to match the boiler capacity in view of green house area and plumbing units.

For smaller range of green house, hot water is used with a water temperature of 82 to 95°C. Large volume of water is required because 1 Btu (British thermal unit) heat unit is obtained from 1 lb of water when it losses 1°C temperature. For circulating the water circulating pipes are required. For this the pipe diameter of 5.1 mm is used. Placement of these pipes in green house is very important. If all the pipes are stacked on the side wall or end wall, undesirable pattern of hot air flow will occur and certain cold spots will be created in the centre. This can be avoided by placing one third pipes across the green house running from end to end. The remaining of two third pipes is placed in stacks along the outer wall. To reduce the cost of pipe, fin pipes should be used. Fin pipes are conventional pipes with numerous thin metal plates radiating out ward which increases its surface area which transfer heat more rapidly and evenly. It is known that 1 foot of fin pipe is equal to 4 feet conventional pipes. However, for better results fin pipes should be alternate with conventional pipes at equal intervals. Pipes in coils can also be used and these coils can be arranged in two styles. For hot water system box style is used whereas for steam, tram borne is used.

Apart from central heating system, localized system is also popular because of low initial price. In this small heaters with self contained fire box system are installed in each green house. But in long term, this system proves costly due to high cost of maintenance than a central system.

Low intensity of infra red radiant heaters can also be used which can save 30% energy over conventional heaters. So far, solar system of heating is found useful only for small or hobby green house.

Heat Conservation. Efforts should be made to conserve the heat of green house so that green house production becomes economical. A sizeable decrease in heat loss can be made possible through green house design and by taking extra care during peak days of winter *e.g.* by double covering (it saves 40% energy). Covering of glass by films, wall insulation, by sealing leaks and thermal screen which are popular among green house in cold season. Thermal screens are curtains of different material such as polyethylene, polyester film, polyesters film strips or polyester cloth. It can be drawn eave to eave or gutter to gutter as well as around inner perimeter of green house. These thermal screens reflect back the heat to plant and soil. Such steps can save the consumption of fuel by 20-60%. Automation can also be done in bigger green house if used on large scale.

By planting green wind break in north and south side at a distance of 2.5 times the height of green house, 5-10% fuel saving can be achieved.

Watering System for Green House Crops

Green house crops should be watered judiciously. Both the conditions of under watering and over watering are harmful and may result in poor growth of plants, less production of quality blooms. The ill effects of under watering are reduced plant growth, less photosynthetic activity, shorter stem internodes, wilting and drying of leaves, smaller flower size, etc. Whereas over watering affects root system of green house crops adversely by inhibiting oxygen supply, water and nutrient uptake which results in several nutrient deficiencies, wilting, hardened growth and overall stunting of plants.

For efficient watering, always use well drained medium of high water holding capacity. Thus good rooting mixture should have coarse texture and high degree of stable structure. These qualities can be achieved by formulating media and not field soil alone.

Each time when water is to be applied, it should be applied thoroughly and it should 10-15% extra so that water from pots or bench run out of the bottom. Due to the extra water, excessive fertilizers and salts will leach out otherwise they become toxic to roots.

The water should be applied when plants need it. This condition can be judged by observing the plants and when plants enter into the early symptoms of water but it should be prior to the plant damage. These symptoms differ from crop to crop. Only experienced manager by observing the crops can predict correctly. By observing the colour and feel of the root medium or by weighing the pots (if pot crops) decision can be made of watering the crop.

The quality of water should be good. It should not have high amount of sodium bicarbonate or sodium chloride or boron or fluoride or any other heavy metal.

Different system used for watering green house crops are as follows :

Hand Watering. It is most common but most uneconomical and expensive system of watering but still is followed where labour is cheap and scale of operation is rather small and automation is not practiced. Hand watering requires considerable time and it is not pleasant work. But still it is followed where crops is at high density like nursery production, seed flats or pots. Where hand watering is practiced care should be taken that force of water should be broken either by using fine rose spray or breaker at the end of pipe so that it does not result in washing of root medium out of pots or bend and disturb structure of root medium surface.

Perimeter Watering System. In this system, pipes run around the perimeters of a bench with a nozzle that sprays water root medium surface below the foliage. Thus this system is most pertinent for production of fresh flowers. The pipes may be of galvanised iron or PVC can be fixed. Rubber pipes if are used they can be rolled if it is not in use. In these pipes, nozzles are fitted at an angle of 45, 90 or 180 degree. Normally pipes of 19 mm diameters is used. Depending upon the length of beds, valve may be provided accordingly.

Tube Watering System. This system is mostly used for watering the pots. Water is carried to each pot by the polyethylene micro tube. These micro tubes are available in different inner diameter varying from 0.9, 1.1, 1.3, 1.5, 1.9mm and above. The number of pots that can be watered from a single 19 mm main water pipe will depend upon inner diameter of micro tube being used *e.g.*, 0.9, 1.1, 1.3, 1.5 mm micro tube will be able to handle 600, 900, 700 and 400 pots, respectively. These micro tubes should be provided with a weight at the end of it so that speed of water can be broken otherwise there is a danger that pipe being thrown out of the pot and will dig a small pit in medium.

Drip Irrigation System. This system is highly suitable for green house cultivation. There is a saving of 50-70% of water in comparison to conventional system of watering. It provides uniform distribution of water, nutrients, pesticides and fungicides without wastage.

Drip irrigation system delivers water to the crop using a net work of main line, sub mains, lateral line with emission point spaced at regular interval along the length side. Each dipper or emitter supplies a measured precisely controlled uniform application of water, nutrients, pesticides, fungicides and growth substances directly into the root zone of plant. Water and nutrient enter soil from emitters moving in root zone of plants through combined forces of gravity and capillary. In this way, plants get moisture and nutrient replenished immediately ensuring plants which never suffer from water stress thus enhancing quality. This results in optimum growth and high yield.

PVC pipe line is used for distribution of water from the main source to sublines. Drip laterals of 12-16 mm are suitably placed in each row of plants which are directly connected to sub lines. Drippers/emitters fitted to the laterals are laid just near the plant for gradual distribution of water upto the root of plants.

There should be regular maintenance of pipes by adding chlorine or other chemicals in drift line to kill bacteria and algae. Treatment with acid is also needed to dissolve calcium carbonate. Care should be taken against rodents to avoid any damage to pipes. Monitor the pipes thoroughly for any leakage which should be done regularly.

Overhead Spray system. Green house crops are most easily and cheaply can be irrigated from over head. In this system pipes are hanged over the plant from 60 cm to 180 cm above the plants. Nozzles are fitted in the pipes at 360° pattern. Care should be taken that the water should be of good quality and properly filtered so that these nozzles are not clogged.

Ebb and Flow System. This is a sub irrigation system for potted and bedding plants. Pot or flats are grown in a leveled bench made of plastic or fiber glass available in various width (1.2-2.0 m) and length is 1 m. These benches are glued to gather to make a bench of desired length and width. These benches can move over a stand. These have channels to drain out irrigation water. There is a water tank below the bench. This tank is covered to avoid dust or algal growth. The area of watering and size of pump should be matched. To ensure watering of pot, water remains on the bench for 10 minutes and collected in tank and reused. Fertilizer may also be applied through this system of watering. Pot with bottom hole is better than side holes.

New Cropping systems

Other than conventional method of green house cultivation, there are few innovations which are being practiced on limited scale which open up new possibilities of crop raising. Soil sterilization or changing the soil of a green house is tedious exercise so to overcome these difficulties new cropping systems have been developed which are getting popular elsewhere, but still they being tried on limited scale. Few of them are described in brief.

***(i)* NFT System.** (Nutrient Film Technique System) It is a specialized form of hydroponics. In this system, the growing of plants in a system where thin film (Few millimeters) of nutrient solution is circulated through a trough that contains the plant roots. This system was outlined in detail by Cooper (1979). This system has a channel laid out on 1% slope. The channel must be level across its width to ensure that the whole floor will be covered with nutrient solution. Such channels may be situated on raised platforms. The width of channel is 23 cm and depth is 5 cm. It may be from concrete, metal or wood. The channel should be water tight. Channels lined with a plastic film sufficiently large to cover the top of channel including propagating blocks. This film should be of 5 mil thickness which will not adhere with the roots and will force nutrient solution to move along with the slope. Covering of channel will check the evaporation, restrict light entry to prevent algal growth and help in controlling room temperature. Nutrient solution is circulated in a closed recirculating system from tank. The flow rate of solution should be sufficient to maintain a film of 3 mm thickness. Cooper used following nutrient solution successfully for more than 50 of species.

Plants for NFT system are propagated in containers such as blocks of rock wool, foam cubes, netlike pots containing soil less media.

The main advantages of this system are : It eliminates, the labour of sterilization. It has tremendous potential of water and nutrient conservation. Whole system is automated so it saves labour.

TABLE 16.6
The Concentrations of Nutrients for NFT System

	Concentration (ppm)
Nitrogen	200
Phosphorus	60
Potassium	300
Calcium	170
Magnesium	50
Iron	12
Manganese	2
Boron	0.3
Copper	0.1
Molybdenum	0.2
Zinc	0.1

***(ii)* Floor Ebb and Flow system.** Potted and bedding plants are grown on the floors. The floor is raised around and drained in the centre so that is can be flooded and drained in the fashion as an ebb and flow bench system. The advantage of floor recirculating system is considerable cheaper, labour cost and expensive benches are not needed. However, floor must be leveled with laser beam. Some difficulty may felt about labour as they have to bend during the cultural operation.

***(iii)* Trough Culture.** It is modification of ebb and flow culture. In this system, single rows of plants are grown in water tight trough. Troughs are sloped from one end to another with a decline of 2.5 cm per 10 m. Nutrient solution is pumped from high end where it slowly trickled down by gravity to supply to each pot on the way. At the low end solution is collected through connected gutters and recalculated again and again.

***(iv)* Rock wool culture.** In this system rock wool slabs are covered with polythene used on a floor. Each slab is used for a single plant and are placed on top of slabs at the desired spacing for the crop involved. Root penetrate into the slabs below the cubes in two or four days. Each slab is connected with water tube. Nutrient solution is applied with each watering. Crops can require 3-10 applications per day. This application will vary with plant size and weather conditions. This method has certain advantages as it does not need pasteurization. Rock wool is an excellent inert substrate for open system. Rock wool is light in weight and self contained. The light weight permits growth of crop on benches. This reduces the cost of production as there is reduction in average growing space.

***

17

Post-Harvest Handling and Storage of Cut Flowers

The flowers are highly perishable and are prone to post harvest losses. Nearly 30 percent flowers perish during handling. The term cut flower is used to define flower which is cut along with portion of stem and leaves and has a comparatively long vase life. These cut flowers are used in decorating interiors, Ikebana and making bouquets. In India as well in international market, the demand of cut blooms is increasing year after year. The major cut flowers which dominate in the market are : rose, carnation, chrysanthemum, orchid, tulip, gerbera, anthurium, lilies, gypsophylla, bird of paradise, alstroemeria, etc.

Cut stem is a living entity. When detached from the plant, the cut stem is deprived of its natural sources of food, water, minerals and hormones and carries out its metabolic functions. Various terms such as vase life, shelf life, longevity, keeping quality etc. are used to express potential life of a cut flower. The termination of vase life is marked by the wilting of individual petals or the whole flower. Before reaching the consumer, flowers pass through many handling steps such as pre-cooling, grading, storage, packing and transit. Various factors -affecting postharvest keeping quality of cut flowers consist of inherent factors, pre-harvest growing conditions, harvesting of flowers and post-harvest handling of flowers.

INHERENT FACTORS

Post-harvest behaviour and lasting quality of flower species and cultivars vary considerably. Differences in flower longevity and quality among the varieties maybe due to anatomical, physiological, physical, biochemical and genetic makeup of the plants. Cut flower longevity depends upon the carbohydrate reserves of flower, osmotic concentration and pressure potential of petal cells, stomatal functioning, differences in number of thick-walled supporting cells in the xylem element and phloem fiber.

HARVESTING OF FLOWERS

Right stage, method and time of harvest are of considerable importance to ensure long vase life of cut flowers. Flowers should always be harvested during cool hours of day. High temperatures lead to faster respiration and excessive water loss. Flowers should, therefore, be harvested either in the morning or in the evening.

In the morning, flowers are fully turgid due to low transpiration loss of water during night. Before packing the flowers, it should be ensured that there are no dew drops on the flowers because high moisture conditions favour the growth of pathogens. Harvesting during evening is also advocated because of high sugar level in the stem due to increased activity of photosynthesis during daytime.

Proper method should be adopted to harvest the flowers. The stems should always be cut with sharp knives or secateurs. While harvesting, care should be taken not to crush the stem. Hardwood stems should always be given slanting cut so as to expose the maximum surface area in order to ensure rapid water absorption. The nature of the cut, however, does not matter much in case of the herbaceous stems because such stems have ability to absorb water through epidermal tissue also.

Specific guidelines are needed to be followed for harvesting of flowers *e.g.*, stems of chrysanthemum should be harvested at least 10 cm above the ground level. The stem close to the soil is harder and lignified which allows lesser water uptake by the stem. The rose stems should be harvested leaving at least two 5 leaflet leaves on the stem. Some flowers such as dahlia, poinsettia and papaver release latex upon cutting. To overcome this problem, such sterns should be given dip in hot water (80-90°C) for few seconds.

Flowers should always be harvested at an optimum maturity stage. Too immature buds do not open properly whereas over mature ones wither quickly. The stage of harvest varies with the species and even with the cultivar, prevailing temperature conditions, distance to the marketplace and requirement of the consumer. In general, for transport to the distant markets, the flower should be harvested at bud stage whereas for the local market, the flower can be harvested at relatively open stage. The optimum stages of harvest for some important flowers are mentioned bin Table 17.1.

Harvesting of flowers at bud stage is always advantageous as the buds continue to open over a long time and hence, possess long vase life. Besides the buds are less sensitive to ethylene, easy to handle during storage and transport, economical to transport and are less prone to damage by diseases and pests.

FLORAL PRESERVATIVES

To enhance the vase life of flowers, floral preservatives are used. Term Floral Preservative is used for any chemical formulation which is used to extend the vase life of flowers. Besides increasing vase life, floral preservatives help to improve flower opening, flower size, shape and color. Many commercial formulations are now available in liquid or powder form. These formulations are to be used as per the guidelines provided by the manufacturer.

TABLE 17.1. Optimum stages of harvesting of important flowers

Flower Name	***State of Harvesting***
Alstroemeria (*lily of the Incas*)	4-5 florets open
Anthurium andreanum	Spadix almost fully developed
Antirrhinum majus (*snapdragon*)	1/3 florets open
Calendula officinalis	Fully open flowers
Chrysanthemum morifolium	Fully open flowers
Pompons and decorative	Flowers open but before shedding of pollens Centre of the flower fully open
Cymbidium species (*orchid*) *Cattleya* spp. (*orchid*)	Open but before central disc florets begin to enlongate
Dendrobium spp. (*orchid*)	3-4 days after opening
Dahlia variabillis	Fully open flowers
Dianthus caryophyllus (*carnation*)	Fully open flowers
Standard	Paint brush stage, half open
Spray	At least 2 flowers fully open
Freesia hybrids	Coloured buds
Gerbera Jamesoni	Flowers open but outer 2 rows show
Gladilous cultivars	Shedding of pollens
Gypsophila paniculata	1-5 florets show color
Iris germanica	Flower just open
Lillium spp. (Lily)	Goose neck stage
Narcissus spp. (daffodil)	
Polianthes tuberosa (tuberose)	Buds fully developed but unopened
Single	Maxium florets open
Double	1-2 petals beginning to unfold
Rosa hydrida (Rose)	First floret opens
Strelitzia reginae (bird of paradise)	1-2 petals beginning to unfold
Tulipa spp. (*Tulip*)	Fully open flowers
Zinnia elegans	

(*a*) Types of Floral Preservatives

Floral preservatives can be applied as (*i*) pulsing solutions (*ii*) bud opening solutions and (*iii*) holding or vase solutions.

Pulsing solutions : Pulsing refers to short duration (16-24 h) pre-shipment or pre-storage treatment. The effect of such a treatment lasts throughout the entire vase life of the flower. Sugar (sucrose) is a main component of the pulsing solution. Since pulsing involves short duration treatment, relatively higher levels of sucrose are used.

Bud opening solutions : Immature buds of many flowers can be made to open in chemical solutions, referred to as bud opening solutions The components of bud opening solutions are essentially the same as those of pulsing solutions, but in case of bud opening solutions, lower concentrations (2 to 5 per cent) of sucrose are used.

Holding or vase solutions : Holding or vase solutions are meant to hold flowers continuously, till the termination of their vase life. The level of sucrose in vase solutions is, therefore, also kept very low (0.5 to 2 per cent), due to long durations for which flowers are kept in the solution. Many commercial vase preservatives are also available for use in many countries as holding solutions and are required to be used as per recommendations of the manufacturer.

(*b*) Constituents of Floral Preservatives

The main constituents of floral preservatives are (*i*) Water (*ii*) Sugar and (*iii*) Biocides besides mineral nutrients and acidifying agents. Anti-ethylene compounds and growth regulators are also used in floral preservatives. The brief account of these constituents is given herewith :

Water : Water is the most important component of floral preservatives. It affects vase life in terms of its quantity as well as quality. Tap water can be harmful due to its high pH (alkalinity), presence of total dissolved solutes (TDS) and toxic ions. High pH of water promotes the microbial growth and hence, reduces vase life. Therefore, acidifying water to low pH 3.0-3.5 is advantageous because it decreases microbial growth and considerably improves water uptake by the stem. If tap water is to be used, it should be boiled and decanted to remove the dissolved salts.

Sugar : Sucrose is the most widely used sugar in floral preservatives. Sugars acts as additional food source and also improves water balance of cut flowers. Besides, sugar also decreases sensitivity to ethylene. The optimum concentration of sugar required, however, varies from species to species and type of the treatment. Since sugar promotes microbial growth, it is usually combined with biocides before use.

Biocides : Biocides are chemical compounds which are used to inhibit, the microbial growth in vase water as well as on the stem surface. Important biocides used are 8-hydroxyquinoline citrate, silver nitrate, aluminum sulphate, citric acid, slow release chlorine compounds and quaternary ammonium compounds.

1. **Hydroxyquinoline citrate (8-HQC)** is very effective broad-spectrum biocide. It acidifies water and also induces partial closure of stomata. This compound is used at the concentration of 200-600 ppm.

2. **Silver nitrate ($AgNO_3$)** is also very effective biocide. Silver Nitrate's lethal concentration of 25 ppm completely inhibits the microbial growth. High concentration (1,000-1,200 ppm) of $AgNO_3$ can also be used as short term (10-15 minutes) treatment. $AgNO_3$ has, however, very low mobility and remains concentrated at the cut surface of the stem where it inhibits the microbial growth. A lot of concern is however, being voiced in the developed countries regarding the use of silver as being a heavy metal, it is considered to be an environmental hazard.
3. **Aluminum sulphate (Al_2SO_4) (100-300 ppm).** It has also been found to be very effective in increasing keeping quality of gladiolus, carnation and roses. It reduces pH of the solution, very effectively inhibits bacterial growth and also causes partial stomatal closure, thereby increasing vase life. Aluminum sulphate is also reported to cause foliage wilting in chrysanthemum, and hence, must be used, very carefully.
4. **Citric acid (50-1,000 ppm).** It has been reported to improve vase life of roses, chrysanthemum, gladiolus and carnation. It acts as an acidifying agent, lowers pH of the solution and also prevents blockage of xylem vessels.

STORAGE OF FLOWERS

Storage of flowers is highly useful to regulate supply of the flowers in the markets to prevent gluts especially, during the periods of high production and low demand when the prices crash down. Storage is also useful to hold the flowers till sufficient quantities are available for shipment.

Factors Affecting Storage Life

Important factors which determine the storage life of the flowers are quality of the flower to be stored, stage of harvest. Temperature and relative humidity during storage light, ethylene production and pathogens. The flowers to be stored should be of good quality. Flowers should be free from any physical injury and incidence of pests or diseases. The injured and diseased tissues are potential source of ethylene production and are also highly prone to infection with pathogens. The flowers must also be cut at an optimum harvest maturity. Some flowers like gladioli, roses, lilies, narcissus and irises can be stored at bud stage as buds of these flowers possess the ability to open after storage.

Temperature management during storage is an important component which determines subsequent vase life of flowers. Temperature fluctuations in the cold room should he avoided. Relative humidity inside the cold room should preferably be kept high (90-95 per cent). Low relative humidity causes excessive water loss from the flowers thereby stimulating their desiccation. Cold rooms should be kept clean to prevent the spread of pathogens as well as ethylene accumulation by the decaying plant tissues and microbes.

Storage Methods

There are three methods for the storage of flowers; Simple Refrigerated Storage (which includes both wet and dry storage), Controlled Atmosphere (CA) Storage and Hypobaric or Low-Pressure Storage (LPS).

1. Simple Refrigerated Storage : Simple Refrigerated storage is the most widely used method for storage of cut flowers. It includes two types of storages *viz.*, wet storage and dry storage. In wet storage, the stems are stored with their basal portions dipping in water or preservative solution. Wet storage is used to hold the flowers for short duration and for their day to day handling. The temperature in wet storage is kept at 2-4°C *i.e.* slightly higher than that in dry storage.

In dry storage, flowers are sealed in plastic bags to prevent the loss of moisture. Dry storage is also referred to as Modified Atmosphere (MA) storage because sealing of flowers in plastic bags leads to reduction in O_2 and increase in CO_2 levels due to respiration of the tissue. Buildup of very high levels of CO_2 (beyond 4-5 per cent) or extremely low levels of O_2 (lower than 0.4 per cent) in the packages. However, may damage the flowers. Therefore, it is very important to identify the suitable films for optimum buildup of both CO_2 and O_2 levels inside the packages. Dry storage is more laborious but can be used to hold the flowers for longer duration. Before dry storage, flowers should be pre-cooled to remove the field heat. Pulsing of flowers before storage also improves their storage life. Most flowers are prone to freezing injury below –2°C but since free water freezes at –2°C, therefore, free water droplets condensing on the petals can also cause injury to the tissues. Therefore, the flowers should be stored at 0.5-1°C. The temperature range for tropical flowers like anthurium, cattleya and poinsettia is 10-15°C and for sub-tropical flowers such as gladiolus and strelitzia is 2-8°C.

2. Controlled atmosphere (CA) Storage : Controlled atmosphere (CA) storage refers to low temperature storage in gas tight chambers under decreased levels of O_2 and increased levels of CO_2. Carbon dioxide levels higher than 4-5 per cent may produce anaerobic conditions. Major limitation in CA storage of flowers is that the optimum levels of O_2 and CO_2 required for storage vary for different flowers and hence, different flowers cannot be stored at the same time in the same chamber.

3. Hypobaric or Low-Pressure Storage (LPS) : Hypobaric or low-pressure storage refers to storage at low atmosphere pressure (*i.e.* 40-60 mm Hg) under continuous ventilation and high relative humidity. Major disadvantage of low-pressure storage is that it causes rapid loss of water and faster removal of gases from the tissues. Low pressure storage, however, holds considerable promise for prolonged storage of flowers but due to high cost of installation, it is of limited use.

The optimum storage temperature and duration of storage varies with each individual flower. As a general rule, flowers originating from temperate regions *e.g.* rose, carnation, lily, cymbidium and chrysanthemum are best stored at a temperature close to 0°C above the freezing point of plant tissues *i.e.* at above –2°C. Flowers originating from tropical regions *e.g.* anthuriurn, vanda, cattleya and bird of paradise are sensitive to chilling injury and hence, have to be stored at higher temperatures. The recommended commercial storage conditions are selected flowers at 90-95% relative humidity are given below :

Storage	*Crop*	*Storage temperature*	*Maximal storage period*
Dry	Carnation	0 to 1°C	16 to 24 weeks
	Chrysanthemum	1°C	3 weeks
		4°C	15 days
	Gerbera	2°C	2 days
	Gladiolus	4°C	2 to 3 weeks
	Rose	0.5 to 2°C	2 weeks
Wet	Anthurium	13°C	2 to 4 weeks
	Carnation	4°C	4 weeks
	Chrysanthemum	4°C	2 weeks
	Dendrobium	5 to 7°C	10 to 14 weeks
	Gerbera	4°C	3 to 4 weeks
	Gladiolus	4°C	19-12 days
	Tuberose	7 to 10°C	3 to 5 days
	Rose	2 to 5°C	5 days

GRADING, PACKING AND TRANSPORT

Grading refers to categorization of flowers on the basis of their quality. For a market, it is important that each bunch is made up to the same size, weight or quality of the bloom. It is very essential to grade the flowers before marketing them. Major parameters on which the grading of flowers is based are appearance, harvest maturity, blemishes or injuries due to diseases, attack of insects or pests, color and size of the bud and straightness, strength as well as length of the stern. The flowers should look harvested at right maturity and free from damage by pests and diseases. The stem should be straight, free from side shoots and should be strong enough to hold the buds erect. The foliage should be dark green and free from the dust or residue of any chemical spray. The stem should be free from physiological disorder such as bent neck (in roses), tip bending (in gladiolus and other flowers), stern break (in gerbera), calyx splitting (in carnation) etc.

In the United States, the Society of American Florists has recommended various grades for some species of cut flowers. Under this grading, the flowers are categorized as blue (fancy), red (special), green (standard) and yellow (utility). The quality of the bloom declines from blue to yellow category. The U.S. standards for hybrid tea rose, gladiolus, carnation (standard) and chrysanthemum (standard) are given under the respective crops. Unfortunately, this type of classification is not being used in India or other countries. Growers have, hence, developed their own grading systems based upon the market requirements.

Bunching of flowers is done in 5, 10, 12 or 20 stems and loosely tied with rubber band. The bunches should be carefully put together to prevent any damage to the blooms and leaves and secured with the rubber band at the base and close to the bunched head. There is a need to take great care at this stage to avoid bruising of the buds or foliage. Flowers such as gerbera, orchids, anthurium and standard chrysanthemum are packed individually. The bases of some tropical flowers such as dendrobium and anthurium are kept moist by putting them in specially designed vials filled with water or in moist wool. The bunches are held in polyethylene sleeves or buds are wrapped in corrugated paper to protect them from mechanical damage. Different flowers, however, require different methods of wrapping or sleeves. Polyethylene sleeves are generally employed for wrapping a wide range of flowers, but such a wrap may tend to retain moisture and encourage fungal attack. Hence, many wholesalers prefer naked bunches. For delicate flowers, clean papers should be used, and the use of newspapers should be avoided because it sometimes leaves the print marks on the buds.

The flowers are held tightly in the boxes so that they are not damaged by the jerk movements during transit. The flowers are generally packed in telescope types boxes made of corrugated fiberboard. The quality board (usually 4-5 ply boxes strong enough to hold the weight of at least eight such completely filled boxes) should he used. The dimensions of the packing boxes are worked out keeping in view the length of the stem, type of the flower to be packed and efficient utilization of the space in the cargo or in refrigerated trucks. Box sizes which are in use are given below :

Flower	*Length (cm)*	*Width (cm)*	*Height (cm)*	*Weight (kg)*
Carnation	100	40	20	13
Chrysanthemum	80	50	23	15
Gladiolus	120	50	15	15
Rose	100	40	30	17

Over packing or under packing both contribute to high chances of excessive damage. A wooden batten should be placed across the center of the box under light pressure and nailed to the sides to prevent the movements of bunches. The flowers such as gladiolus snapdragon and some other species which show geotropic bending during transport should be transported in vertical position. A new packaging system, the Proconas system which uses plastic container and a cardboard sleeve and allows transport of flowers upright in water is very useful to keep flowers fresh during transport.

Flowers are usually transported by air and by refrigerated vans/trucks. But when the distances are short, non-refrigerated insulated trucks can also be put to use for transportation of flowers. Temperature management during transportation is very essential. The flowers should be pre-cooled before transport and also kept cool during transit.

Speedy dispatch of flowers from the farm to the airport and final destination is essential. Postharvest problems of flowers mostly arise while loading the local vehicle and storage at the airport. Before loading, it is to be seen that the truck or van is pre-cooled to the ideal temperature. Before loading, all boxes are also required to be pre-cooled to the optimum transportation temperature. To avoid the box becoming warm, loading is to be carried out as quickly as possible.

CAUSES OF DECLINE IN VASE LIFE

1. Depletion of carbohydrates. The intact stem on the plant continues to get supply of nutrients, water, carbohydrates and growth hormones from different parts of the plant. Once the stem is harvested, it becomes an independent entity, which means it must survive on the basis of what it contains. Carbohydrates in the form of reserves determine the vase life of flowers. The stems that are deficient in sugar will have shorter vase life. During respiration sugar is required which provides energy to the stem so if less sugar is available then the vase life will also be shorter.

2. Inability of stem to absorb water due to xylem blockage. Generally, 90% water is present in the stem. In a cut stem, loss of water is a continuous process from leaves. Vase life of any flower depends upon how much water a stem can absorb and retain. If water loss will be more than water absorption, then a stem will develop the water stress conditions which cause wilting of flowers. Decrease in water uptake by the stem is mainly due to blockage of xylem vessels caused by bacteria, fungi and other microorganisms which grow in vase water or on the dipped portion of the stem. The microorganisms also secrete certain enzymes which dissolve the cell wall components. The material released after the degradation of the cell all wall also blocks the xylem vessels. The continuous water loss primarily through leaves during storage or transit also creates empty columns in the xylem and to fill this empty space, air enters into the stem. The air bubbles block the xylem vessels and render them non-functional. The air can, however, be displaced from the stems by putting them in tepid water and placing at cold temperature as well as by re-cutting the stem under water. Physical blockages of xylem vessels also lead to decrease in uptake by the stem, thereby decreasing vase life. The microbial blockage of xylem vessel can be minimized by the use of biocides in vase water.

3. Excessive water loss from cut flowers. Excessive water loss from the flowers leads to wilting and reduction of flower quality and vase life. Immediately after harvesting, field heat should he removed from cut stems and refrigerated as soon as possible. Removal of excessive leaves should be done, and flowers should be kept in water and stored in cold storage after harvesting.

4. Ethylene. It is a gaseous hormone which plays an important role in senescence. All flowers are not sensitive to ethylene. In many species, pollination-induced ethylene production serves as a signal for initiating the process of petal senescence. In ethylene sensitive flowers like carnation, pollination induced production of ethylene is sequentially trans located through stigma, style and ovary to the petals where it regulates genes for ethylene biosynthesis which induces autocatalytic production of ethylene. The senescence of flowers is marked by the loss of dry matter, primarily due to the hydrolysis of macromolecules such as starch, sugars, proteins and nucleic acids. The starch is broken down into sugars which cause bluing of petals in red cultivars of rose. Senescence is also marked by changes in the property of membranes such as increase in micro viscosity and loss of phospholipids as well as permeability which results in leakage of pigments, sugars, mineral ions and other metabolites. These lead to death of the tissues. It has been observed that flowers show two peaks of respiration; first when the bud starts opening and the second when the senescence take place. This is followed by decreases in respiration rate which has been ascribed to the depletion of sugars or inability of mitochondria to utilize available sugars.

5. Pests and diseases. They cause physical damage to the plants in the form of some visible symptoms. Crop should be protected right from nursery stage or even in green house when plants are not in flowering. These pests and diseases cause physical damage to the plants and reduce photosynthesis. These infested plants also invite secondary infection and produces more ethylene which is responsible for senescence of flowers.

6. Pollutants. The green house areas or areas for flower cultivation should he away from the industries or polluted areas which release toxic gases such as sulphur dioxide, nitric oxide and fluorides in the atmosphere. Fluoride in the atmosphere or irrigation water also causes damage to flowers like gladiolus and tulip. These gases damage the foliage as well as petals of the flowers. Tip burn of leaves caused by fluoride toxicity has been commonly observed in gladiolus under Punjab conditions.

18

Flower Seed Production

The Commercial flower seed production of F_1 hybrids and open pollinated is considered a profitable venture and, hence, it is popular amongst farmers on limited scale. Earlier, seed production was being done on limited scale in Sri Nagar and plains of North India and in other localities due to limited demand. Mr. Man Mohan Attawar of M/s Indo American Hybrid Seeds, (India) Pvt. Ltd., Bengaluru has started producing F_1 hybrid seeds of Petunia for 100% export during mid sixties. However, production of seeds of open pollinated flower crops was revolutionized by Mr. Avtar Singh, M/s Beauscape Farms, Sangrur, Punjab who started flower seed production involving farmers on large scale. Now many companies have started producing seed on large scale for export to Holland, U.K., USA, France, Germany, Japan etc. At present in India, the area under flower seed production is about 600-800 ha. The main areas of flower seed production in India are : Punjab (Sangrur, Patiala, Ludhiana); Haryana (Panipat, Sirsa); Karnataka (Bengaluru, Rani Banur); Himachal Pradesh (Kullu Valley); J & K (Sri Nagar Valley); and West Bengal (Kalimpong). The cost of seed production per ha varies from ₹ 10,000 to 15,000 and generate a net profit of ₹ 25,000 to 75,000. Farmers have started using innovative techniques for harvesting seed as well as their cleaning which made the seed production easy and profitable. However, flower seed production is labour intensive. It has been observed that for wheat and paddy the labour requirement is 50 men days/ acre whereas for flower seed production the labour requirement varies between 120-366 men days per crop. In a survey conducted in Ludhiana Distt. it has been observed that farmers mostly prefer those crops which are easy to harvest, clean and whose seed size is bigger but this preference seems to have no value as they have to grow those crops which the companies suggest.

Climatic requirement for seed production

Seasonals as term indicate are divided in to three seasons—Summer, Rainy and Winter season. Summer and Rainy seasonals in India are available in limited number whereas Winter annuals are rich in kinds and available in large range of colours and height and bring riot of colours in the garden and, hence, very popular amongst gardeners. Thus from seed production view point also winter annuals cover large area. Ideal climate condition for seed production is long duration of cool and dry season which helps in good seed setting of bold size. While excessive hot and dry season hampers seed setting of summer annuals in north Indian plains. Excessive rain at the time of flowering washes away pollen grains resulting in poor seed set.

According to climàtic requirements, the production of flower seed is divided in following climatic zones.

1. Mild climate (Kashmir Valley, Kullu Valley etc.)—Delphinium, Giant Pansy, Zinnia, etc.
2. Sub-Topical area—Antirrhinum, Anchusa, Ageratum, calendala, Brachycome, Lineria, Californian poppy, Candytuft, Carnation, Dianthus, Daisy, Dimorphotheca, Nasturtium, Petunia, Portulaca, etc.
3. Tropical—Tagetes, Salvia, Ipomea, etc.

Mode of Pollination

For successful seed production of flowers, it is necessary to know whether the particular flower is self, often cross or cross pollinated crop. To find out whether a particular flower is self or cross pollinated, the very simple method of bagging or growing in complete isolation may be followed. A number of plants are grown in flower bed and when flowers are about to open some of them may be covered with muslin bags and tied properly at the base so that no insect can enter. If the plants or flowers which were covered with muslin bags produce seeds, then it is obvious that variety or species is self pollinated. If the bagged flowers do not produce seeds, it is an indication that flowers have to be hand pollinated. Some time this is not true of certain species or varieties. (Table 18.1)

TABLE 18.1
Mode of Pollination in different flower crops

Self Pollinated	*Often Cross pollinated*	*Preferential outbreeds*	*Cross Pollinated*
Balsam	Antirrhinum	Ageratum	Alyssum
Clianthus	Aster	Corn Flower	Arctotis
Lupin	Dahlia	Dephinium	Calendula
Sweet Pea	Salvia	Marigold	Cineraria
	Wall Flower	Verbena	Gazania
	Linum	Chrysanthemum	Stock
	Linaria	Nemasia	Zininia

For production of pure seeds numerous devices are adopted for cross pollination. Most common method is to grow them in isolation. (Table 18.1) This means to grow two species/variety or strain of the same species at certain distance from one another. Other methods such as bagging flowers, growing them under nets or in glass houses, emasculation or special methods *e.g.* collection of pollen grains with vacuum pump in petunia, use of floor guns etc. or use of male sterile line for seed production are followed to produce seeds in cross pollinated crops.

Types of Flower Seeds

In the market different types of seeds are marketed depending upon the derivation of seeds. They are :

***(i)* Open pollinated variety.** In India more than 90% seed of open pollinated type is produced and exported. These are true breeding cultivars and can be used for all genetic improvement. However, amongst flower crops, these are used for landscape design for main effects and creating colour patterns. These represent true self pollinated lines and produce uniform plant both genetically and phenotypically and produce the uniform colour and of same size. The seed of often or cross pollinated type is produced by providing proper isolation distance to avoid foreign intimidation of pollen grains.

***(ii)* F_1 hybrids.** The F_1 hybrid seeds are very popular amongst many flowers due to superior in plant vigour, earliness, increased flower size, and numbers of flowers. These are produced by crossing two inbred lines/variety. Various method are adopted like hand pollination by following the procedure of emasculation and pollination or by use of male strility. For example in petunia or pansy emasculation and pollination method is used whereas in marigold, zinnia, male sterile system is used. On selfing these do not breed true to type and produce mixture of progeny due to genetic segregation and recombination.

***(iii)* F_2 seed.** In flowers like pansy, Antirrhinum, F_2 seeds are also very common in trade due to wide variation in flower colours, but flower size may be little reduced. It results from self fertilization of F_1 hybrids.

***(iv)* Mixture.** In flower seed trade, mixture of seed is sold by blending the seed of different colour in certain proportion thus field looks more colourful in certain flowers colours. It has been observed that in many flowers, certain flower colour tend to produce more seeds than others do. Therefore such mixtures are preferred.

Seed Production Techniques (Plate 7)

Selection of Soil. For successful seed production, soil should be preferably loam. Other soils like sandy loam to clay are also good. The soil should be well fertile and free from water stagnation and have 6.5 to 7.5. Since most the crops have tendency to shatter the seeds in the field and hence, the same crop may not be selected to grow year after year to avoid contamination.

Raising of seedlings. Generally seeds of annuals are sown in controlled space for better handling owing to small size of seeds. In general, 10-15 cm raised nursery seeds of 1 m width and 2-3 m long are prepared. The soil should be well prepared by mixing 10-15 kg/m^2 well rotten farm yard manure. To check the soil borne diseases in nursery, soil should be drenched with 0.2% brassicol or captan or soil can be sterilized by drench with 2% formaline and covering with polythene. For better handling of small seeds, these are mixed with bulk material like sand or ash. Seeds are sown by hand in line 5-6 cm apart and .5 cm deep. The depth can be increased with the increase in size of seed. A seed sowing machine has been developed by Beauscape Farms, Sangrur. The seeds sown from this machine are sown at uniform depth and distance. After sowing seeds are covered with well sieved mixture of farm yard manure and soil. Watering should be done twice or thrice a day or use sprinkler system. The seeds are covered with newspapers or bed sheet to create darkness which helps in better germination. The seeds start germinating within 2-3 days and covering is removed. It takes about one month to grow the seedling sufficiently and are ready for transplanting when seedlings have 3-4 true leaves. For raising one acre nursery 8-10 beds of 5 × 1 m are required. The quantity of seed for raising in nursery will depend upon size and weight of seeds. Following amongst of seed is required for raising seedlings for keeping 3 seeds required for raising one seedling : Marigold (African)—400-500 g, Petunia—200 g; Island Poppy—150 g; Oenothera—200 g; Nemophylla—500 g; *Eschscholtzia californica*—200 g; *Bellis perernnis*—100 g.

Direct sowing of seeds. Many flowers are sown directly. For this 15-20 cm wide ridges are prepared by manual or tractors. Three to 4 seeds are sown directly by hand at 15-20 cm apart. After complete sowing light watering is done. Seeds start germinating after 4-5 days.

Preparation of land and transplanting of seedlings. Before laying out the field, the land should be well prepared by ploughing by harrowing and mixing 15-20 tonnes well rotten farm yard manure/acre. It has been observed that instead of flat planting raised bed method of planting is more useful. Therefore, raised beds with tractor are prepared and keeping the spread of plant, seedlings are transplanted on both sides of raised bed or in center.

In general 50 kg urea, 250 kg of single super phosphate and 60 kg of muriate of potash should be well incorporated in the soil. The application of remaining 100 kg urea should be applied in to two equal splits *i.e.* one and two months of transplanting of seedlings. It is advised that before transplanting light irrigation should be applied which helps in better establishment of seedlings. The distance for tall annuals like helichrysum; delphinium; gaillardia is kept 60 × 45 cm, whereas for medium annual, (petunia, verbena, phlox, eschscholtzia, etc.) 45 × 45 cm and for dwarf annuals (pansy, mesembryanthemum, alyssum, daisy) 25 × 25 cm is kept. The seedling should be transplanted in cool hours of evening and should be watered lightly immediately.

Isolation requirement. To produce genetically pure seeds, crop should be isolated with other varieties of same crop. (Table 18.2)

TABLE 18.2
Isolation distance for different flower crops

Self Pollinated	*Often cross pollinated*	*Cross Pollinated*
No distance	50-100 m	1000 m

Generally no isolation distance is required for strictly self pollinated crops, whereas a distance of 25 m is beneficial. For often cross pollinated and cross pollinated crops the isolation distance of 100 and 1000 m is kept, respectively. Seed producing companies those are busy in custom seed production, select the farmers in different villages and are given a programme of seed production of one line of single type. This helps in better isolation.

Management of annuals. Annuals are quite tender and require proper attention through out their life. Regular water supply is essential for successful raising which should be done according to the requirement of the crop. Sufficient moisture is required till harvesting of seed is done. Depending upon season and soil type frequency of irrigation is decided. In rainy season generally irrigation is not required except during dry spell. During winter season, irrigation is required at 10-12 days where during summer season it should be done at 4-5 days. Regular weeding and hoeing is essential for the development of seedling in to healthy plants.

Rogueing. Seed producer should have a detective eye to observe off type or improved type in the field. He should keep constant vigil from the beginning of the crop to the maturity. It is utmost important to remove and destroy the off types whereas if any plant exhibits superior trait in terms of vigour earliness, colour, or size of flower, etc., the seed of such plant should be preserved for further testing and utilization.

Insects, pests and diseases. The annuals are commonly attacked by *Heliothis* sp., aphids, leaf miner, etc. Protective control measure should be taken well in advance to avoid any considerable loss. The heliothis sp. attacks mostly *Antirrhinum,* coreopsis, caranation, Dianthus, etc. can be controlled by spraying Chlorpyrifos @ 1 ml/l of water at an early stage of caterpillar. Aphids commonly attack alyssum, stock, nemesia, calendula etc. Spraying of Imidacloprid 200 SL 0.4 @ ml/l or Thiamethoxam 25 WG @ 0.4 g/l of water is effective to control it. Different fungi attack many annuals at different stages such as in nursery or in the field. Seed dressing with Bavistan (l g/kg) or captan (3 g/kg) or drenching with the infected nursery beds with 0.2% Brassicol or Bavistan (0.1%) is quite effective. The seed virus can not be controlled by using chemicals and effective control is destroying of the plants or by controlling the vectors by using effective insecticides.

Harvesting and storage of seeds. After vegetative phase, flowering occurs in succession which may last 30-60 days or even more resulting in maturity of individual flowers in succession. Therefore pods are harvested individually before the splitting occurs.

Whereas in many cases to save labour single harvesting with sickle is done at the cost of seed yield *e.g.* nasturtium, mesembryanthemum, *Coreopsis tinctoria, Verbena tennusecta* Phlox, Oenonenthra, Clarkia, etc. The seeds of Nasturtium are very bold and thus it is collected by brooming the field. Similarly the seeds of *Chrysanthemum paludosum* are collected by shaking the plant over sheet of cloth. The harvested seeds are spread over the tarpauline under shade or in ventilated room for week but reshuffled daily. In many cases seed along flower stock is plucked which provides moisture and nutrient to maturing seed *e.g.* daisy. Many big seed producers use green netlon 12′ high to provide shade for drying the seeds.

The seeds are cleaned and sieved with different types of seed machine and seeds are finally cleaned by hand winnowing or using table fan to separate light seeds. In certain machines brushes are attached to remove the hairs or appendages. The seeds after proper cleaning are packed in muslin bags H.D.P.E. bags with silica gel and are stored in cool and well ventilated room. The techniques of seed production of few important annuals are discussed below.

The average yield per acre, no. of seeds per gramme and stage of collection of commonly grown annuals are given in tables 18.3 & 18.4.

Marigold. (*Tagetes erecta*). It belongs to family composite and is most commonly grown commercial flower and quality seed is always in demand. It is highly cross pollinated crop, hence an isolation distance of 1 km is required amongst varieties. However, different species do not cross pollinate with each other. For seed production, under North Indian conditions, August planting has been found most promising. The planting is done at a spacing of 60 × 40 cm and plants start blooming after seventy days of transplanting. Seeds start maturing from November onward and should be collected when cypsela (seed crop) has dried up. After drying individual flower should be hold in the hand and dried petal should be rotated to obtain the clean seed. Seeds are also removed by crushing the cups. The seed yield of 312-375 kg/ha is obtained.

Gaillardia pulchella. It belongs to family compositae and is highly cross pollinated crop. The plant attains the height at 60-90 cm. Flowers are produced abundantly in many colours *viz.* yellow, orange, scarlet, copper colours or combinations of these. There are two types *i.e.* picta type—single type and large flowers, and lorenziana type—flowers are double having quilled petals with split tips. Thus an isolation distance 1000 m is required. Plant are planted at 60 × 60 cm distance during November and seeds are harvested from May to July. Seeds are borne on capitulum and have appendages. Therefore heads are harvested individually and are cleaned in special machines which removes it. Yield of 500-625 kg/ha is obtained.

Coreopsis lanceolata. It belongs to family compositae and being a highly cross pollinated and isolation distance of 1000 m is required for pure seed production. The crop can be transplanted as late in November to mid December. The plants attain a

TABLE 18.3

Seed yield of annual flowers and approximate number of seeds per gram

Name	Kg/acre	No. of seeds/g
Alyssum maritimum (Dwarfwhite)	70-80	350
Antirrhinum majus (Tall)	120-130	8000
Antirrhinum majus (Dwarf)	60-70	8000
Calendula officinalis (Pacifica mixed)	400-425	140
Callistephus chinensis (Ostrich Plume)	100-110	568
Celosia plumosa (Tall)	150-160	1600
Celolsia plumosa (Dwarf to medium)	40-50	1900
Centaurea cyanus (Tall)	250-300	325
Centaurea cyanus (Dwarf)	70-80	325
Centaurea moschata	150-160	250
Chrysanthemum paludosum	80-90	1800
Chrysanthemum multicaule	100-110	600
Clarkia elegans	300-320	2800
Coreopsis lanceolata	400-425	—
Coreopsis tinctoria	250-260	—
Cosmos sulphureus	150-160	178
Delphinium consolida	250-260	450
Dimorphotheca aurantiaca	150-160	4195
Eschscholtzia californica	250-270	850
Gaillardia pulchella	150-200	535
Gomphrena globosa (Tall)	150-160	359
Gomphrena globosa (Dwarf)	30-40	359
Helichrysum bracteatum (Tall)	150-160	1560
Helichrysum bracteatum (Dwarf)	30-40	1560
Helianthus annus (Sun burster type)	250-260	25
Impatiens balsamina	300-320	170
Limonium sinuatum	40-50	460
Lupinus hartwegii (Tall)	200-220	56
(Pixie delight type)	80-100	—
Mesembryanthemum crinifolium	150-160	6600
Molucella laevis	150-160	—
Papaver rhoeas	110-110	4563
Pansy-viola type	250-260	800
Petunia hybrida (Open pollinated)	60-70	11,000
Phlox drummondii	120-140	677
Tagetes erecta	80-100	350
Tropaeolum majus (Single type)	400-450	10
(Double type)	60-80	10
Verbena hybrida (Ideal floristmix)	120-130	673
Verbena tenusecta	150-160	—

height of 100-120 cm. Flowers are of sulfur yellow in colour. The ideal distance of transplanting is 60 × 40 cm. Flowers mature in April-May and seeds shatter when are too dry therefore collect heads individually when are partially dry. It yields about 200-250 kg/acre. *Coreopsis tinctoria* produces flowers of yellow to yellow with maroon making and produces 625-750 kg/ha.

Helichrysum bracteatum. It belongs to family compositae and is a cross pollinated crop. A large number of varieties are available in different colour and height. Seedlings are transplanted in first week of November at a distance of 60 × 40 cm and seed is ready for harvesting in March-April. Individual flower heads are harvested when the colour shows the signs of maturity and head become fuzzy. A good yield 160-200 kg/ha is harvested.

Petunia hybrida. It belongs to family solanaceae and flowers vary from *i.e.* single to double in different colours like white, cream, pink, purple, magenta, mauve, salmon or bicoloured are available. The height of plant varies from 50-70 cm. The seedlings are transplanted in first week of November at 45 × 45 cm distance. Seeds are ready for harvesting from April to mid June. Harvest the pods as they began drying otherwise shattering will occur.

Ammobium alatum. It belongs to family compositae and flowers are bicoloured (disc is yellow and ray florets are cream coloured). After transplanted in November at 45 × 30 cm, plants remain spreading and dwarf in nature up to 90 days and afterward length of main stem increase till anthesis. Due to its papery florets and long stem, it is useful for drying purposes. The branches are cut when flowers are dried in May and June. Seeds are cleaned and yield about 125-150 kg/ha is obtained.

Chrysanthemum paludosum. It belongs to family compositae and flowers are single, white in colour and are produced in great profusion. Plants attain ultimate height of 30-45 cm and has spreading in nature. Seedlings are transplanted at 60 × 60 cm distance in November. Seeds are ready for harvest during March-April. Since there are numerous flowers and due to small size of seed, seeds are collected by shaking over sheet of cloth or chajj 4-5 times. Seed yield of 200-225 kg/ha is obtained.

Monarda citriodora. It belongs to family Labiatae and needs cross pollination for good seed set. The plants attain height of 70-100 cm. The crop is transplanted in November-December at 60 × 45 cm. Flowers dry during May to July and harvested singly. It yields about 200-225 kg/ha.

Gazania splendes. It belongs to family compositae and highly cross pollinated crop. The plants attain ultimate height of 30 cm and produce daisy like showy flowers in red, brown, white, yellow and pink colour. The seedlings are transplanted in first week of November at 30 × 30 cm. Flowering occurs during March and harvesting pods is done when dry fully before seed shatters. By this method yield of 125-150 kg/ha is obtained from a good crop. However, due to employment of more labour farmers are

TABLE 18.4
The stage of collection of different flowers

S.No.	Plant	Stage of Collection
1.	*Alyssum maritimum* (Sweet Alyssum)	The seeds shatter easily. Remove pods when just about to dry.
2.	*Antirrhinum majus* (Snap dragon)	Cut when just about to dry, Spikes mature from lower branches onwards.
3.	*Arctotis stoechadifolia* (African daisy)	Cut the whole plant when maximum amount of seed matures and dry then on canvas.
4.	*Calendula officinalis* (Pot marigold)	Seeds shatter when too dry. Collect heads when partially dry.
5.	*Companulla* (Bell flower)	Whole plants may be collected.
6.	*Celosia* sp.	Collect the heads when dry on the plant. Protect drying heads from rain.
7.	*Chrysanthemum coronarium* (Annual chrysanthemum)	Cut when almost all the flower heads dry.
8.	*Clarkia elegans*	Remove seeds as they begin drying.
9.	*Clianthus*	Collect seeds as pods dry.
10.	*Cosmos bipinnatus*	Collect the seeds as the heads dry on the plants.
11.	*Dahlia variabilis*	Collect heads of flowers as they dry. Take out tubers when plant almost dry. Store in dry and cool place.
12.	*Delphinium-Consolida* (Larkspur)	Cut the whole plant when the lower capsules begin to dry and dry in shade.
13.	*Dianthus* (Sweet Williams, Carnations and Pinks)	Seeds may shatter if allowed to dry too much on the plant. Collect individual heads as they begin to dry.
14.	*Dimorphotheca aurianticeca* (African daisy)	Cut the whole plant when the maximum amount of seed is mature.
15.	*Gaillardia pulchella* (Blanket flower)	Cut the entire plant when the maximum amount of seed is mature and dry on canvas.
16.	*Gazania splendens* (Treasure flower)	Cut the entire plant when the maximum amount of seed is mature and dry on canvas.
17.	*Godetia grandiflora*	When lower capsules open and begin drying, cut the entire plant and dry in shade.
18.	*Gomphrena globosa* (Globe Amaranth)	When the heads dry, collect individually.
19.	*Gypsophila elegans*	When the majority of the capsules have turned brown, cut whole plant and dry on canvas.
20.	*Helianthus annus*	When the flower heads dry, collect individually.
21.	*Iberis amara* (Candytuft)	Cut the whole plant and dry in sun set at the first signs of seeds becoming dry (self incompatible).

22.	*Helichrysum bracteatum*	When the heads become fuzzy, collect individually.
23.	*Impatiens balsamina* (Balsam)	Cut pods burst at the slightest touch. Collect individually when turning yellow brown. Place in a box the half matured seeds to dry.
24.	*Impomea* sp.	Cut the entire plant when the maximum amount of seed is mature and dry on the canvas.
25.	*Kochia scoparia*	Cut the entire plant when the maximum amount of seed is mature and spread on canvas to dry.
26.	*Lathyrus adoratus* (Sweet pea)	When the lower pods commence drying remove the entire plant and dry in shade.
27.	*Limonium sinuatum*	Cut the whole plant when the maximum amount of seed is mature and spread on canvas to dry.
28.	*Linaria bipartita*	When lower pods begin drying, remove the entire plant and dry in shade.
29.	*Linum grandiflora*	Cut the entire plant when the maximum amount of seed is mature and dry on canvas.
30.	*Lobelia erinus*	Cut and collect when all seed pods have dried.
31.	*Lupinus hartwegii*	Remove individual seed pods as they dry. If allowed to dry in excess on the plants, they would burst.
32.	*Mathiola incana* (Stock)	Remove the plant when seed pods begin drying. Dry in sun or shade. Single flower seeds produce 50% or more double flowering plants.
33.	*Mesembryanthemum cirmifolium* (Ice plant)	Whole plant should be harvested and seed should be extracted.
34.	*Molucella laevis* (Bells of Ireland)	Whole plant should be harvested and seeds should be extracted by beating the plants with stick.
35.	*Nemesia strumosa*	When the flower head dries seed is collected by rubbing in flower heads.
36.	*Nicotiana alata*	Collect the seed capsules as they dry. If allowed to dry too much, they are likely to shatter.
37.	*Niglla damascena*	Cut the entire plant when the maximum amount of seed is mature and dry in the canvas.
38.	*Papaver roheas* (Poppy)	Remove seed pods when just about to open after drying (self incompatible).
39.	*Petunia hybrida*	Remove seed pods as they begin drying. Double flowers require hand pollination.
40.	*Phlox drummondii*	Remove seeds when just about to dry to prevent shattering.
41.	*Pimpinella monoica*	Harvest umbles when completely dry and collect seeds by thrashing.
42.	*Portulaca grandiflora*	Collect when capsules begin to dry.

43.	*Rudbeckia bicolour*	When flower heads fuzzy collect them and dry in shade.
44.	*Salvia splendens*	When seed cap dries, remove plant and dry in shade.
45.	*Tagetes* (Marigold)	Collect flower-heads as they dry. Plants of dwarf species should be removed when all the flower heads are dry.
46.	*Tithonia speciosa*	Collect the flower heads as they become fuzzy and dry in shade.
47.	*Tropaeolum majus* (Nsturtium)	Collect seeds as they dry. If allowed too long on the plant they fall off.
48.	*Viola wittrockiana* (Pansy)	Collect seed pods when just about to dry. They shatter when overdried on the plant.
49.	*Venidium fastuosum*	When flower heads become fuzzy collect them and dry in shade.
50.	*Verbena hybrida*	Remove seeds as they dry.
51.	*Zinnia elegans*	Cut flower heads as they dry. Dwarf varieties may be removed when all the flower heads dryout.

harvesting a crop as a single cut when 75% crops is mature may result in reduction of marginal yield.

Phlox drummondii. It belongs to family polemoniaceae and is very popular in all gardens due to its easy culture free flowering and wide range of colours with long blooming period. The flowers are produced on clusters at the end of branches well above the foliage. The individual flowers are small about 2 cm across with round or pointed petals. The plants attain a height of 50-60 cm and seedlings are transplanted at 45 × 30 cm in November and seeds are harvested from April to June when the pod colour changes from green to brown. Harvesting is repeated 3-4 times for securing good yield of 300-325 kg/ha.

Cosmos sulphureus. It belongs to family compositae and flowers are highly cross pollinated and hence, an isolation distance of 1000 m should be provided to maintain purity. It is transplanted in July at a distance of 40 × 30 cm. It flowers during August upto end of September. Seeds maturity are of yellow or orange colours and are very liable to shattering and thus seeds are individually collected daily. A good yield of 150-175 kg/ha is obtained. *Cosmos bipinnatus* has more range of flower colours *i.e.,* white, pink, salmon, violet, magenta etc. It is transplanted in August and flowering continues up to mid-November. A yield of 175-200 kg/ha is obtained easily.

Verbena hybrida. It belongs to family verbenaceae and small colourful flowers are produced in different colour like white, red, blue, violet, salmon and intermediary colours on the Capitulum. It is preferentially out breeds but self pollination occurs. For seed production an isolation distance of 500 m is kept amongst different varieties. For seed production it is planted in first week of November at a distance of 45 × 30 cm and seeds are harvested during March-April. A good seed crop yields about 250-375 kg/ha.

Viola wittorckiana. It belongs to family violaceae. Flowers are very pretty due to shape and marking of contrast colour on blotches and size varies from 5 to 10-13 cm across. Flowers are of self colour or contrast colour. The pansy varieties are grouped into two groups *i.e.* small or large sized flowers. Small types produce flowers in great number and produce abundant seeds (62-75 kg/ha) where as large types flowers varieties produce less number of flowers but of big size and produce very less seed in plains of North India and requires hand pollination. These varieties produce good quality seeds in Sri Nagar valley of J & K.

Calendula officinalis. It belongs to family compositae and flowers are highly cross pollinated. Proper isolation distance of 1000 m helps in maintaining the purity of varieties. The seedlings are transplanted 45 × 30 cm during November and seeds are harvested during March-April. Individual dried flowers are harvested. A yield of 1000-1060 kg/ha is obtained from Pacifica mixed.

Mesembryanthemum crinifolium. It belongs to family Aizoaceace and plants are dwarf with thick flashy leaves. Single flowers in different colours are produced and flowers open in the sun which are very shining and attractive. Different colours are maintained by 1000 m isolation distance. Seeds are extracted by beating the mature fruits when they get completely dried. A yield of 375-400 kg/ha is obtained.

Tropaeolum majus. It belongs to family Tropaeolaceae. Plants are of trailing type and flowers have large long spur. Highly cross pollinated crop and it needs an isolation distance of 1000 m for maintaining purity. The crop is sown directly. Seeds are collected by brooming when crop is completely dried in April-May. An average yield of single type is about 1000-1250 kg/ha whereas of dwarf type is 150-200 kg/ha.

Alyssum maritimum. It belongs to family Cruciferae and very small flowers of white and purple colour are produced in great profusion. Plants are very dwarf and hardly grow about 20 cm. Seedlings are transplanted in last week of October to first week of November at spacing of 30 × 20 cm and flowering occurs during January to March. Seeds are harvested in April as a single cut and good yield of 125-150 kg/ha is obtained very easily.

***

19

Hedges, Edges and Topiary

HEDGES

When shrub is planted on boundary for fencing, it is called as hedge. It may be ornamental or protective or both. Hedges also become an important feature of formal layout to serve various functions as screening of the area, out buildings, tennis courts, vegetable gardens, hiding of unwanted places, etc. All the shrubs cannot be planted for making an ideal hedge due to different nature of growth. Only some selected shrubs are used for making an effective hedge.

For selecting an ideal shrub for a hedge, it should have following characteristics;

(*a*) It should have thick texture and quick growth.
(*b*) It should stand trimming to shape.
(*c*) It should be easily propagated through seeds or cuttings.
(*d*) It should withstand drought conditions.
(*e*) It should not attract reptiles or other animals.

Classification of Hedges

Hedges are planted to protect the area to avoid the trespassing by man or animals or to beautify the boundary of different. Therefore, according to purpose, hedge can be classified as :

1. Tall protective hedges. The height is about 1-3 m and growth is very dense alongwith thorns. The plants that can be selected are : *Inga dulcis, Acacia fernesiana, Carrisa carandus, Bougainvillea* sp., etc.

2. Dwarf protective hedges. Dwarf shrub grows about 1 m and have thorns and are protective in nature *e.g. Euphorbia bojori, Opuntia* sp., *Agave* sp., *Pedilanthus* sp., etc.

3. Tall ornamental hedges. Shrubs grow 1-3 m tall and have attractive foliage. Plants produce colourful flowers also *e.g. Lawsonia alba, Duranta plumieri, Dodonea* *viscosa, Casuarina equisetifolia, Murraya panniculata, Thevetia peruviana, Hibiscus rosasinensis, Stenolobium stans, Hamelia patens* etc.

4. Dwarf ornamental hedges. Height of dwarf hedge is about 1 m and plants are very attractive *e.g. Acalypha* sp., *clerodendron inerme, Thunbergia erecta, Daedalacanthus nervosus, Lantana* sp., etc.

Growing of Hedges

Preparation of soil. Hedge plants can be grown in various kinds of soil. For better growth of plants soil should be deep, well drained and fertile. If the soil is not so good, it should be changed upto the upper 45 cm profile with fertile garden soil. Hedges are planted in the garden for various purposes and persist for many years. Therefore, planning and preparation of ground should be done carefully. For making a good growth of hedge, a trench about 60-75 cm deep and 30-60 cm wide should be dug up and left exposed for a fortnight or so before planting, in order to destroy harmful microflora and insects by scorching sun. Then it should be refilled with soil and sufficient amount of well rotten farmyard manure.

Planting time. The most suitable time of planting hedge is rainy season. In the beginning of monsoon weather conditions are hot and humid. Therefore, during August, when temperature cools down, planting can be done safely. Planting can also be done during February-March depending upon the availability of regular supply of water and planting material.

Planting Method. Rooted cuttings or seeds are planted. The distance of planting for tall hedge should be about 60-90 cm and for dwarf hedge it should be about 20-30 cm. The planting should be done by triangular system so that it makes a good dense hedge.

Maintenance

Irrigation. Water requirement of hedges depends upon season and soil types. During rainy season generally no irrigation is required except during drought conditions. During winter season, watering is required once in 10-15 days and at weekly intervals during summer season. In light soil more frequent watering is requiring than in heavy soils.

Weeding. Weeds should not be allowed to grow to compete with hedge plants specially in early stages and hence weeds should be removed as they make their appearance.

Trimming. When the hedge plants attain a height of 15 cm they should be topped back to 10 cm height with the garden shears. This will result into production of several side growths on each plant. If such topping is not done and plants are allowed to grow unchecked, hedges will not be dense and look untidy. Topping is done many times till desired height is achieved. Later on hedges can be trimmed to get different shapes. The top of hedges can be kept flat, wavy, blocks of square or any shape. The frequency of trimming depends upon season and type of plants. In rainy reason frequent trimming is required because plants make faster growth than winter or summer season. Faster growing plants need frequent trimming than the slow growing ones.

EDGES

When low growing perennial plants are grown on the border of plots or beds, they are called as edge plant or an edge. These plants hardly grow upto 20-30 cm. Like hedges, they also become a part of formal or informal landscape designs. In garden, edges are also planted around rockery, big trees, alongside walks, pathways and to divide the area. Their culture is similar to that of hedges. The edge plants are usually propagated through terminal cutting during rainy season and stand against trimming. There are only few plants suitable for making perennial edge *e.g. Eupatorium cannabium, Justicea* sp., *Iresine lindenii* and *Alternanthera amoena* and *A. versicolour.*

TOPIARY

It is an art of training plants into different shapes, *i.e.*, of birds, animals, domes, umbrellas, etc. It is an old art and now-a-days it is becoming common in city parks to provide passive recreation to the visitors specially children. In the houses, miniature topiary is also practised in the pots. For making a good topiary all the shrubs and trees are not ideal. The ideal plants should be quick growing with small foliage and stands against trimming which is done frequently. The most common plant used for making topiary is *Clerodendron inerme* and other plants for topiary work such as *Duranta plumieri, Sesbania aegyptiaca* (for temporary topiary) and trees like *Inga dulcis, Ficus benjamina, Callistemon salignus* (**Plate 8**) and *Acacia modesta*, etc.

Training frames are employed for making topiary. These frames are generally made from soft steel rods or galvanized wire frame. Making frames true to shape is highly technical and artistic for true depiction of figures. Hence, good artist should be employed which will make good curves of shapes. The figures of lady with a pitcher, elephant, girraffe, camel, ox, monkey, birds, peacock, farmer and bullocks and umbrella etc. are liked by everybody. For picturesque topiary, there should be a broad base or an platform of plants.

Planting of selected plants at selected place is done at an early stage. Stopping of branches is done for dense growth and then frame is placed.

For large topiary the desired frame is first erected and number of plants are planted depending upon the size of the frame. For large animals 2-4 plants are planted at different points. The desired shape can be achieved by training, tying and constant trimming of branches.

Maintenance of topiary plants need special care. Plants should be liberally watered and fertilized so that plants make a dense and colourful growth.

***

20

Shrubs

Shrubs are very important in the garden as flowering shrubs produce beautiful flowers at eye level and fragrant shrubs emit fragrance at nose level. They display their varied beauty by way of richly coloured flowers, handsome foliage and their shape. Variegated shrubs especially with silver or dark grey foliage can also produce pleasing effects if planted in appropriate schemes as they provide most suitable contrasts in the garden. They become an integral part of the garden, being permanent characters. They form part of the framework of the garden and create very pleasing picturesque effect if selected and planted carefully. Shrubs are very popular amongst gardeners on account of their great diversity in colour of flowers, height, ease in propagation, being grown easily in varied climatic conditions and in good well drained soil.

There are four chief purposes of growing shrubs in the garden :

1. They enhance the beauty of surroundings.
2. They take the place of garden boundary wall and provide liveliness to the garden.
3. They are planted for screening purposes to hide unwanted places of the garden like manure pits, dust bins, etc.
4. They are also planted to divide the area of the garden such as ladies corner or children's corner.

The shrubs can be classified easily on the beauty of plant parts.

1. Flowering Shrubs

These shrubs produce very attractive and spectacular flower which are grown for mass effect and carpetting purposes. Suitable shrubs are : *Hibiscus rosa-sinensis, Hibiscus mutabilis, Caesalpinia pulcherrima, Bouganinvillea* sp., *Ervatamia divaricata, Malavaviscus arboreus, Nerium oleander, Thunbergia errecta, Stachyterfeta indica, Ixora parviflora, Calliandra* sp., etc.

2. For Foliage

These shrubs are grown in the garden for handsome and richly variegated foliage *e.g. Acalypha tricolour, Nandina domestica, Codium variegata, Manihot variegata,* etc.

3. For Flowers and Foliage

Some shrubs have handsome foliage and produce good flowers too. They add to the beauty in the garden tremendously *e.g. Bougainvillea,* (Thimma and Dr. Rao), *Buddleia asiatica, Buddleia madagascariensis, Hamelia patens,* etc.

4. For Fragrant Flowers

These shrubs produce fragrant flowers *e.g. Cestrum nocturnum, Cestrum diurnum, Murraya panniculata, Jasminum sambac, Jasminum auriculatum*, etc.

Shrubs in the garden are planted in the following ways :

1. Specimen shrubs. These shrubs are planted singly or in group in the beds on the lawn to create a picturesque effect. These shrubs have elegant foliage and produce beautiful flowers. For the best effect different shrubs can be selected for different landscaping setting. Suitable shrubs are : *Bougainvillea, Hibiscus rosa-sinensis, Hamelia patens*, *Thevetia peruviana*, etc.

2. Standard shrubs. Any shrub can be trained as standard shrub. In case of full standard shrub, main stem is clean upto 100 cm whereas in half standard shrub stem is clean upto 50 cm.

3. Shrubbery border. The area of the garden devoted exclusively to shrubs is called as shrubbery border. In the garden, to secure privacy and provide partition, formal hedges are used. The labour and cost of maintaining hedges is much more than keeping a well planned shrubbery. Therefore, the shrubbery is preferred which also adds beauty to the garden. In planting shrubberies, consideration of foliage and colour effect should be kept in mind before planting. Prior to laying out shrubbery border, the following points should be kept in mind.

1. By keeping shrubbery in front of the trees, a very pleasing effect will be obtained.
2. Best result is achieved by facing east or south.
3. Allow greater width where greater height is desired. Greater height is necessary when shrubbery is away from the house. But when it is near to the house, the low height is needed.
4. The shade of trees should be avoided as far as possible.
5. Shrubbery is also used to limit the gardens especially in case of large gardens.

Planning a shrubbery border. The general principles are described earlier which should be kept in mind before planting a shrubbery.

After deciding site for shrubbery, first of all a paper plan should be prepared which will help in final planting.

Arrangement of Shrubs

1. According to height. Shrubs are divided into three groups *i.e.*, tall, medium and dwarf shrubs. Tall shrubs should be planted 30-45 cm away from the edge of the bed. On very large area two or more *shrubs* may be planted in a group for better results. The *intermediate* or *medium* shrubs should be planted in group of 3-4 plants and 30 cm away from the inner edge of the bed. The *dwarf shrubs* should be planted in group of 4-6 plants.

Double face shrubbery border. Just to get the pleasing effect of the shrubbery border from both the sides, double face shrubbery border is preferred. In this type, tall shrubs are planted in the center followed by medium and dwarf shrubs both the ways.

2. According to colour. The flowering shrubs produce a nice colour effect if planted carefully. Before actually planting, consideration of colour scheme should also be kept. It depends upon gardener to use colour contrast scheme or dominance of one colour.

Growing of Shrubs

Soil. Any garden soil is suitable for growing shrubs. Height and rate of growth depend upon to some extent on type of soils. If soil is bad, their top layer upto 45 cm should be removed and refilled with good fertile and well drained soil. The shrubs can be grown in varied pH of soil. If the soil is highly alkaline or acidic in nature, it should reclaimed by using soil amendments. Following is the list of shrubs which can tolerate high pH of the soil. Salt tolerant shrubs are : *Bougainvillea* sp., *Russelia equistifolia, Thevetia peruviana, Nerium indicum* and *Caesalpinia pulcherrima*. Salt sensitive shrubs are : *Buddleia asiatica, Lagerstroemia indica, Barleria cristata,* and *Hamelia patens*.

Preparation of the ground. After the alignment has been done for the planting of shrubs, soil should be prepared thoroughly atleast 15 days prior to planting. Prepare the pits of 60 cm deep and expose the soil to the sun for atleast 2 weeks. Refill the pits with the mixture of soil and farmyard manure (2-4 baskets/pit). Fill the pits 10-15 cm higher than the general level of the field to avoid the formation of a depression after the first irrigation. Prepare the rest of the ground by digging it about 30 cm deep and removing all the weeds. Chlorpyrifos @ 5 ml/l of water should also be mixed with the soil to prevent the damage by termites.

Propagation of shrubs. The shrubs are propagated by seeds, cuttings and layering.

By seeds. Seeds are collected when they are fully ripe. They are dried in the shade and stored in airtight bottles in dry places. In rainy season, nursery is prepared and seeds are sown. Seedlings are transplanted in prepared ground. The shrubs which are mainly propagated by seeds are : *Stenolobium stans, Thevetia peruviana, Caesalpinia pulcherrima, Calliandra* sp. etc.

By cuttings. Majority of the shrubs are propagated by cuttings. The best season for propagation by cuttings is rainy season, although cuttings can be made in February-

March if water supply is not limited. The cuttings are made 15-20 cm long and planted in pots or beds containing ample of sand. The plants raised from cuttings are planted in the next planting season. The shrubs propagated by cuttings are; *Hibiscus rosa-sinensis, Achania malvaviscus, Jasminum sambac, Hamelia patens, Bougainvillea, Cestrum diurnum, Cestrum nocturnum*, etc.

By layering. The shrubs which are not easily propagated by seeds or cuttings are propagated by this method. In rainy season, layering is done by removing the bark of 2.5-3.0 cm long and wrapping by garden soil and sand mixture or sphagnum moss. The shrubs like *Bougainvillea, Ixora* sp., *Jasminum sambac* and *Jasminum multiflorum* (ground layering) etc. are propagated by layering.

Planting of shrubs. The best planting season of shrubs is rainy season. After the onset of monsoon, planting can be done. Planting can also be done successfully in the month of February-March depending upon the water facilities during summer months.

The planting distance of different shrubs is variable according to their height and spread. The tall shrubs may be planted at the distance of 150-180 cm. In very large shrubbery two or more shrubs may be planted in groups for better results. The medium shrubs should be planted in one or more lines between tall and dwarf shrubs. They are planted at a distance of 90-120 cm. The dwarf shrubs in front line should be planted atleast 30 cm away from the inner edge. The distance of planting is 45-90 cm depending upon the shrub. For the best effect they should be planted in a group of 4-6 plants in the length of 3.0-4.5 m.

Watering. Immediately after planting, watering should be done frequently till they are established.

Filling of gaps. Newly planted shrubberies need attention in the beginning specially for filling the gaps. As soon as mortality occurs, new plants should be planted.

Irrigation. The shrubbery should be irrigated at regular intervals. Frequency of irrigation depends upon soil texture, season etc. Under Punjab conditions, during rainy season no irrigation is required. From September onwards, irrigation should be done at monthly intervals upto March. After that irrigation should be done at fortnightly intervals. In sandy soils, more frequent irrigation may be required.

Weeding. Weeds should be remove regularly otherwise they compete with main plants for nutrients and moisture etc. During rainy season, one or two weedings are necessary and after that occasional weeding will serve the purpose.

Pruning of established shrubs. Shrubs need careful pruning. Winter flowering shrubs *e.g., Poinsettia pulcherrima, Holmskioldia* sp., and *Doembeya* sp. are pruned at the end of summer season whereas others are pruned at the close of winter season. Cutting back the growth of the shrubs is very important and the performance of the plants is improved. Trimming and method of their pruning depend upon the form and age. All the superfluous growth, old or weak branches are best cut at ground level or as low as possible without leaving the stub.

Description of Shrubs

There is almost an unlimited varieties of plants falling within a broad category of shrubs. But for excellent results and picturesque effect, the small number of shrubs should be planted in the garden. The characteristics of important shrubs are described here under.

Acalypha (Euphorbiaceae)

It is native of East Indies. This shrub is famous for its copper coloured and variegated foliage which is highly desirable for contrast effect in landscaping of garden, buildings etc. It is a medium shrub growing upto 1.2-1.5 m, prefers semi-shady situation flowers are red and long like catkin. The shrub produces the flowers throughout the year but profusely from February to July, and is very easily propagated by cuttings. Following are the species which can be grown very successfully : *A. hispida, A. wilkesiana, A. colorata.*

Malvaviscus arboreus (Malvaceae)

A medium shrub growing upto 1.2-1.5 m tall. This shrub can be trained as standard bush shrub or can be planted closely and trimmed to make an attractive hedge. The foliage is dull green but flowers are bright scarlet, pink half-open, producing throughout the year. Easily propagated by semi-hardwood or hardwood cuttings.

Barleria cristata (Acanthaceae)

This shrub is native to India. It is a dwarf shrub growing upto 0.6-1.2 m and produces numerous flowering shoots from the ground level. The shoots are cylindrical and quite hardy. Violet blue flowers appear during March-April. This is very easily propagated by cuttings. There are several species producing white, rose, orange and red flowers.

Brunfelsia americana (Solanaceae)

It is native of West Indies and commonly called as Lady of the Night. It is an erect, tall, spreading shrub which produces yellow flowers during summer months. It is propagated through terminal cuttings. It is highly suitable for group planting or on the mounds.

Brunfelsia pauciflora (Solanaceae)

It is commonly called as yesterday, today and tomorrow and is bushy nature. Leaves are leathery, lanceolate and glossy dark green. It produces purple flower from spring to summer which fades away almost to white.

Buddleia asiatica (Scrophulareaceae)

It is native to South East Asia and Indonesia. An attractive shrub, growing upto 1.5-2.4 m. Leaves are lanceolate green upper side and whitish underneath. Flowers are white, fragrant produced on long drooping recemes, abundantly from February to April. Easily propagated by cuttings.

Buddleia madagascariensis (Scrophulearaceae)

It is native of Madagascar. This is a very vigorous shrub which can be planted on the periphery of the gardens. The leaves are comparatively longer, dark green, and whitish underneath. Flowers are whitish yellow, highly scented and produced freely in months of February-March. Propagation is done by cuttings. Very severe pruning should be done after flowering.

Senna sulpurea **Syn.** ***Cassia glauca*** (Leguminosae)

It is native of tropical America. It is tall, handsome and quick growing shrub suitable for shrubberies and group planting. It produces yellow flowers in September-October. It is easily propagated from seeds. (**Plate 9**)

Senna pallida **Syn.** ***Cassia biflora*** (Leguminosae)

It is ever green medium shrub with spreading branches. Leaflets are small, dense and of dark green coloured. It bears yellow flower during March-April and October-November. It is propagated through seeds.

In gardens it is used as a specimen or for group planting. It has become very popular in Indian gardens in very short duration due to its elegant beauty.

Calliandra sp. (Leguminosae)

The name Calliandra means "beautiful stamens", and several of the approximately 150 shrubs and small trees belonging to this genus are grown for their dense heads of pink or white silky stamens that appear on short stalks along the upper sides of the branches.

Calliandra brevipes. It is native to South Eastern Brazil, Uruguay, and northern Argentina. It is a small shrub with double-compound leaves. The number of leaflets is much larger than that in *C. haematocephala.* Leaflets are very closely spaced and are often 'closed up'. It rarely grows taller than 1.8 m. Flowers are a beautiful power-puff of erect stamens which are white towards the base and pink above.

Calliandra haematocephala. The species name haematocephala comes from the greek for with blood-red head. Flowers are 2-3 inches, big red or white puffs, appearing almost throughout the year. This large, multiple trunked, low-branching, evergreen shrub has silky leaflets which are glossy copper when new, turning to a dark metallic green. It can be used as a specimen, hedge and is often recommended for buffer strips around parking lots or for median strip plantings in the highway. It grows best in sandy soils and full sun. It responds favourably to regular watering while young but does not need special care after establishment except occasional pruning to keep it confined. It can grow to a small tree of about 15 feet height. Pinching the new growth increases branch number and produces more flowers on a more compact plant and should be trimmed after it finishes blooming.

Calliandra tergemina* var. *emarginata* Syn. *Calliandra emarginata. This is a medium-sized shrub which looks spectacular when in bloom. The flowers are brilliant red and give a sharp contrast to leaves. The foliage is also attractive with light green kidney shaped leaflets, appearing in pairs. It can grow up to six to seven feet. The flowers attract hordes of butterflies. These plants flower all year round, but the best blooming is in spring and summer.

Ceasalpinia pulcherrima (Leguminosae)

It is native of tropical America and commonly called as Peacock flower. It produces scarlet to yellow flowers in great profusion from May to August. It is quick growing shrub suitable for shrubbery and inner roads. It is easily propagated through seeds.

Cestrum diurnum (Solanaceae)

This is native of tropical America and commonly called as 'Din Ka Raja' and very popular for sweet scented flowers. It grows upto height of 1.5-1.8 m. Leaves are lanceolate and lush green. Flowers are white borne in clusters and open in the day. Flowers are produced throughout the year. Propagation is done by cuttings.

Cestrum nocturnum (Solanaceae)

It is native of tropical America and popularly known as 'Raat ki Rani' and is very common shrub in the Indian gardens. Plant grows upto 1.5-1.8 m high. Leaves are dark green. Shrub is valued for highly fragrant flowers which opens after dusk, making night fragrant. Flowers are pale in colour, produced in clusters twice or thrice in a year. This shrub is easily propagated by cuttings.

Dombeya spectabilis (Malvaceae)

This Madagascar native shrub has large maple-shaped leaves and dense heads of little pinkish white flowers. It blooms from November-January. Its flowers are lighter in color, not as pink as the *D. wallichii* and the stalks from which the flower bunches hung are much shorter. Most importantly, the leaves are very much lobed like a maple, whereas in the *D. wallichii*, the leaves are heart-shaped. Leaves are shallowly 3-lobed, serrate with rusty pubescent underneath. It is commonly grown as a large roadside shrub.

Dombeya wallichii (Malvaceae)

It is known by the common names pinkball, pinkball tree, and tropical hydrangea. Its native range includes India, East Africa, and Madagascar. It can grow to height of 20 to 30 feet tall, spread of up of 25 feet making a small tree if not regularly pruned. The alternately arranged leaves are heart-shaped with serrated edges. The hanging flower clusters are pink, showy, and fragrant. The shrub maintains a rounded form and makes a beautiful specimen plant. Growing rapidly in full sun to partial shade on a wide variety of soils, it needs protection from frost in initial years and is not salt tolerant. The dried flower heads should be removed regularly for a neat appearance of this otherwise very low-maintenance plant. Plants need regular watering until established and once established, can withstand prolonged dry periods.

Eranthemum pulchellum Syn. ***Daedalacanthus nervosus*** (Acanthaceae)

It is native of east parts of India. This is a hardy, dwarf shrub growing upto 60-90 cm tall. Flowers are blue in colour, borne on spikes. Flowers are produced in March. This is very suitable for hedge. Easily propagated by cuttings.

Ervatamia divaricata (Apocynaceae)

It is commonly called as 'Chandni' or Malti in Hindi on account of its coverage with pure white flowers in profusion. It is very common in Indian gardens as flowers are used in worship. A tall, evergreen shrub, growing upto a height of 1.8-2.4 m. The leaves are smooth and glossy green which make shrub very attractive. Flowers are white, single, produced in great profusion throughout the year. They are sweetly fragrant. It is easily propagated by cuttings. It has another form which produces double flowers.

Very recently, natural dwarf form of *Ervatamia divaricata* has been isolated and getting popular amongst gardeners. It can be mostly used as dwarf bush plant in shrubbery or used as pot culture.

Eunymous japonicus (Celastraceae)

It is native of Japan and commonly called as Privet or Japanese Spindle. It is an ever green, dwarf bush with variegated leaves. It is commonly propagated through terminal cuttings. In gardens it is used as dwarf hedge or edge or pot plant.

Euphorbia cotinifolia (Euphorbiaceae)

In gardens it is planted for its coppery coloured leaves which makes good contrast with other green shrubs. It is winter deciduous under North Indian conditions. It is commonly propagated through cuttings. It is highly suitable for shrubbery border or pots.

Ficus panda (Moraceae)

It is most showy hardy plant which has dominated in Indian landscape in very short period. Its new emerging leaves are light green coloured and older leaves are of green colour which makes it more beautiful. It stands to trimming very well and hence it is used garden as specimen, group planting, topiary, hedge, etc. It is commonly propagated through terminal cuttings during rainy season with 85-90% success.

Galphimia gracilis (Malpighiaceae)

It is native of tropical America. The shrub is medium in height and gives bushy appearance. It is very attractive shrub on account of its neat yellow flowers which appear on comparatively long stalks arranged on terminal recemes. Golden yellow flowers make a very effective contrast against its dark green foliage. Flowers are produced almost throughout the year but very profusely during July-November. Propagation is done by seeds.

Hamelia patens (Rubiaceae)

It is native of tropical and sub-tropical America. This shrub is very famous amongst gardeners for attractive foliage and scarlet flowers which are produced throughout the year. It is a tall shrub growing about 2.4-3.0 m high. Due to profuse branching it shapes well and appears elegant. Foliage is green and greenish bronze which stands clippings and trimming to any shape. This shrub can serve very well either as standard, specimen or hedge in the garden. It can be very easily propagated by cuttings.

Recently, natural dwarf form has been selected which is getting very popular in the garden. It is mostly used as dwarf bush in shrubbery, edging plant or as a dwarf specimen bush plant or ground cover or shrub for larger pots. It is very floriferous and is liked most for its bright scarlet coloured flowers.

Pseudocaryopteris bicolor* Syn. *Clerodendrum odoratum (Verbenaceae)

It is a tall shrub of about 2.4-3.0 m. Leaves are light green in colour which provide freshness and coolness in summer. Flowers are mauve coloured and appear in March. They are borne on terminal corymbs. It should be pruned in end of June to maintain the shape of the bush.

Hibiscus rosa-sinensis (Malvaceae)

It is commonly known as "China shoe" flower and in Hindi as 'Gurhal' and is native of China. This shrub is tall growing about 1.8-2.4 m and spreading in nature. Leaves are dark green and shining. Flowers are brilliant rosescarlet and produced throughout the year. Flowers are big and prominent and due to their bright colour and can be seen from a distance. There are large number of hybrids available *e.g.*, Alipore Beauty, Daffodil, Australian Rose, Sweet Heart etc. The height of the bush can be maintained by pruning. Highly suitable for specimen, standard and ornamental hedge. Easily propagated by cuttings.

Hibiscus mutabilis (Malvaceae)

It is native of China. It is a tall shrub growing about 1.8-2.4 m. Leaves are light green and very rough in look. Flowers are double and produced in end of October to November and they change the colour from creamy white to deep rose and blue with the advancement of day. Pruning should be done in the month of April. It is propagated by cuttings.

Holmskioldia sanguinea (Verbenaceae)

It is commonly called as cup and saucer on account of its lower shape and is native to India. It is evergreen and handsome shrub bearing brick red to orange flowers from November to February. It is easily propagated by cuttings. *H. acuminata* is another handsome species producing yellow flowers.

***Ixora* sp.** (Rubiaceae)

In Hindi it is called as "Rukmini" and is native of tropical Africa. It is a tall shrub growing about 1.5-2.5 m. Leaves are smooth, dark green. Flowers are borne on terminal trusses throughout the year but in profusion in the hot and rainy season. They are liked by gardeners due to variation in colour, long cut life and natural flower arrangement. The Ixoras are propagated easily by layering. The pruning should be done after flowering season. There are several species but common are : *I. coccinea*—bright red flowers, *I. parviflora*—pure white, *I. singaporensis*—terra cotta coloured flowers.

Jatropha multifida (Euphorbiaceae)

It is native of West Indies. A tall shrub growing upto 1.8-3.0 m and is very common in gardens. It looks better if kept bushy by pruning the branches after flowering. Flowers are small, scarlet and borne at the end of branches in hot and rainy season. It is easily propagated by seeds and cuttings.

Justicia carnea (Acanthaceae)

It is an evergreen erect dwarf bush and scarcely branched. It bears purplish pink flowers from late summer to autumn. It is commonly propagated through terminal cuttings. In garden it is commonly used for edging or shrubbery or pot plant.

Lagerstroemia indica (Lythraceae)

It is native of China and is commonly called as "Shravani" because its attractive bloom might have been observed in the month of "Sharavan" but it blooms from May-August. It is a tall shrub growing about 1.8-3.0 m and winter deciduous in nature. Leaves are small and dark green. It is very popular in the gardens on account of its being easily grown in any garden soil. Flowers are very attractive, soft, fringed, arranged in long erect sprays produced in white, purple and rose colour on shoots of current year's growth. Pruning should be done in February. It is easily propagated by cuttings and seeds.

Lantana camara (Verbenaceae)

It is native of Jamacia. It is very hardy shrub and popular in the gardens as it appears pretty when in bloom. It grows upto a height of 1.0-1.5 m and has spreading habit. Foliage is dark green and rough in texture. Flowers are produced in different colours *i.e.* yellow, white, red, orange, lilac, mauve etc. It is easily propagated by seeds or cuttings. Highly suitable for ornamental hedge.

Murraya panniculata (Rutaceae)

It is native of South Asia and is commonly called as 'Kamini'. It is most popular shrub in the garden. It is a tall shrub growing upto a height of 2.4-3.0 m. The foliage is dark and shining comprised of deep glossy green pinnate leaves. The shrub is highly suitable for specimen shrub or Topiary or for ornamental hedge because it stands trimming well. Flowers are white, produced in clusters and emit strong fragrance in rainy season. Propagation is done by seeds or cuttings.

Mussaenda luteola (Rubiaceae)

It is native of tropical America and medium shrub. Flowers are yellow which are produced from May to August but yellow foliaceous bract is very attractive. It is highly suitable for specimen and group planting. Easily propagated by cuttings. *M. philipica* is beautiful species which is prized for its pink bracts.

Nerium aleander (Apocynaceae)

It is called as 'Lal Kaner', and is native of tropical and sub-tropical Asia. It is tall shrub growing upto a height of 2.4-3.0 m. Stems arise from ground level. Leaves are dull green, hard and rough. The flowers are produced in clusters throughout the year. There is a great variation in flower colour *i.e.* from white, cream, pink, rose, red, crimson, etc. Flowers may be single or double. It is easily propagated by cuttings.

Nyctanthes arbortristis (Oleaceae)

It is native of India and is popularly called as "Har Shingar" or "Tree of Sadness" because flowers are shed like tears by falling of morning sun rays. It is a tall shrub. If left unpruned it grows to a height of dwarf tree. Leaves are dull green and rough in texture. The flowers are small, numerous, white with orange coloured centre, very fragrant and star shape borne in large terminal clusters. Flowering season is from August to December. The pruning should be done in winter. It is easily propagated by seeds or cuttings.

Pentas carnea (Rubiaceae)

It is commonly known as star cluster. It is a woody dwarf, bushy with attractive foliage. Star shaped flowers of pink, magenta, red and crimson are produced from March to November. It is very easily propagated through terminal cuttings. It is highly suitable for pot culture and shrubbery border.

Preynia disticha Syn. ***Phyllanthus nivosus*** (Euphorbiaceae)

It is native of tropical forest of Asia and Australia and commonly called as 'Snow Bush'. Leaves are round and upper leaves of branches are variegated. New emerging leaves during April-May become almost snowy white and hence 'Snow Bush' name has been given. It is dwarf bush in frost prone area whereas it grows to its medium height in tropical climate. *P. nivosus compacta* is a much dwarf kind. *P. nivosus roseopicta* is another handsome kind in which upper leaves of branches turned into deep pink and get mottled. It is easily propagated through separation or through soft wood cuttings.

Plumbago auriculata (Plumbaginaceae)

It is commonly called as "Chitra" and is native of South Africa. It is an evergreen small shrub growing upto a height of 1-1.2 m. Numerous branches are produced from the ground level. Leaves are small and light green in colour. The light blue flowers are produced throughout the year but in profusion during July-October and December to

LIST OF IMPORTANT SHRUBS

S.No.	Name of shrub	Colour of flowers
Tall Shrubs		
1.	*Buddleia asiatica*	White
2.	*Buddleia madagascariensis*	Whitish Yellow
3.	*Cassia glauca*	Yellow
4.	*Caesalpinia pulcherrima*	Red, Yellow
5.	*Doembeya wallichii*	Pink
6.	*Ervatamia divaricata*	White
7.	*Hamelia patens*	Red
8.	*Pseudocaryopteris bicolor*	Mauve
9.	*Hibiscus rosa-sinensis*	Red
10.	*Hibiscus mutabilis*	White changing to deep Pink to Blue
11.	*Holmskioldia snaguinea*	Red
12.	*Jatropha multifida*	Red
13.	*Lagerstroemia indica*	White, Red, Purple
14.	*Murraya panniculata*	White
15.	*Nyctanthes arbor-tristis*	White with orange colour
16.	*Nerium indicum*	White, Pink, Red
17.	*Punica granatum*	Red
18.	*Poinsettia pulcherrima*	Red, Yellow
19.	*Stenolobium stans*	Yellow
20.	*Thevetia peruviana*	Yellow
Medium Shrubs		
1.	*Acalypha* sp.	Colourful foliage
2.	*Malvaviscus arboreus*	Red
3.	*Cestrum nocturnum*	Creamish white
4.	*Cestrum diurnum*	White
5.	*Jasminum multiflorum*	White
6.	*Lantana* sp.	White, Red, Yellow
7.	*Mussaenda luteola*	Yellow
8.	*Stachytarpheta indica*	Violet
9.	*Tecoma capensis*	Red, Yellow
10.	*Thunbergia erecta*	Violet
Dwarf Shrubs		
1.	*Barleria cristata*	Violet-blue
2.	*Daedalacanthus nervosus*	Blue
3.	*Galphimia gracilis*	Yellow
4.	*Jasminum sambac*	White
5.	*Lantana sellowiana*	Blue
6.	*Pumbago auriculata*	Blue
7.	*Russelia equistiformis*	Red
8.	*Russelia floribunda*	Deep red

March. Highly suitable for ornamental hedge, carpetting or as a pot plant. It is easily propagated by suckers and cuttings.

Euphorbia pulcherrima Syn. **_Poinsettia pulcherrima_** (Euphorbiaceae)

This is a very attractive shrub and highly suitable for bedding purpose. This grows upto a height of 2.4-3.0 m. The branches are cylindrical smooth. Leaves are alternate, smooth and glabrous. Inflorescence is terminal. This shrub is highly valued for showy bracts in different colours *i.e.* white, pink and crimson. They are produced in December-January. It is easily propagated by cuttings. This shrub should be cut back drastically in the month of February-March.

Punica granatum (Lythraceae)

It is commonly called as "Anar". It is a hardy, tall ornamental shrub growing upto a height of 1.8-2.0 m. The flowers are bring scarlet in colour which make a very good contrast against dark green foliage. They are produced in rainy season. It is easily propagated by cuttings.

Ravenia spectabilis Syn. _Lemonia spectabilis_ (Rutaceae)

It is a native of Caribbean region. Leaves are divided into three leaflets which are elliptic, dark green and glossy. It bears bright dark pink flowers, prefers semi-shade conditions and regular watering. The flowers have a flattened appearance and petals have a grainy, hard texture. It is a very good plant for attracting butterflies. It can grow to a height of 3-5 meters but is mostly maintained as a shrub. It is especially attractive when grown in shrubby borders or groups. It blooms throughout summer months in North India.

Russelia equisetiformis (Scrophulariaceae)

It is native of India. It is dwarf shrub with thin scattered fine leaves. Flowers are scarlet which are produced round the year. It is easily propagated by division. *R. floribunda* produce dark red flowers.

Stachytarpheta indica (Verbenaceae)

It is native of West Indies. It is a hardy medium grown shrub. The branches arise from close to the ground. Leaves are green and have neat venation. Flowers of deep violet colour, produced on long spikes throughout the year. It is very easily propagated by seeds or cuttings.

Tecoma stans Syn. _Stenolobium stans_ (Bignoniaceae)

It is native of America and is very common in Indian gardens. It is a tall shrub growing upto a height of 2.4-3.0 m. The leaves are pinnate with 5-7 leaflets. It is a very hardy and quick growing, thus needs annual prunning to keep in shape. Highly suitable for specimen or ornamental hedge. The flowers are golden in colour, trumpet in shape and are produced in terminal clusters throughout the year, propagated by seeds.

***Tecoma castanifolia* Syn. *Tecoma gaudichaudii* (Bignoniaceae)**

It looks quite similar to *T. stans* but the leaves are usually simple, can have up to 3 leaflets, toothed, up to 12 cm long and bears flowers in bunches. It blooms almost throughout the year.

***Tecoma Capensis* (Bignoniaceae)**

It is commonly called cape Honey Suckle and native of South and East Africa. It is an evergreen, medium in height growing about 1.5-2 m and behaves like rambler and hence, it is kept in shape by trimming. The leaves are glossy green, pinnate and leaflet are opposite in pairs of oval to round serrated leaflets. Flowers are produced in orange and yellow colour in terminal clusters.

It grows happily in full sun or in partial shade in well drained, fertile, rich soil and adaptable to various conditions. It can tolerate moderate drought conditions. It is easily propagated by seed or semi hard wood cuttings.

***Cascabela thevetia* Syn. *Thevetia peruviana* (Apocynaceae)**

It is native of Peru and is commonly called as 'Pili Kaner' and very popular in Indian gardens as flowers are used in offering worship to 'Lord Shiva'. It is a tall shrub growing upto a height of 2.4-3.0 m. The foliage is lush green and are very attractive. Flowers are yellow, orange or white funnel shaped and are produced in terminal cymes throughout the year. It is highly used for specimen, screening or as an ornamental hedge. Propagated by seeds.

***Thunbergia erecta* (Acanthaceae)**

It is native of tropical Africa. It is a medium shrub growing upto a height of 1-1.2 m. It is hardy shrub producing numerous erect branches close to ground. Flowers are violet, funnel shaped and look very attractive when they peep out from foliage during summer and rainy season. Highly suitable for shrubbery and for ornamental hedge.

***

21

Trees

Trees are perennial, tall, with marked trunk and grow for several years, bear flowers and fruits. On this planet, trees are growing for the last 50 million years. In different biotic zones, different types of trees dominate which have developed through evolutionary process and ecological succession. With the spread of human civilization and increase in population pressure, forests have been cleared for cereal production. Unjudicious felling of trees resulted into ecological imbalance, change in climate, desertification and deterioration of soil conditions.

Trees provide us fruits, shelter, shade, fuel wood, and timber wood for making furniture, ships, paper etc. Other economic products like rubber, *broza*, gum and resins are also obtained from trees. From medicinal viewpoint, parts of trees are also used to make medicines. The value of trees in terms of service given to the society is very high according to calculation done by T.M. Das. Fifty year old tree weighing about 50 tonnes, would produce for the society in the next fifty years value worth of ₹ 1.57 m for different services, for instance value for oxygen—0.25 m, prevention of air pollution and cleaning of atmosphere—0.50 m; rain water and regulation of humidity in atmosphere—0.30 m; protection against soil erosion and silting of dams, reservoirs and rivers—0.25 m, shelter to birds—0.25 m and protein for cattle from tree leaves—0.02 m.

Trees play important role in improving total living of this planet. They maintain balance between oxygen and carbon dioxide which is essential for the life of fauna. Trees serve as natural air conditioner specially in hot and dry climates by increasing humidity around them through transpiration, lower down temperature and thus change microclimate of surroundings. Trees regulate the rainfall and check the spread of desert. They are also the guardian of soil and check soil erosion by binding the soil particles with their roots. They improve fertility of soil by adding organic manure in the form of leaves every year. To check air pollution due to the gases, dust, smoke, planting of trees is highly advantageous.

Planting of trees alongside the roads has a long history and in early days large planting was done by Buddha Bhikshu. Emperor Ashoka and Sher Shah Suri were well known for planting trees along side the roads. Mogul kings did not lag behind in planting of trees. Now-a-days with the increase in vehicular traffic, it has become much more compulsory to plant more trees.

Apart from all these virtues, trees have immense beauty from aesthetic viewpoint. They bring the change in skyline on account of variation in their height: shapes, foliage texture and flower colour. They are used in landscape plan for aesthetic and functional purposes. Trees are used in garden as specimen, avenue plantation, wind breaks, and screening.

A judicious planting of trees enhances the beauty of surroundings. As the trees are the framework of the garden, they should be easy growing, hardy and requiring little attention.

SELECTION OF TREES

This depends upon many factors which should be kept in mind when selection is being done.

1. Purpose of Growing

(a) Specimen tree. Such trees are planted singly for their attractive shape, beautiful foliage, flowers or for drooping branches which reflect humbleness, for example : *Araucaria cookii, Callistemon lanceolatus, Salix babylonica, Cassia fistula, Magnolia grandiflora, Plumeria alba, P. acutifolia, Cassia nodosa, Ficus elastica* var. *Decora or Black Prince*, etc.

(b) Shady trees. Such trees have mostly round canopy or umbrella crown. Leaves are large and dense so that no or very little sun is allowed underneath them. Common trees are *Azadirachta indica, Kigelia pinnata, Alstonia scholaris, Ficus religiosa, F. benghalensis, F. infectoria, F. benjamina, Mimuspos elengi, Pterospermum acerifolium, Morus* sp., *Eugenia jambolina, Mangifera indica, Melia azedarach, Milletia ovalifolia, Pongamia pinnata, Swietenia mahagoni, Swietenia macrophylla*, etc.

(c) Flowering trees. These trees produce colourful flowers and are planted for their beautiful flowers *e.g., Bauhinia variegata, B. blakeana, Callistemon lanceolatus, Cassia fistula, Cassia nodosa, Cassia siamea, Delonix regia, Erythrina variegata* var. *Orientalis, Jacaranda acutifolia, Plumeria alba, Plumeria acutifolia, Butea monosperma, Chorisia speciosa, Heterophragma adenophyllum, Spathodea campanulata, Thespesia populnea, Lagerstroemia speciosa*, etc.

(d) Trees for avenue or road side. Trees planted along side avenues or roads are generally for shade or for flowers. Commonly planted trees are *Cassia fistula, Grevillea* *robusta, Jacaranda acutifolia, Heterophragma adinophyllum, Ficus infectoria, Milletia ovalfolia, Kigelia pinnata, Pterosperomum acerifolium,* etc.

(*e*) Screening purpose. When tall upright trees are planted very close to give an ultimate look of curtain or screen. Such trees are planted to hide some objects or sides. For example, *Grevillea robusta, Eucalyptus* sp., *Poplar* sp., *Polyalthia longifolia, P. pendula,* etc.

(*f*) For fragrant flowers. For this purpose trees like Kanak Champa (*Pterospermum acerifolium*), Pagoda tree (*Plumeria* sp.), Bara Champa (*Magnolia grandiflora*), Swarn Champa (*Michelia champaka*), Molsari (*Mimuspos elengi*), Akas neem (*Millingtonia hortensis*), etc. are suitable.

(*g*) For checking air pollution. Industries are the major source of pollution and hence, in such areas deciduous trees or trees having thick shining leaves will be more successful *e.g. Morus* sp., *Poplar hybrida, Plumeria acutifolia, Ficus infectoria,* etc.

2. Climatic Factors

Selection of trees will depend upon prevailing climatic conditions of the locality and trees of other climatic conditions may or may not do well or later on decline. This depends upon adaptability of individual types. Trees of tropical climate like *Delonix regia, Neonauclea purpurea* decline later on in subtropical climate. Similarly, trees of temperate climate do not grow well in tropical regions.

3. Soil Factors

In general trees are adaptable to varying soil conditions but specific trees excel in particular conditions. In alkaline and saline soils, the trees which can be grown successfully are : *Casurina equistiefolia, Cassia fistula, Parkinsonia aculeata,* etc.

4. Availability and Cost Factors

The selection of trees for general planting should be done which are easily available at low price; however, specimen trees may be exception to this. For example, pants of *Araucaria* sp. or *Ficus elastica* var. Decora or Black Prince are sold at exorbitant prices but still these are in demand due to their beauty and rarity.

Planting Schemes for Roadside

There are different types of roads depending upon width and vehicular traffic. Thus, the selection of trees differs from road to road. For wider roads—tall trees, for medium roads—medium trees and for narrow roads-dwarf trees are selected. For national highways tall shady trees like *Azadirachta indica, Ficus infectoria, Terminalia arjuna, Pterospermum acerifolium, Kigelia pinnata, Millingtonia hortensis, Polyalthia longifolia,* etc. should be selected. For economic reasons, earlier most of national highways were planted with *Dalbergia sisoo*; Mango seedlings (*Mangifera indica*); Jamun (*Sygizium cumini*), etc. but now fast growing tree *i.e. Eucalyptus* sp. has been planted. For inner roads of cities/ colonies and roads of big campus, mostly colourful trees are preferred.

Planting of Trees

Preparation of ground. Trees are perennial and will live for several years. Since the preparation of ground is done just once so it is worthwhile to follow proper method of planting. According to the landscape plan, pits should be dug of 60 × 60 × 60 cm size. Soil should be well mixed with 10-15 kg well rotten farmyard manure and Chlorpyrifos @ 5 ml/l of water to check the attack of termites. Refilling of pit should be done which is followed by immediate watering to settle down the soil. After settling down the soil, planting of trees is done in the centre of pit.

Planting time. For mass plantation best planting time is rainy season. After 2-3 good showers planting should be done so that water does not become a limiting factor. Plants establish easily and grow luxuriantly. Planting can also be done in February-March depending upon the availability of water.

Maintenance and Care of Trees

After planting the trees, constant care is necessary to get the desired effect. Different necessary operations are :

Staking. Staking is necessary to keep the trunk of trees straight otherwise trunk would bend and plants make abnormal growth. Stake should be such that it suits the size of the tree and should be strong enough to provide the support. For this mostly, straight bamboos or iron bars can be used. It should be placed about 5 cm away from the main trunk so that it does not harm root system.

Fencing and tree guards. Young trees need protection against grazing by animals, human carelessness and adverse weather conditions until they are fully established.

The type of fencing depends upon factors like cost involved and availability of material. Most common and cheap fencing is obtained by putting thorny bushes around the tree. Other materials like bricks, perforated bitumen drum and wire netting are also used. These all materials make the poor aesthetic effect and hence, cemented architectural structures or fabricated iron tree guards though little costly can be used.

Weeding. Weeding is important to prevent the shadowing and suffocation of young trees by vigorous growing weeds until these trees outgrow surrounding vegetation. Two to three weedings are sufficient in a year.

Irrigation. Irrigation in first few years must be done regularly and according to requirement. Watering during hot and dry period is necessary.

Gap filling. Replacement of dead trees is also very important and replanting should be done at the earliest with the tree of same age so that it does not disturb the uniformity of the trees.

Training of young trees. All dead wood, misshapen or dangerous branches hanging over roads, buildings, etc. should be cut. Damaged or decayed branches should be cut away. All cuts more than 5 cm should be painted with Bordeaux mixture against the attack of pathogens.

Description of Trees

Acacia sp. (Leguminosae)

This genus is native of arid zone of central Australia and has large number of species. The trees of this genus grow luxuriantly in dry climate. Profuse flowering and heavy seed set help in their propagation. *Acacia arabica*, *A. fernesiana* and *A. modesta* are used to plant in unproductive land to protect soil from wind erosion. Animals are grazed on their leaves.

Acacia auriculiformis (Australia wattle) is another species which is planted in the garden for its attractive shape, foliage and shade. It is a dwarf tree and produce typically spiral legume shaped fruits.

A. cyanophylla commonly called as Golden acacia, is another handsome species which bears golden flowers. It is also dwarf but possess brittle branches.

Albizzia lebbek (Leguminosae)

It is native of tropical Asia and Africa and is commonly called as Siris. This is quick growing tall tree having pinnate leaves of light green colour. Trees make growth of an umbrella shape. Flowering occurs in April-May and flowers are pale, greenish yellow in colour and mildly fragrant. This is a good tree for roadside plantation and its pods of straw yellow colour hang on tree for long time which makes it more attractive. It can be easily propagated by seeds.

Albizzia procera (Leguminosae)

This is native of Indo-Malaya region and is large quick growing tree. Trunk is tall and colour is greenish yellow white. Leaves are bipinnate of light green colour. Tree produces profuse flowers in July-August. Flowers are small heads of yellowish white in colour. It is good tree for roadside plantation and for big parks. It can also be easily propagated by seeds.

Alstonia scholaris (Apocynaceae)

It is commonly called as Devil's tree or Sapt parni or Satpattia and is native of Indo-Malayan region. In the beginning it is slow growing tree but later on it grows fast. It is a tall tree. Leaves are shining green and produced in a whorl of 5-7. Flowers are produced in October-November in clusters which are greenish white. It is suitable for planting alongside the roads, groups or as a specimen in parks for shade purpose.

Very recently, it has been observed that the trees of *Alstonia scholaris* have been attacked by leaf gall forming psylids (*Microcerposyla nigra* and *Aerocercops astaurota*) which causes serious damage on newly developing leaves, developed leaves and epidermis of tender shoots. The trees look very ugly if not controlled. It can be controlled by spraying 2-3 ml of Monocrotophos or 3-4 ml of Metasystox or Rogor in 1 L of water. However, keeping the size of the tree in view it will not be possible to control it. Hence, this tree should not be planted any more.

Araucaria cookii (Coniferae)

It is commonly called as monkey puzzle and native of South America. This genus is most handsome of evergreen gymnospers. Now-a-days, it is a matter of pride to have this tree in the gardens. There is always a great demand due to its attractive shape and soft foliage. The branches which are horizontal, and well placed on the main trunk are covered with needle like soft, dark green leaves. In Punjab, plants are brought by nurseıymen from Kolkata and Bengaluru and are sold at prices which depend on plant height. The plants need protection from severe cold in winter and from scorching heat during summer. Artificial rain by sprinkling water helps proper growth.

Azadirachta indica (Meliaceae)

It is commonly known as neem and is native of Indo-Burma region. Tree is tall with brown stem. Leaves are pinnate, dark green and bitter in taste. Flowers are small, white and sweet scented which appear in the end of April to middle May. Fruits are drupe, oval in shape and golden yellow in colour at maturity. Propagation is done from seeds. This is a medicinal tree and new emerging leaves are eaten for keeping away from many blood diseases. Leaves are hanged in the house and rooms of patients to repel harmful insects. Semi hardwood shoots are used as dattuns to clean teeth. Margosa oil is prepared from fruits and seeds which are used in making soaps or medicines. Neem cake is used as a fertilizer which kills harmful insects, termites and nematodes. From landscaping viewpoint, it is a good tree for roadside plantation and in big compounds for shade and in hospitals as a medicinal plant as well as for shade.

Madhuca longifolia* var. *latifolia* syn. *Madhuca indica (Sapotaceae)

It is an indigenous tree and commonly known as Mahua. It is a tall growing tree with a rounded crown. Leaves are large which fall in April and reappear in bright red colour in May. Flowering occurs in end of April to first fortnight of May. The flowers are light yellow in colour which are not very showy. They bloom in night and fall in morning. Fruits are consumed in tribal areas of central India and used to make a delicious brew. It is easily propagated by seeds. It should be planted in groups or singly in parks for shade.

Barringtonia acutangula (Lecythidaceae)

It is native to coastal wetlands in southern Asia and northern Australasia, from Afghanistan east to the Philippines and Queensland. It's an evergreen tree 5-8 m tall with rough fissured dark grey bark. Leaves are obovate and red flowers are produced on pendulous racemes about 20 cm long. Four sided fruits are produced periodically throughout the year. It can tolerate heavy clay soils with poor drainage and can grow in a wide variety of soils. It is a very good roadside tree and forms umbella shaped, low canopy without much training or pruning. It can also be used in parks for shade and as a specimen tree.

Bauhinia (Leguminosae)

The genus is comprising of flowering trees and most of them are native of India. The generic name has been given after twin brother, John and Casper Bauhin—famous botanists of 16th century. Plants of the genus are easily identified on account of peculiar leaves which are composed of two oval shaped leaflets united in the centre. Flower buds are cooked as vegetable. Easily propagated through seeds which are produced in abundance. Following species are common :

B. variegata. This is a tall tree producing flowers of rose coloured variegated with red and yellow which appear in mid of March. Most ideally suited for planting as specimen or in groups in gardens and big parks and as avenue tree alongside the road.

B. variegata* var. *candida* syn. *B. alba. This is a medium tree bearing white flowers during middle of March. It is suitable for specimen or group planting.

B. purpurea. It is medium tall tree producing purple colour flowers in mid of November. It is good for roadside plantation and group planting.

B. blakeana. This species is recently introduced from Hong Kong and is very handsome flowering tree. It is a medium tall tree with round top producing large purple flowers in end of December or February. It is propagated through air layering. It is highly suitable for specimen and group planting.

Bauhinia tomentosa (Leguminosae)

It is a slow growing dwarf to medium tree which produces yellow flowers during rainy season.

Koelreutaria elegans (Sapindaceae)

It is commonly called as golden rain tree and is native of Korea and China. It is a winter deciduous, medium sized and spreading tree. The emerging leaves are coppery-pink colour. It produces small yellow flowers on long panicles in late summer and is followed by development of ball/bladder like red flushed fruit capsules. It is easily propagated through seeds.

It is highly suitable for planting along side the road for shade or flowers or in group planting in large parks.

Bombax ceiba (Bombaceae)

It is native of Indo-Burma region and popularly known as Semal, Simbal or Red cotton silk. It is considered as a sacred tree and is associated with Goddess Kali. It is a fast growing, tall, stately upright tree which is deciduous during February-March. Trunk has spines. Leaves are compound. Flowers are large, bright red or yellowish red in colour which appear in mid of March. Fruits contain silky floss which is used in stuffing pillows and cushions. It is easily propagated through seeds. It is highly suitable for boundary plantation of bigger compounds, national highways and in parks.

Butea monosperma (Leguminosae)

It is commonly called as Flame of Forests or Dhak or Palas or Tesu and is native of India. It is a tall tree but does not look attractive due to crooked and distorted stem. Leaves are large, trifoliate and are used to prepare plates (*Doonas*) for community functions. It bursts into blooms in deciduous condition during end of March and whole tree is clothed with scarlet orange crescent shaped flowers. When planted in groups, from distance it gives appearance as if there were fire and hence, it is called as 'Flame of the Forest'. Lac insects are reared on its branches. Temporary dye is extracted from its flowers and in ancient days during holi festival this colour was used to play.

Callistemon lanceolatus (Myrtaceae)

It is commonly called as bottle brush on account of appearance of red colour flowers at the end of branches, which resembles brush used for cleaning bottles. Though it is native of Australia, it has adapted well under Indian conditions and being commonly planted at the gate of house, in corner of lawn as a specimen tree and near water bodies. It is also highly suitable for grouping, screening and to be planted in the shrubbery. Its drooping branches are symbol of welcome on which dark green, lanceolate leaves are nicely laced. Flowers of bright red colour are produced on young growth of branches in end of March to middle of April. Commonly propagated through seeds. *Callistemon polandii* is another species which produce yellow flowers.

Callistemon salignus (Myrtaceae)

It is commonly called as white bottle brush or willow bottle brush and is native of Australia. It is a very handsome erect to spreading type of small tree or having thin branches and fragrant, narrowly lanceolated pale green leaves. It produces white flowers in the end of March to middle of April which attracts lot of honey bees. It is propagated through seeds or air layering.

It is very suitable for small compounds gardens as a specimen or group planting.

Cassia (Leguminosae)

This is another important genus which has many colourful species and beautify the surroundings with spectacular flowers in leafless condition. Following are important species.

C. fistula. It is native of India and commonly called as Amaltas or Golden laburnum which is considered as tree of prosperity by traders. It is a tall tree having white stem. Leaves are pinnate leaflets are larger than other species. Colour of emerging leaves vary from copper. red to light yellow. Colour of flowers in earlier type is deep yellow while in latter colour of flowers is light yellow, respectively. Flowers are well arranged on long drooping rachis which appear in middle of May when tree is in leafless condition and flowering continues upto July. Since it flowers in hot months and its beauty is enjoyed early in morning or in evening. It is an excellent specimen tree, and for avenue, or group planting. Fruits are used to prepare medicine. It is propagated through seeds.

C. nodosa. It is commonly called as Pink Cassia and native of Indo-Malaya region. Tree is of medium height and having spreading branches. Leaves are compound and light green in colour. Tree sheds leaves during April and May and it is clothed with pinkish white flowers. Flowering starts in middle of May and continues upto July. It needs protection against frost in initial stages during winter. It is an ideal specimen tree, suitable for grouping and roadside plantation.

C. javanica. It is native of Java, Malayasia, Sumatra and is commonly called as Java cassia. It is a medium tree with spreading branches and leaves are compound and green. It is also deciduous during April-May and flowering starts from middle of May which continues upto July. Flowers are of deep pink to rose pink. Tree is suitable for specimen, group and roadside planting.

Senna siamea* Syn. *C. siamea. It is native of Indo-Malaya region and is commonly called as Kassod tree. A quick growing, evergreen tree, size medium to large with columnar to spreading crown. Leaves are compound and of olive green colour. Yellow flowers start appearing in July and continue upto January. It is easily propagated through seeds. It is good for boundary plantation and for demarcation of large areas. It is also suitable for road side planting. Branches are, however, brittle and breakaway very easily with storm.

Casuarina equsietifolia (Casurinaceae)

It is native of Australia, Burma, Malaya and Pacific Islands and is commonly called as beefwood, Jangli saru, Farash or Jor-Tor on account of cord like leaves which are easily separated at the node and can be temporarily fixed without knowing breaking point. It is a tall, upright, evergreen and fast growing tree. It has rough bark and spreading branches. Trees are dioceous producing male and female cones separately. Easily propagated through seed. It is tolerant to saline soils. It is planted for screening purpose and can be planted closely and trimmed as tall hedge.

Toona ciliata Syn. *Cedrella toona* (Meliaceae)

It is commonly called the toona or Indian Mahagoni tree and is native of Indo-Malaya region and Australia. It is a large, quick growing tree having shining compound leaves. Flowers are small, white and appear in middle of April. Easily propagated by seeds. It is planted in the garden in groups for shade and also suitable for roadside plantation.

Ceiba speciosa Syn. *Chorisia speciosa* (Malvaceae) (Plate 10)

It is native of Mexico and Brazil and commonly called as Mexican Semal. It is an erect tree with well placed branches. Stem is waxy green with spines. Leaves are palmate in shape, of light green colour. Flowers are pink with creamish white centre which appear in October-November when tree is in leafless condition. It flowers when other trees are not in bloom. It is planted in the garden as specimen.

Cupressus sempervirens (Coniferae)

It is commonly called as Saru and native of West Asia and South Europe. It is considered as tree of tomb on account of its dull green leaves. It is an erect tree without branches (sometime branches appear which should be removed) and make columnar growth. It is a very common tree of Persian and Mogul gardens and planted in rows along the side of water canal. It is also preferred in the home landscapes. It is easily propagated through seeds.

Delonix regia Syn. _Poinciana regia_ (Leguminosae)

It is native of Medagascar and is commonly known as Gulmohar, Peacock or Flamboyant flower. It is a fast growing large tree. Limbs are spreading and form an umbrella. Leaves are compound and leaflets are small with round apices. Showy flowers or orange, red, scarlet to salmon colour are produced in May-June. Easily propagated through seeds. It is an ideal tree for big parks for beautiful flowers as well as for shade. Under North Indian conditions it starts declining after 10-12 years. It has shallow root system and hence grass does not grow under this tree.

Erythrina stricta var. _suberosa_ (Leguminosae)

It is commonly known as Pangra tree or Parrot flower on account of its flowers resembling beak of parrot and is native of India. It is a tall tree with spines on the trunk. Leaves are trifoliate. Tree sheds leaves during February-March and flowering occurs in first fortnight of April in leafless condition. Tree is clothed with scarlet flowers which are produced at the end of branches. It is easily propagated by cuttings. It is suitable as specimen, or group planting.

Erythrina blakei (Leguminosae)

It is native of India. It is small tree, grown for its beautiful scarlet flowers which are produced in flushes from March to November.

Eucalyptus sp. (Myrtaceae)

It is native of Australia and is commonly called as Safeda or Blue gum. It is a tall upright and fast growing tree. It attains the conical shape to round crown. Bark is smooth and of shining colour which peels off in long flakes. Leaves are simple lense shaped whose upper side is shining and lower side dull. Leaves are rich in oil. Flowers are greenish white produced abundantly in April and provide ample pollens for honey bees. It is planted for quick effect of green vegetation on boundaries and alongside roads. It is highly suitable for areas where water table is high. Its stem and branches are used in making paper, as fuel or pole and hence, fetches remunerative price. It has become popular now-a-days amongst absentee landlords and big land owners. The important species are : *E. citridora, E. hybrida, E. robusta* and *E. camadulensis.*

Ficus sp. (Moraceae)

The genus is comprising of very important trees which are associated with Indian life from ancient period. The trees of this genus are still being worshipped which started with the arrival of Aryans in India. The important species are :

F. religiosa. It is commonly called as Pipal or Bodhi tree and native of India. It is associated with three deities of Hindu religion *i.e.* Brahma, Vishnu and Shiva and also with Lord Buddha, hence, it is considered to be a sacred tree and is worshipped. It is a huge tree with spreading branches. Leaves are cordate in shape and shining. Flowers are very small hidden in synconium which appear in April-May. Easily propagated through seeds and cuttings. It is commonly planted in villages near ponds for shade and also in big parks. Birds take shelter on this tree, eat fruit and disperse the seeds.

F. benghalensis. It is commonly called as Bor, Bargad or Banyan tree and is native of India. This tree is also sacred tree according to Hindu mythology and is worshipped especially by ladies for getting child, being, the symbol of fertility. It is very huge tree and is commonly planted in villages near ponds for shade. It also provides shelter and food for birds. It can be planted in big parks for shade.

F. infectoria. It is commonly known as Pakar or Pilkan and native of Indo-Burma region. It is a tall tree with spreading branches and round crown. Leaves are simple dark green with clear veins and shed in end of March. New leaves appear in first fortnight of April which are coppery red and look attractive. Easily propagated through cuttings. It is good for group planting and roadside plantation for shade.

F. racemosa* Syn. *F. glomerata (Moraceae). Popularly known as the Cluster Fig or Gular, this is native to Australia, Malaysia, South-East Asia and the India Subcontinent. The most distinctive aspect of this tree is the red, furry figs in short clusters, which grow directly out of the trunk of the tree, termed cauliflory. It has evergreen leaves, if it is close to a water source; otherwise it sheds its leaves for a short time. It is a good shade tree and can be used in large parks and open spaces.

F. microcarpa* Syn. *F. retusa. It is commonly called as chilkan and is native of Indo-Malaya region and China. It is a large, evergreen tree with round crown. Leaves are simple, oval, and shining green. It is useful tree for shade in parks and alongside the roads.

F. benjamina. It is native of Indo-Malaya region. It is medium to large, an evergreen tree and is planted for its attractive foliage. Branches are drooping and leaves are shining green. Propagated through seeds. It is suitable for planting in parks as specimen, groups and alongside the roads for shade.

Ficus benghalensis* var. *krishnae. It is native of India and is commonly called as Makhan Katora or 'Krishna' butter cup due to its folded leaves giving shape of a cup and is known to have association with Lord Krishna. It is a natural mutant of *F. benghalensis*. It is a medium evergreen tree and good for specimen as novelty or in groups.

F. elastica

It is commonly called as rubber tree and is native of Indo-Malaya region. It is a tall tree. Leaves are large, deep green and shining. The growing tip is covered with red. It is propagated through air layering. It is planted as specimen. Other varieties are variegata, Decora and Black Prince.

Ficus lyrata (Moraceae). It is commonly called as Bingo or Fiddle leaf fig and a native of western and central Africa. It is an open, evergreen tree with leathery, glossy, fiddle shaped dark green leaves. It bears spherical figs which are produced singly or in pairs only on the mature trees. It is very easily propagated through cuttings.

It is suitable for planting in parts as a specimen for group planting and alongside the roads to provide shade.

Grevillea robusta (Proteaceae)

It is commonly called as silver oak and is native of Australia. It is an upright, quick growing tall tree producing columnar shape of growth. Bark is brown and rough. Leaves are pinnately divided, green above and silvery below. Flowers of golden yellow colour are produced in first fortnight of April. It is propagated from seeds which are collected in July and sown immediately thereafter. It is good for boundary planting, specimen, group planting, screening purpose and alongside the roads.

Haplophragma adenophyllum (Bignoniaceae)

It is native of Indo-Malaya region and is commonly called as maror phali or monkey tail due to its long curved fruits capsules. It is a tall, evergreen, and quick growing tree. Leaves are compound having large leaflets. Flowers are large, yellow in colour produced in the month of February-March. It is propagated from seeds. It is suitable for roadside plantations.

Pithecellobium dulce Syn. *Inga dulcis* (Leguminosae)

It is popularly known as Jangal jalebi due to its curvaceous fruits. It is quick growing evergreen tree having thorns. Leaves are compound and of dark green colour. Flowers are not very showy which appear in the month of March-April and easily propagated through seeds. It is highly suitable for boundary plantation and tall protective hedge.

Jacaranda acutifolia (Bignoniaceae)

It is native of Brazil and is commonly called as Nili Gulmohar. It is medium in height, quick growing and winter deciduous tree. Leaves are compound and leaflets are small with acute apex. Flowers are large, trumpet shaped and blue in colour which appear in first fortnight of April. Easily propagated through seeds. It is most ideally suitable for group planting and alongside roads.

Juniperus chinensis (Conifereae) (**Plate 11**)

It is a conifer but very well grown in the gardens. It is slow growing mostly unbranched and makes a clean shape of conifer and preferred in the gardens for its elegant shape. It is being ever green, does not shed the leaves and keeps garden clean. It is ideally suited for formal planting, specimen and group planting. It is easily propagated by seeds or stem cuttings.

Kigelia africana Syn. *K. pinnata* (Bignoniaceae)

It is native of tropical Africa and called as Sausage tree or balam khira on account of its long fruits. It is a tall and evergreen tree. Leaves are compound and leaflets are large and rough. Flowers are big, trumpet shaped of crimson colour which appear in May-June on long pendulous rachis. It can be easily propagated through seeds. It is ideally suitable for shade, to be planted in large parks and alongside roads.

Lagerstroemia speciosa (Lythraceae)

L. floribunda is another important species producing white and purple flowers, Pride of India or Jarul. It is a medium sized tree with light green smooth bark. Leaves are large, elliptic, which is winter deciduous. Flowers of mauve purple or mauve pink appear in middle of April and flowering continues upto August. Easily propagated through seeds. Most ideally suitable as a specimen tree, group planting and alongside the roads.

L. floribunda is another important species bearing pink flowers which fade to white. It blooms from May to August and is suitable for specimen or group planting. *L. parviflora* is a medium tree which produces white colour flower in May-June and looks very attractive.

Magnolia grandiflora (Magnoliaceae)

It is native of United States of America and is commonly called is Bara Champa on account of its big sized flowers which are highly fragrant. There are several species but this species is common in Indian gardens. It is a medium tree with upright growth. Leaves are simple, large and have shining greenish red upper side and lower side is dull yellow. Flowers are white, fragrant and appear in the month of April-May. It is suitable as specimen tree.

Melia azedarch (Meliaceae)

It is native of Persia and is commonly called as Persian Lilac or Dharek. It is fast growing tree having grey bark. Trunk is straight and branches spreading upward and forms round crown. It is winter deciduous. Leaves are compound and of dark green colour. Flowers of lilac colour appear in first fortnight of April and later on tree is laden with golden coloured fruits. It is easily propagated through seed. It is suitable for boundary planting and for shade.

Magnolia champaca Syn. *Michelia champaca* (Magnoliaceae)

It is native of India and is commonly called as Swaran Champa or Sone Champa. The tree is tall with brown bark. Leaves are simple, large and of light green colour.

Fragrant flowers of light yellow colour appear in second week of May. Easily propagated through seeds which should be sown immediately after collection. It is ideally suitable for shade as well as for fragrant flowers.

Millingtonia hortensis (Bignoniaceae)

It is native of Burma-Malaya region and popularly called as Akash neem or cork tree or tree Jasmine. It is fast growing, tall, upright tree and looks attractive. Trunk is straight and branches emerge vertically. Leaves are compound, handsome of dark green foliage. Flowers are white and are highly fragrant which appear in April-June. Easily propagated through root suckers, cuttings and fresh seeds. It is suitable for boundary plantation and roadside plantation.

Millettia peguensis* Syn. *Millettia ovalifolia (Leguminosae)

It is native of Burma (Mayanmar) and is commonly called as The Moluemain Rosewood. It is a fast growing tall tree with spreading crown. Leaves are trifoliate, leaflets are of oval shape and light green colour. Leaves shed in the month of February and light purple coloured flowers appear in the middle of April when tree is in leafless condition. It is easily propagated through seeds. It is ideally suitable for boundary and roadside planting.

Mimusops elengi (Sapotaceae)

It is commonly called as Molsari and is native of Indo-Malaya region and Sri Lanka. It is slow growing, evergreen, tall, umbrella shaped tree. It has deeply fissured bark of brown colour. Leaves are simple, dark green and shining. Flowers are of dirty white colour and highly fragrant which appear from May to August. It's propagation is done by seeds. It is highly suitable for big parks for shade. Flowers are used for making garlands and essential oil is also extracted. It needs training to single stem to form an attractive crown.

Morus sp. (Moraceae)

It is native of Western Asia, Afghanistan, Europe and China and is commonly called as sehtoot or toot or mulberry. Tree is a quick growing, winter deciduous, tall with upright branches forming spreading crown. It's bark is brown, leaves are simple, cordate with serrated leaf margin which sheds in December. New leaves emerge in March. It is easily propagated through cuttings. Flowers are catkins which are not attractive and appear in April. It is planted in the garden for shade and for fruit.

Neolamarckia cadamba* Syn. *Anthocephalus cadamba (Rubiaceae)

It is native of Indo-China region and Sri Lanka. It is commonly called as Kadam tree and is associated with Lord Krishna who used to play and swing under this tree. It is tall, evergreen tree having straight and long trunk. Bark is of light grey colour with longitudinal fissures. Leaves are simple, large, leathery and shining above. Ball shaped flowers are produced on a round pedicel in the month of August-September. In the beginning spherical buds are of green colour and change to golden colour in due course of time. It is propagated through seeds. It is a good shady tree for parks and can be planted on roadsides also. Under Punjab conditions, it shows decline.

Parkinsonia aculeata (Leguminosae)

It is native of America and commonly called as Jerusalem thorn or vadi-Bathal. It is medium tree and has thorns. Leaves are compound. Leaflets are thin, numerous, oblanceolate and shortly. Yellow flowers are produced in April-May. It is highly suited to alkaline soil, arid zones and drought conditions. It can be easily propagated through seeds. It is ideally suited for roadside plantation.

Plumeria species (Apocynaceae)

This genus comprises of most handsome trees which are very common in gardens, homes and are liked for their architectural shapes and fragrant flowers. This genus is native of Mexico and Guatemala and commonly called as pagoda tree or temple tree or frangi pani or gulachin. These are easily propagated through cuttings. The common species are :

***Plumeria rubra* f. *acutifolia* Syn. *P. acutifolia*.** It is winter deciduous tree of medium height with round top. Branches are dichomatous or trichomatous. Leaves are simple, large, dark glossy green with acute apex. Fragrant flowers of light yellow colour are produced from May to September. It is most ideally suited as specimen tree or alongside of inner roads.

***P. rubra*.** The characters are similar to P. acutifolia except flower colour is red and flowering is not as profuse as in earlier species.

***P. obtusa*.** This is another evergreen species, medium in growth. Leaves are shining, dark green with obtuse apex. Flowers are waxy white and produced round the year.

Plumeria pudica

It is very recently introduced in Indian gardens and commonly called as Bridal Bouquet or white Frangipani. It is native to central and south America (Columbia, Venezuela and Curacao). Dwarf tree growing upto 3-4.5 m and 1-1.5 m in width having multiple stems which arise from close to the ground and have an upright shape. Stems are thin, weak and liable to break very easily due to high winds. Leaves are large, spoon shaped, sessile and of dark green colour during spring, summer and autumn whereas it becomes yellow green during winter. Tree is evergreen only in warm and tropical conditions but becomes leafless during severe winter and frosty situations. Young and old plants may die to severe frost.

Flowers are showy, single of white with yellow or light green throat. Flowers are waxy but not fragrant like other frangipani which appear during spring season continue to flower during summer and autumn. Fruit is a two part follicle and seeds are winged which germinate easily. Trees are propagated through stem cuttings.

It is recommended as an accent tree and are suitable for containers in frost fre locations and sunny situations.

Polyalthia longifolia (Anonaceae)

It is native of Sri Lanka and commonly called as Ashok. It is an evergreen, tall tre and producing conical crown. Branches spread horizontally form straight trunk. Bark i of brown colour and split longitudinally. Leaves are simple alternate shining green wit wavy margin. In hot summer shining green leaves provide coolness. Flowers are produce in April which are of pale green colour. It can be easily propagated through seeds. It i graceful tree mostly planted in temples and ideally suitable for approach roads an roadside plantations. It needs protection from frost in its early stages of growth.

***P. longifolia* var. *pendula*.** It is commonly called as pendula or weeping Ashok It has long trunk with drooping branches so it gives stately appearance. It is also ideall suitable for approach roads of buildings and in front of tall buildings to reduce th tallness of buildings. It needs adequate staking for proper shape.

Pongamia pinnata (Leguminosae)

It is native of Burma-Indonesia region and is commonly called as Karanj or *Suk Chain*. It is a medium tree having spreading habit. Leaves are compound and leaflets ar round and shining green but mostly attacked by insects and develop brown spots. Thes attacked leaves look very ugly. Foliage is densely crowded. Flowers are of purple colou and produced profusely in May. It is planted for shade.

Pterygota alata* Syn. *Sterculia alata (Sterculiaceae)

This India native is known as Buddha Coconut. It's a tall tree, which gets its nam from its coconut like fruit. Ovate-heart shaped leaved are carried on long stalks, crowde towards the ends of branches. Flowers have no petals; sepals are fleshy, densely pubescen outside, sparsely pubescent and purple with red streaks within. It can be used as a ver good avenue tree because of its columnar growth.

Pterospermum acerifolium (Sterculiaceae)

It is native of Bengal, Assam and Burma (Mayanmar) and is commonly called a Kanak Champa. It is a tall tree with conical crown. Bark is of brown colour. Leaves ar large, peltate, round, of dark green above and whitish or rust coloured beneath. Leave are densely crowded. Flowers are of yellow colour with long petals which appear in firs fortnight of April and continue upto June. Flowers are highly scented. Easily propagate through seeds. It is ideally suitable for group planting, shade and roadside plantation

Putranjiva roxburghii (Euphorbiaceae)

It is native of Indo-Burma and Sri Lanka and is commonly known as Jiva Putra It is an evergreen tree, growing slowly in beginning but finally attains the medium to ta height. It has small trunk with brown stem. Branches are slender and drooping whic makes it attractive. Leave are simple, short, petioled, ovate, lanceolate smooth, shinin and dark green with wavy margin. Flowers are light yellow, not very showy which appear in the month of March-April. It is easily propagated by seeds. It is an ideal tree as specimen, good for avenues, alongwith paths and very effective for ponds, or lake planting.

Ravenala madagascariensis (Scitmineae)

It is native of Madagascar and commonly called as travellers tree. This tree has a unique shape resembling like a dancing peacock. It is tall and evergreen tree whose leaves resemble like banana leaves which are arranged or thick succulent stem in one plane giving the appearance of gigantic fan. There are as many as 20-24 leaves having about 2 meter long leaf stalks. It is propagated through seeds or suckers which are produced under mature trees. To create picturesque effect with this tree, plant as a specimen keeping a direction of crown towards east-west. It does well in tropical climate.

Salix babylonica (Salicaceae)

It is native of China and is commonly called as weeping willow. It is tall winter deciduous with brown bark. Branches are thin and drooping. Leaves are shed in December-January and again appear in March. Flowers are not very showy. Easily propagated by cuttings which are prepared in winter. It is ideally suited as specimen specially near lakes or ponds and alongside walks and paths in park. Tree requires high moistures content in soil.

Saraca asoca (Leguminosae)

It is native of Sri Lanka and is commonly called as Sita Ashoka. It is associated with Goddess Sita who was kept in captivity under this tree in gardens of Sri Lanka. It is medium to tall, evergreen tree with spreading shining green leaves. Scarlet red flowers are produced in clusters in the month of April and plant become very attractive. It is planted in parks for shade.

Schleichera oleosa* Syn. *S. trijuga (Sapindaceae)

It is native of Indo-Burma region and Sri Lanka and is commonly called as Kusum. It is slow growing tree but ultimately attains tall height with round to conical crown. It sheds leaves in February-March and again new bright red leaves appear in April. Small green of white coloured flowers appear in April. It is propagated by seeds. It is ideally suited as specimen, for shade and as avenue tree.

Senna spectabilis* Syn. *Cassia spectabilis (Leguminosae)

This is a medium to large tree native to tropical America and can reach 60 feet in height. The pinnately compound leaflets have fuzzy undersides and are three inches long. The bright yellow flowers are 1.5 inches wide but appear in dense racemes up to two feet long. It has delicate foliage covered with yellow flowers and blooms for about two months in early winters. This tree is best used in an open, sunny setting in parks and open spaces.

Spathodea campanulata (Bignoniaceae)

It is commonly called as fountain tree and native of tropical Africa. It is evergreen, tall tree with upright growth. Leaves are compound with dark green leaflets which are elliptical in shape. Bell shaped flowers of scarlet red colour appear at the end of branches in late April and looks very attractive. It is propagated through seeds and root suckers. It is very good as avenue tree but shows decline under Punjab conditions.

Swietenia mahagoni (Meliaceae)

It is native of tropical America and is called small Mahagoni. It is tall evergreen tree with spreading crown. Bark is blackish brown with deep fissures. Leaves are compound and leaflets are dark green and comparatively small. It is easily propagated by seeds. It is suitable for shade in the park and along the roadside. *S. macrophylla* is another good species which has comparatively larger leaflets.

Tabebuia rosea (Bignoniaceae)

It is native of Mexico and is commonly called mauve tabebuia. It is a medium sized tree. Leaves are compound with 3-5 leaflets which are elliptic oblong. Flowers of mauve colour are produced in October-November, which are funnel shaped and arranged on a peduncle. It is easily propagated through seed. It is ideally suited as specimen or group plantings in landscape plan. Other important species is *T. spectabilis* which produce flowers of rose colour in the month of February-March.

Tabebuia aurea* Syn. *Tecoma argentea (Bignoniceae)

It is a medium sized tree with open top. It is very handsome when it produces sulphur yellow trumpet like flowers during end of March to middle of April. It is easily propagated through seeds.

It is highly suitable as a specimen or for group planting in parks.

Tecomella undulata (Bignoniaceae)

It is native of India. It is a dwarf tree with bushy appearance. Flowers are large, bell shaped of crimson colour which are produced in April and continue upto July-August. It is suitable for small parks.

Tectona grandis (Verbenaceae)

It is native of Indo-Burma region and commonly called as Sagwan or teak. It is a tall, quick growing tree. Bark is light grey in colour which peels off easily in longitudinal flakes. Leaves are large, green in colour, rough above and light yellow underneath. White small flowers appear in March-April. It is generally propagated through seeds. It is a timber tree and the wood is in great demand for furniture industry. It can be planted on boundary and alongside roads.

Terminalia arjuna (Combretaceae)

It is native of India and Sri Lanka and is commonly called as Arjun tree. It is an evergreen, tall tree making columnar growth. Bark is whitish or pinkish grey. Leaves are opposite and oblong-elliptic which are shed in March-April and again appear in the month of May. Flowers are yellow in colour which appear in May and continue upto July. Fruits are five winged. It is propagated through seeds. Trees are suitable for roadside plantation and in groups for shade.

Terminalia catappa (Combretaceae)

It is native of India and is commonly called as Badam or Indian almond. It is a tall tree with columnar habit of growth. Bark is greyish brown. Leaves are small, rough and turn red in February and September. White flowers appear in May. It is propagated through seeds. It is suitable for group planting and as avenue tree.

Terminalia belerica (Combretaceae)

It is also native of India and Sri Lanka and is commonly called as Bahera. It is suitable for roadside plantation and fruits are used for medicinal purposes in the Ayurvedic system.

Thespesia populnea (Malvaceae)

It is native of India and is commonly called as *Pahari Pipal*. It is medium in height with round crown. Bark is light grey in colour with long fissures. Leaves are cordate in shape and shining. Flowers are of yellow colour changing to cinnamon red which appear round the year. It can be easily propagated through seeds. It is suitable for roadside planting.

Triadica sebifera* Syn. *Sapium sebiferum (Euphorbiaceae)

Commonly called Chinese tallow, it is native to eastern China, Taiwan, and Japan. It is a winter deciduous tree reaching to a maximum height of 15 m. The simple, heart shaped leaves have an extended tail often resembling the Peepal tree. The leaves are light green in color and slightly paler underneath turning coppery red in autumn. It produces cluster of greenish-yellow and white flowers in spike-like inflorescences in spring. Chinese tallow tolerates almost any soil and can grow 5 ft (1.5 m) tall in its first year. It can be grown as a shade tree and for its brilliant fall foliage. However, the plant is now known to be extremely invasive in some areas.

***

22

Climbers

Climbers are the group of plants which have weak stems and ability to climb up the support with the help of modified organs for sun light and air. Twiners differ from climbers in the way that they don't possess such modified organs but twine around the support, cover it and reach the top. Nature has equipped these climbers with different modified organs which help in climbing of the support. They are tendrils. (*e.g. Antigonon leptopus, Bignonia gracilis, Pyrostegia venusta, Clematis panniculata*), thorns (*e.g. Bougainvillea* sp.; climbing Roses); roots and rootlets—(*e.g. Campsis grandiflora, Ficus repens*) by secreting sticky substances from growing points—*Ficus repens.*

Climbers and twiners are important group of plants which add beauty, colour in striking way or fragrance in gardens and artificial structures like wall, arches, pergola, pillars, cascades, topiary, etc. are well decorated with the help of climbers. Trees are also used to train the climbers and this enhances the beauty of garden manyfold.

Selection of Climbers

There are different situations in the garden like sunny, partial shade and different structures according to which suitable types can be selected from an available wide choice. The climbers for different situations are as follows :

***(i)* For sunny situation.** *Pyrostegia venusta, Quisqualis indica, Antigonon leptopus, Adenocalymma alliaceum, Bougainvillea* sp., *Campsis grandiflora* etc.

***(ii)* For Partial shade**. *Clerodendrum splendens, Petrea volubilis, Lonicera japonica, Trachelospermum jasminoides.*

***(iii)* Heavy climbers.** These climbers produce luxuriant vegetative growth and grow very vigorously. They cover large area and are suitable for covering bigger area *e.g. Antigonon leptopus, Bignonia magnifica, Beaumontia grandiflora, Bougainvillea* sp. *Hiptage benghalensis, Quisqualis indica, Wisteria sinensis, Pyrostegia venusta, Clerodendrum splendens.*

***(iv)* Light climbers.** These climbers make sparse vegetative growth and remain light in spread. These climbers are suitable for limited spaces *e.g. Lonicera japonica, Trachelospermum jasminoides, Clitoria ternatea.*

***(v)* Climbers for fragrant flowers.** *Jasminum grandiflorum, J. officinale, Trachelospermum jasminoides, Hiptage benghalensis, Clematis panniculata.*

***(vi)* Climbers for pots.** *Bougainvillea, Clitoria ternatea.*

***(vii)* Making hedge.** *Clerodendrum inerme* and *Bougainivillea.*

***(viii)* Indoor decoration.** Shade loving climbers like *Pothos, Monstera deliciosa, Philodendron* sp., *Asparagus* sp., etc. can be grown in pots and can be kept indoor also.

***(ix)* For screening.** *Vernonia elegnaefolia, Pyrostegia venusta.*

***(x)* Annual climbers.** These are climbers which are seasonal in nature and can be grown temporarily. The examples are : Sweet pea, trailing nasturtium. There are woody climbers which can also be grown as annual like *Ipomea lobata, Clitoria ternatea.*

Planting of Climbers

These climbers thrive well in any garden soil. The soil should be fertile, deep and good in water holding capacity and water drainage should be proper. For planting the climber, a pit measuring 60 cm × 60 cm × 60 cm should be dug and refilled with 10-15 kg of well rotten farmyard manure and Chlorpyrifos @ 5 ml/l of water. The main planting season for evergreen climbers is July-September and February-March whereas deciduous climbers are planted during January-February. After planting regular watering should be done. Proper support should be provided at the time of planting. General training in first year is necessary to have the desired effect depending upon the kind and shape of support.

Maintenance

After planting the climbers, regular watering, weeding and hoeing are required which should be done. In case of any mortality, the gap filling should be done. During subsequent years, pruning of these climbers is essential to keep them in limit and desired shape. General pruning consists of removal of dead wood and excessive growth. Heavy climbers need drastic pruning whereas light climbers require light pruning. General care should also be taken to keep the climbers healthy and free from insects, pests or diseases.

Description of Important Climbers

Mansoa alliacea Syn. *Adenocalymma alliaceum* (Bignoniaceae)

It is native of Brazil. It is an evergreen climber and makes profuse growth. Leaves are alternate and compound and smell like garlic. Leaflets are 2 or 3, ovate, shining green and leathery. Flowers are bell shaped and of lavender in colour which are produced profusely in November. Easily propagated by cuttings and layering. It is highly suitable for planting against wall, arches, pergola, etc. The pruning should be done in end of January to keep it in limits.

Antigonon leptopus (Polygonaceae)

It is native of South America and commonly called as Coral vine or loves chain. It is an evergreen, tuberous rooted and quick growing climber. Leaves are cordate in shape, rough and dull green. Flowers are produced on a long receme in white, rosy-red and red colour in September to November. The climber is highly suitable for arches and pergola. It should be pruned in January. It can be easily propagated by seeds.

Aristolochia grandiflora (Aristolochiaceae)

It is native of South America and is commonly called as Duck flower. It is an evergreen climber. Leaves are simple, alternate and cordate in shape. Flowers are big sized, white in colour with purple blotches which appear in September. Flowers are liked for its peculiar shape but has offensive odour. It is easily propagated through seeds. It is ideally suited for trellis, screen, against walls, pergola, etc. It should be pruned in the end of January.

Bauhinia vahlii (Leguminosae)

It is native of India and common in Western Ghats. It is a luxuriant grower and makes very fast growth. Leaves are big and are bilobed. Flowers are creamish white and appear in April-May. It is also easily propagated by seed. It is useful for bigger parks and can be trained on trees or large stumps.

Beaumontia grandiflora (Apocynaceae)

It is native of Eastern India and is commonly called as Nepal Trumpet creeper. It grows luxuriantly and climbs up the support upto 15-20 m. Leaves are large, opposite and oblong ovate in shape. Flowers are big, bell shaped and white in colour which are produced in the month of March. Propagation is done by air layering. It is highly suitable for arches, pergola and to cover large area. It should be pruned in the end of June.

Banisteriopsis laevifolia* Syn. *Benisteria laevifolia (Malpighiaceae)

It is native of tropical America. It is an evergreen climber and makes luxuriant growth. Leaves are big opposite, elliptic and of dark green colour.

Flowers are of showy yellow colour which are produced on trichotomous umbellate panicles during hot and rainy season. It is easily propagated by cutting.

Bignonia magnifica (Begnoniaceae)

It is a native of Columbia and South America. It is an evergreen climber making profuse growth. Leaves are opposite, compound consisting of 2 leaflets. Flowers are large, funnel shaped, purplish violet in colour which appear in October-November. It is propagated through cuttings and layering. Suitable for planting against wall and arches. It should be pruned in January.

Dolichandra unguis-cati Syn. *Bignonia gracilis* (Bignoniaceae)

It is a native of South America. It is an evergreen climber and makes impressive growth against high walls. It produces yellow coloured trumpet flowers of large size in March-April and in July-August. It is easily propagated by seeds and layerings.

Bougainvillea sp. (Nyctaginaceae)

Bougainvillea is a versatile plant and rich in its varietal wealth which can be used in different ways like bush, standard shrub, climbers, hedges, pot plant, bonsai, ground cover for sloppy lands and to make the garden colourful for most of the year. The flowers are surrounded by colourful bracts which are single or multiple and have wide colour range that varies from white, mauve, purple, deep purple, orange, red, scarlet, yellow to crimson with various intermediate shades. In many varieties like Thimma, Dr. Rao, Vaisakha Archana, Surekha, Odissi, Jaya Lakshmi variegata, Arjuna, Lady Hudson of Ceylon variegata and Silver Top, have variegated leaves which also add to the beauty of the garden.

Bougainvillea is a native of Brazil introduced in India in early 20th century. It has been acclimatized in various agroclimatic zones of India in such a way as it is native of this country. It has dominated in Indian gardens from Northern hilly to Southern parts of the country and from East to West in short span of time due to its floriferous nature, recurrent blooming, easy propagation and free from insect pests and diseases.

There are four main species of Bougainvillea *viz. B. peruviana, B. glabra, B. spectabilis, B. xbuttiana,* which have played major role in evolving varieties.

There are about 150 varieties of Bougainvillea, available in India and suitable varieties can be selected for planting in the garden for various situations. These varieties can be planted in such a way that garden is colourful in most parts of the year. Now-a-days, few gardens have been planted exclusively with Bougainvillea and it has been used in all possible ways. In Northern India, two flushes of bloom appear in end of November to first week of December and during March-April. The varieties like Shubra, Partha, Glabra, bloom during summer also.

There is a vast choice of varieties for different purposes in the garden and suitable varieties should be selected for the desired effect. It has been observed that varieties like Partha, Formosa, Sanderiana are highly suitable for making good hedge whereas for a dwarf shrub varieties like Cypheri, H.B. Singh, Formosa, Golden Glow, Sanderiana can be selected. For training Bougainvillea as a tall shrub, there is a wide choice in different colours and varieties like Jayalakshmi, Meera, Scarlet Ohara, Tomato Red, Partha, Mary Palmer, Lady Mary Baring, Thimma, Shubra, Mrs. Manila and Louis Wathen are suitable whereas for climbing purposes varieties like Mary Palmer, Mrs. H.C. Buck, Mahara, Louis Wathen, Lady Mary Baring can be used. For pot culture varieties like Shubra, Los-Banos Beauty, Begum Sikandar, Wajad Ali Shah, Thimma, Mary Palmer, Poultoni Special, H.B. Singh and Mrs. H.C. Buck are highly suitable. For planting on slopes varieties like Partha,

Snow Queen, Thimma and Dr. R.R. Pal can be effectively planted whereas for arches and pergola or on trees varieties should be selected from Magnifica, Mrs. H.C. Buck, Mary Palmer and Lady Mary Baring. For making good bonsai varieties like Meera, Ruarka, Jayalakshmi, Padmini, Elizabeth, Partha, Thimma, Mary Palmer, and Shubra are ideally suited. Flowers of varieties like Mahara, Mary Palmer, Shubra and Splendens can be used as cut flowers which have good keeping quality in air conditioned rooms. Torch Glow is of recent introduction in Indian gardens. It has dense foliage with showy reddish-pink bracts, flowers appear like a torch a top the branches.

Majority of the varieties are propagated through hardwood cuttings in July-August or in January-February and planted in polythene bags. Varieties which are difficult to root like Formosa, Latereitia, Speciosa, Thimma, Mary Palmer, Mahara are propagated through air layering which is done in August.

Campsis grandiflora (Bignoniaceae)

It is commonly called as trumpet climber and is native of China and Japan. It is winter deciduous and new leaves start emerging from the end of March to first fortnight of April. Branches produce rootlets at nodes with which it climb up the support. Leaves are compound and leaflets have serrated margins and are of ovate to obovate shape. This climber is laden with big sized trumpet shaped orange flowers from May and remains in continuous bloom upto October-November. It is highly suitable for training against wall arches and pergola etc. It is easily propagated through cuttings. It should be pruned in January.

Clematis panniculata (Ranunculaceae)

It is native of Himalayan region and is commonly called as virgin bower. It is another good climber which blooms in July-August. Leaves are opposite, compound, leaflets are 3-5 and ovate-cordate in shape. It should be pruned in January. Flowers are white, fragrant and produced in panicles. Easily propagated through cuttings. It is ideally suitable for trellis.

Clerodendrum splendens (Verbenaceae)

It is native of tropical Africa. It is an evergreen climber. Leaves are simple, opposite, big size having crinkled surface and wavy margin. Flowers of deep red colour are produced in clusters from December to March and make very attractive contrast with deep green foliage. It is highly suitable for semi-shady situations and may be planted in north side. It is easily propagated through root cuttings. *Clerodendron inerme* is another good climber highly suitable for making good hedge. It produces white fragrant flowers from July to September and can be multiplied through hardwood cuttings.

Clitoria ternatea (Leguminosae)

It is commonly called as Butter fly pea creeper and is native of India. It is a light evergreen climber. Flowers are attractive in blue, white and purple colours which are produced in three months. It can be propagated through seeds.

Ficus repens (Moraceae)

It is native of India and is commonly called Indian Ivy. It is an evergreen climber and is planted for its foliage. It climbs up the support with help of rootlets as well as by producing sticky substances. It should be planted against the wall. It prefers semi-shady situation. It can be trimmed into different shapes also.

Hiptage benghalensis (Malpighiaceae)

It is native of India and is commonly called as Madhvilata. It is a heavy climber and produces scandent branches like rambler. Leaves are large, simple and dark green in colour. Flowers appear in February-March which are creamish white and scented. It is easily propagated through seeds or layering. The pruning should be done in the end of June to give proper shape.

Ipomoea cairica* Syn. *Ipomoea palmata (Convolvulaceae)

It is commonly called as railway creeper and is native of tropics of both hemispheres. It is a quick growing climber and in a short span of time covers large area. Leaves are palmate and of light green colour. Flowers are large, round, have large corolla tube and purple in colour which are produced in abundance during rainy season and round the year. It is highly suitable for growing in garden on boundaries covering large areas as ground cover. There are many other beautiful species of this genus.

Lonicera Japonica (Caprifoliaceae)

It is native of Japan and China and is commonly called as Japanese honey suckle. It is an evergreen and quick growing climber. Leaves are opposite, ovate or oblong ovate. Flowers are white changing to yellow and highly fragrant and are produced in May-June.

L. sempervirens is another good species which produces scarlet flowers which are not fragrant. Honey suckles are good for trellis. It is rapidly propagated through cuttings or layering.

Passiflora laurifolia (Passifloraceae)

It is native of South America and has peculiar shape of flower which resembles wrist watch. It is commonly called as *Phoolghari*. It grows luxuriantly and is a heavy climber. Leaves are alternate and divided into 3 deeply cut lobes. Flowers are solitary terminal or axillary of purple colour. It is easily propagated through seeds or layers.

Petrea volubilis (Verbenaceae)

It is native of tropical America and is commonly called as 'Purple Wreath'. It is an evergreen climber with greyish bark. Leaves are opposite elliptic or ovate of light green colour but highly rough. Flowers appear in April-May on long spikes of star shaped blue colour. It is easily propagated through layering and suckers. Highly suitable for training against walls or to be planted in shrubbery.

Pyrostegia venusta (Bignoniaceae)

It is native of Brazil and is commonly called as 'golden shower'. It is an evergreen and woody climber. Leaves are compound of deep green colour. There are two leaflets and terminal leaflet oftenly converts into tendrils. Rich orange coloured tubular flowers produced as drooping cymes which cover the foliage completely during February-March. It is most beautiful climber when in bloom adorns most buildings, houses, compounds, etc.

It is propagated through ground layering. It is ideally suited for planting in south side against walls, pergola, trellis, boundary wall, screening and to cover large area. The pruning should be done in the end of June.

Porana paniculata (Convolvulaceae)

It is native of India and is commonly called as Bridal bouquet. It is an evergreen climber which puts a luxuriant growth. Leaves are alternate, cordate, glabrous of light green colour. Flowers are small funnel shaped, scented and produced in sprays during months. Ideally suitable for training against walls, pergola and trellis. It is propagated through cuttings.

Solanum seaforthianum (Solanaceae)

It is commonly called as 'Blue potato creeper' and is native of West Indies. It is a light, woody and evergreen climber. Leaves are alternate, compound with 3 leaflets. Purple blue flowers resembling potato and are produced in several flushes in a year on axillary dropping panicles. It is highly suitable for planting against wall.

Combretum indicum* Syn. *Quisqualis indica (Combretaceae)

It is native of Malaya and is commonly called as 'Rangoon creeper' or 'Jhoomeka Bel'. It is very luxuriant grower, evergreen, hardy and very commonly used for planting. Leaves are opposite, oblong or elliptic. Flowers are produced terminally, axillary on pendulous recemes in three flushes which cover almost the year. It is very floriferous and flowers are at first white, later on change to pink and red. It is easily propagated through cuttings and root suckers. It is highly suited against walls, on trellis, pergola, trees etc. It should be pruned in the end of January to keep it in limits.

Thunbergia grandiflora (Acanthaceae)

It is native of Eastern India and is commonly called as 'sky flower'. It is vigorous growing climber that can cover large area in short time. Leaves are opposite, ovate-cordate, toothed or lobed. Flowers are of blue colour which are produced solitary or in small recemes during February-March and July-September. It is easily propagated through seed, cuttings and layering. It is suitable for trellis, pergola, trees etc. *T. grandiflora alba* is another beautiful species producing white flowers.

Trachelospermum jasminoides (Apocynaceae)

It is native of China and Japan and is commonly called as 'star jasmine'. It is an evergreen climber. Leaves are opposite and elliptic or oblong in shape. Flowers are small, star shaped, white and fragrant which are produced in May-June. It can be easily propagated through semi hardwood cuttings. It prefers partial shade. It should be pruned in the end of January.

***Tarlmounia elliptica* Syn. *Vernonia elaegnifolia* (Compositae)**

It is native of South America. It is evergreen climber which is highly prized for its elegant drooping branches and light green foliage. Leaves are simple, opposite and of light green colour. It is easily propagated through cuttings. It is most ideally suited for creating weeping effect on porch, balcony, boundary wall, against pillars, walls and for screening purpose. It should be pruned according to need.

***Wisteria sinensis* (Leguminoseae)**

It is native of China and Japan and is commonly called as 'grape flower vine'. It is a winter deciduous, woody climber which makes very fast growth in short time. Leaves are alternate and imperipinnate. Leaflets are 7-9 and ovate in shape. Flowers of purple colour are produced in March when plant is still in deciduous condition and covered with hanging flowers. *Wisteria alba* is another pretty climber with white flowers. It is easily propagated through cuttings. It is highly suitable for pergola, trellis or trees. It should be pruned in the end of January to give it proper shape and to keep it within limits.

***

23 Bulbous Plants

In nature, there are certain plants which have modified underground stems in which food material is stored to overcome the unfavourable season. Botanically, plants propagating themselves through modified underground stems are grouped into four groups *viz.* (*i*) Tubers-Dahlia, (*ii*) Corms-gladiolus, Freesia, (*iii*) Bulbs-Amaryllis, Tuberose, Zephyranthes, Crinum and (*iv*) Rhizome-Canna and Iris. But horticulturally, plants of all these four groups are called as bulbous plants. These plants are prized for their magnificent flowers and for being grown for mass effect (*e.g.* Dahlia, Canna, Zephyranthes), cut flowers (*e.g.*, Gladiolus, Tuberose, Amaryllis), pot plants (*e.g.*, Amaryllis, Dahlia), for extracting essential oils (*e.g.*, Tuberose). They are suitable for various situations and are fitted in landscape plants.

These bulbous plants are divided into two categories *i.e.*, hardy and tender bulbs. The hardy types can be left in the ground and are separated after 2-3 years for their multiplication (*e.g.* Amaryllis, tuberose, canna, zephyranthes, crinum etc.). On the contrary, the tender types cannot be left in the ground after flowering and are to be lifted from the ground on maturity. After their treatment against diseases, these are to be stored in cold storage or cool places during off season for next year planting *e.g.,* Gladiolus, narcissus, Daffodils, Freesia, Dahlia, etc.

Climate

In general, cool weather, high atmospheric humidity and moist soil conditions are congenial for the manufacturing of food, better production of high quality flowers and bulbs. According to climatic requirement of these bulbs, they are grouped into two categories *i.e.*, Warm climate—Canna, Dahlia, Tuberose, crinum, Zephyranthes, etc., and cool climate—Gladiolus, Narcissus, Daffodils, Freesia, etc. Modified environmental conditions are to be provided during winter and summer seasons for the production of high quality blooms.

Soil

Soil should be sandy loam or loamy sand in texture for better production of bulbs. It should be moderately fertile with good water holding capacity, and with pH ranging between 7.0-8.0. For pot culture the ideal pot mixture should contain 2 parts sand : 2 parts leaf mould : 2 parts well rotten farmyard manure : 1 part soil and ½ part charcoal.

Propagation

The following are the common methods employed for propagation of bulbous plants :

(i) **Off-sets, cormlets or bulblets.** Mother bulbs produce many bulblets, which are separated and planted for raising new crop. The intensity of bulblet production is a varietal character. Some varieties are very prolific bulblet producer whereas some produce sparse bulblets. It takes about 1-2 years for growing a bulblet into a full bulb.

(ii) **Division.** It is followed in tubers, rhizomes and corms which are divided carefully (in such a way that each piece contains at least one vegetative bud) and planted in the soil.

(iii) **Terminal cuttings.** These cuttings are mostly employed to propagate Dahlia and Rex begonia.

(iv) **Seed.** This method is employed by the breeders to create new varieties. *e.g.* Gladiolus, Dahlia, Freesia, Amaryllis, Lilies, etc.

Manuring

Well rotten farmyard manure @ 5 kg/sq m should be incorporated at the time of soil preparation. The application of 20 g each of P_2O_5 and K_2O/sq m has also been found beneficial at the time of planting bulbs. Nitrogen @ 40 g/sq m into two-three splits should be applied.

Planting of Bulbs

The planting time for different bulbs is different. In North India plains the planting of Gladiolus, Narcissus, Daffodils, Iris, Dahlia, etc. can be done from mid of September to mid of November. If planting is done in succession the time of flower production, will also be longer. Amaryllis is separated in November and is planted. Tuberose and Zephyranthes are separated and planted in February-March. Canna is planted in July with the onset of rains.

The depth of planting of bulbs also differs according to the type of bulb. Generally, bulbs are planted at 7-10 cm depth except few lilies which are planted at 25-30 cm depth. By planting the bulbs at different depths, the flowering period can be extended. The distance between rows and plants are varied according to the purpose of growing. Generally distance of about 30 × 20 cm is maintained.

Intercultural Operations

Weeding and hoeing. Weeds should not be allowed to grow and should be removed as soon as they appear otherwise they compete for nutrients and water with the main crop. Hoeing helps in aeration of soil and better development of bulbs. In early stages of growth frequent weeding and hoeing are required.

Irrigation. Liberal application of water at all stages of development is highly beneficial. Generally 7-10 days interval is appropriate in winter whereas in summer it should be about 5-7 days.

Staking. Bulbous plants like Gladiolus, Dahlia, Lilies, etc. produce flowers on long stem and thus need support. Otherwise spikes may bend down and break with strong winds. Bamboo stakes can be provided.

Harvesting of Flowers. The stage of harvesting of flowers depends upon the distance of travel. It varies with the kind of flower. For local market, gladiolus and tuberose should be cut when lower 2-3 florets have opened. Whereas for distant market, flowers are cut when basal florets show colour. Narcissus and Daffodil are cut at goose neck stage. Before harvesting the flowers, plants should be irrigated.

Lifting of bulbs and their storage. Tender bulbs like Gladiolus, Dahlia, Narcissus, Daffodils, etc. are to be dug from the soil 10-12 weeks after flowering has been finished. Before lifting the bulbs, water is withheld. After digging, bulbs are dried in shade for few days. Then these are treated with 0.2% Bavistin solution for 30 minutes and thereafter stored. Hardy bulbs are separated after 2-3 years and are again planted in the planting season.

DESCRIPTION OF IMPORTANT BULBS

Amaryllis (Amaryllidaceae)

It is commonly called as Belladonna lily, King or Star lily. It is native of South Africa. It is an excellent for bedding, pot, window boxes, green houses, borders, etc. Flowers are sold in the market as cut flowers. The plants are about 75-90 cm tall and leaves are long, dark green and strap shaped. The flowers are large (15-20 cm across), trumpet shaped and are borne in umbels having 3-7 florets in small flowered varieties whereas 2-4 florets of giant size in Dutch hybrids. The colour of flower varies from white, deep pink, dark red, scarlet, crimson, salmon to orange. In some varieties flowers also bear stripes on white with red or white stripes on solid colours. There are numerous varieties available. The important varieties are; Snow white (2 flower-white), Taurus (2 flower-pale pinkish red), Leo (2 flower-pale carmine), Jupiter (3 flower-bright scarlet), James (3 flower-orange scarlet), Picture (3 flower-scarlet), Beautiful (3 flower-deep orange scarlet), Apollo (4 flower-deep vermillion), Star of India (4 flower-deep scarlet), Mount Everest (4 flower-pure white), Black Prince (4 flower-crimson). Other important varieties are Pink Indira,

Silver Lining, White Queen, Snow Queen, Royal Ruby, Red Majesty, Red Master, Christiana Day, Christiana Joy, etc.

***Begonia* sp.** (Begoniaceae)

It is native of Mexico, Central and South America, Asia and South Africa. The genus includes about 1200 species and numerous hybrids, mutants and chance seedlings. The genus has been named by Charles Plumier (1960) after his patron Michel Begon. Begonias are grouped into three main classes according to root types *i.e.*, Rhizomatous (*Begonia red*), tuberous (*B. tuberhybrida*) and fibrous (*B. semperflorens*). Out of three types rhizomatous and fibrous rooted are grown in the plains whereas tuberous rooted can be grown successfully in hills.

Begonias are ideally suited for pot growing, hanging baskets, window gardens, bed and rock garden. They are prized for their attractive leaves and flowers. In the plains, rhizomatous and fibrous types can be planted in October-November and flowers are produced in February-March. If care is not taken, plants die during June-July. Different methods of propagation are employed for different types of begonia. Division of tubers and rhizomes is done for further propagation. Leaf cutting method is also employed for rhizomatous types whereas seeds are used for fibrous types. Frequent application of liquid manure is beneficial for growth and flowering.

The important varieties of rhizomatous Begonia are : Peace Majesty, Silver Queen, Emperor Mikado, Can Can, Crimson Glory, Black Knight, Curly Star Dust, Dew Drop, etc.

The important varieties of fibrous rooted are : Goldilock, Bo-Peep, Little Gem, Ballet, Lady Frances, Pink Jewel, Flamengo, Cindrella, Charm, Silver Star, Pink Camellia, etc.

The tuberous rooted Begonias are valued for their attractive flowers and are grouped into Rose form, Camellia flowered, Carnation flowered, Daffodil flowered or Picotee.

Canna indica (Cannaceae)

It is native of tropical America and is commonly called as 'Indian shot'. Plants are perennial, herbaceous, tall growing about 1 m and with green stem. Leaves are 20-60 cm long, lanceolate, green and in some cases variegated and bronze red. Modern varieties of canna are divided into two classes.

***(i)* Orchid flowering.** In this group flowers do not open together but open one by one *e.g.* Golden Glory, Salmon Yellow, Carmine King, Copper Giant, Golden Wedding, King Humbert, etc.

***(ii)* Truss flowering.** In this group, flowers mostly open together and form large head. Most of the newer creations are in this class which are used for impressive display of colour.

Present day canna varieties are mainly classified according to their height. Tall varieties which grow about 1.0-1.5 m tall and are very useful for big parks and large areas. Important varieties are *Red colour*—Statue of liberty, The President, The Ambassador, Carmine King, Cleopatra; *Orange colour*—American Beauty, Copper Giant, Sunset, Coles; *Pink*—Alipur Beauty, Candilabra, Apricot King, Dorris, Prince of Wales; *Yellow*—Master Piece, Enchantress, Buttercup, Aurora Borealis; *White*—White Queen and Radio. Dwarf varieties grow 0.5 to 1 m. Important varieties are : *Orange*—Orange Glory, Orange Plume, Fairly, Cleopatra; *Pink*—Hungaria, Orange King, Departing Day; *Yellow*—Apricot and Star Dust.

There are number of varieties which have spotted or streaked flowers of different colours. These varieties are preferred over self-coloured varieties. Important varieties are : *Red with yellow border*—Queen Charolette; *Yellow spotted red*—Mikado, Yellow King Humbert, Percy Lancaster, Gladiator, Mrs. Lancaster; *Deep orange with orange yellow border*—Rosamund coles.

In some varieties foliage is also attractive being different from common green leaves. Bronze leaved varieties are preferred for its ornamental foliage beauty. Important varieties are : Black knight, La Globe, King Humbert. Variety with variegated leaves is *Trinacria variegata.*

Canna is a rhizomatous plant and is very easily propagated by dividing the rhizomes into 10-15 cm pieces during end of June. These pieces are kept in cooler place for rest before planting. For evolving new varieties, sexual method of propagation is used. Seeds are black in colour and have hard seed coats. Scarification is done for quick germination of seeds.

Before planting the Canna in first fortnight of July, the beds are prepared thoroughly by digging 60 cm deep and applying 5-20 kg well rotten farmyard manure per sq m. Divided clumps are planted 15 cm deep at the distance of 60 cm between rows and plants.

Regular weeding and hoeing is done. It takes about 4 months to flower after planting and after that flowering continues.

Crinum bulbispermum (Amaryllidaceae)

It is commonly called as Sudarshan or Sukhdarshan or St. John lily or cape lily and is native of South Africa. It is ideally suited for bigger pots, alongside the paths, on the edge of a pool. Bulbs are planted in February-March. Plant has strap shaped long leaves. Flowers of white coloured flushes with red are produced in a scape on long and stout stem. Other important species are : *C. moorii, C. powelli, C. latifolium and C. zeylanicum.*

Freesia-Freesia refrecta (Iridaceae)

It is a native of South Africa. It is highly suitable for pot, window gardens and cut flowers. It can be propagated through corms which are planted in September-October in plains of India. Plant grows about 30 cm and leaves are linear. Flowers are very fragrant.

funnel shaped and are produced on wiry spikes during March. Flower colour varies from white, creamy white, lilac blue, orange, pink, red, mauve and bronze. Modern varieties are hybrids of species *F. refrecta* and *F. armstrongii*. Important varieties in separate colours are : White Swan (white), Pink Giant (Pink with white throat), Gold Coast (orange-yellow), Margaret (purple), Sapphire (Blue), Rynveld's Golden yellow (yellow). Cultural aspects are quite similar to those mentioned for gladiolus.

Glorosia superba (Liliaceae)

It is commonly called as Glory Lily and is native of tropical Asia (India and Ceylon). It is suitable for growing in ground and pot and is a good cut flower. The tubers of pencil thickness are planted horizontally in March-April. It is a tall, twining plant which blooms during late rainy season *i.e.*, September-October. Flowers are smaller with narrow but crisped petals, first green, changing to yellow and finally to orange red. Another species, *G. superba lutea* produces entirely yellow flowers. Other important species are *G. carsonii*, *G. simplex*, *G. rothschildiana* and *G. verschurri*.

Haemanthes multiflorus (Amaryllidaceae)

It is native of tropical Africa and is commonly known as Blood flower or Red cape lily. It is commonly grown in pots. Plants are medium tall with oblong leaves. Flowers are puffy of scarlet red colour which are produced on long stem (30 cm) in spring and summer months. Plants are dormant in winter months. The bulbs are separated in February-March for further plantation.

Hedychium-Alpinia speciosa (Zingiberaceae)

It is commonly called as Shell Ginger and is a native of Sikkim and Bhutan. Plants are 1.0-1.5 m tall and leaves of about 30-40 cm length are produced on long canes. Flowers of white flushed with pink colour are produced in February-March. Plants are best grown in semi-shade.

Hemerocallis fulva (Liliaceae)

It is popularly called as Day lily and is native of Siberia-Japan. It is highly suitable for foundation planting, or near pools or in shrubbery. It is a vigourous perennial herb with long linear leaves. Flowers are produced during April-June on long (60-90 cm) stalks which are showy campanulate. Flower colour varies from light yellow, dark yellow, orange, purple to red. Propagation is done through division of rhizomes during the month of March.

Spider lily—Hymenocallis littoralis Syn. *Pancratium littorale* (Amaryllidaceae)

It is native of South America and has been naturalized in India. It is highly suitable for pot, growing near lilypools, or alongside walks or paths on a large rockery under big trees. The plants are tall 60-90 cm having strap like long leaves of light green colour. White flowers are produced in umbels of 9-10 flowers. At a time 2-3 flowers open. The flowers have six long, narrow segments united at the base by a thin membranous cup or corona. It is commercially grown around Mumbai where flowers are sold in the market. Individual flowers are used in making garlands.

***Iris* sp.** (Iridaceae)

It is a native of South East Asia and Southern Europe and is known as Flag flower. It is suitable for group planting near lily pools, rockey and in shady situations. Leaves of light green colour are pointed at the apex. Different species of Iris are grouped into two main classes *i.e.*, tuberous and bulbous types.

The species is native to Himalaya like *I. kashmeriana, I. nepalensis* and *I. kamaonensis* are rhizomatous which are comparatively easy growing. The rhizomes are planted in October and plants flower in March. Every year fresh planting is not necessary and plants can be left in the ground. The colour of *I. nepalensis* is pale lilac whereas that of *I. kamaonensis* and *I. kashmeriana* is white and purple mottled. Other important species are *I. cristata, I. susiana, I. siberica, I. kaempferi* (Japanese iris) etc. The commercial varieties of German Iris (rhizomatous) are King Midas, Petrea, Black Panther, Sahara etc.

The bulbous Iris is native of Holland, Spain and England. These are highly suitable for group planting, borders and cut flowers. It is available in wide range of colours. The bulbs are planted in September-October about 10 cm deep. Flowering occurs during February-March. The important species of this group are : *I. tuberosa, I. hollandica, I. danfordiae, I. reticulata*, etc. The important varieties of Dutch Iris are; Blue Champion, Harmony, Lemon Queen, King Mauve, White Excelsior, Bronze Beauty whereas Cajanus, Enchamress, King of Blues, Le Mogol, Summer Time are common varieties of Spanish Iris.

***Narcissus* and *daffodils*—*Narcissus* sp.** (Amaryllidaceae)

It is commonly called as Nargis and is native of Asia, North Africa and Europe. Botanically both belong to same genus but differ in the size of cup or corona of the flower. In daffodils the cup or corona is large whereas in Narcissus it is small. They are highly suitable for growing in beds, pots, under the deciduous trees, window gardens, etc. They are very good cut flower and are sold in the market. There are several species but common species are *N. triandrus, N. bulbocodium, N. poeticus, N. aurantiacus, N. albus, N. tazetta, N. tazetta papyraceus, N. odorus, N. jonquilla, N. cyclamineus, N. obvallaris.*

In the plains of north India, the bulbs are planted in October. Soil should be well fertile, sandy loam. Planting is done 20 cm apart and 7-10 cm deep. Flowering occurs during February-March. After flowering water is withheld. Leaves dry and bulbs can be left in soil for 2-3 years as such without disturbing them. After three years, bulbs are lifted from the ground and separated during April-May and are stored at cool place. Again fresh planting is done.

The standard classification of *Narcissus* sp., and varieties is done into eleven groups but broadly it can be divided into two conventional groups *i.e.*, Daffodils and Narcissus. The important varieties of Daffodils are : King Alfred (Yellow trumpet), Queen of Bicolours (Bicolour trumpet), Mount Hood, Beersheba (White trumpet), Golden Harvest (Yellow), Empress of Ireland (White), Flower Record (Creamy white), Golden Yellow (Crown edged bright red). Music Hall (Petals creamy white with golden yellow trumpet).

The common varieties of Narcissus are : Geranium (White petals, cup deep orange red), Poeticus Red Run (White petal and cup with yellow shading and a broad red rim), Poeticus Snow King (Glistening white petals and cup with yellow centre and edge of red), Paper White (Small white fragrant flowers).

Tulip (Tulipa hybrida) : Liliaceae (**Plate 19**)

Tulip ranks first amongst bulbous crop in popularity and occupies highest acerage and cut flower production in the world. Although tulip is a national flower of Iran and Turkey but tulip and Holland are synonyms for each other. In the vicinity of Leiden (Holland) where tulip was found growing in 16th century, now billions of tulip bulbs are being grown every year not only for bulb production but for cut flower production too. The cut flowers are in great demand during November-April otherwise flowers are being produced most of the time of the year. Holland commands highest bulb and cut flower production and export to almost all the countries of the world.

Tulip name comes from the turban of Iran and Turkish's custom due to resemblance of unopened tulip flower with the turban. It is most showy flower comprising of three colourful sepals and 3 petals each of same colour, shape and size in single form. Tulips are available in most of colours including white, yellow, orange, red, salmon, black, purple and more except true blue. The shape of flowers varies from elongated to cup shaped, bowl shaped and plain petals to crested or flathery or crisp or parrot type. This is the flower which is most widely embroidered or painted.

It is native of southern Europe, North Africa, Asia from Anatolia and Iran to east to North of China and Japan. The centre of diversity of this genus is in the Pamir, Hind Kush and Kazakishtan. Tulip is a genus of about 100 species and *Tulipa stellata* is a native of Kumaon hills of Himalaya region of India. In the 16th century, tulips are brought to Holland from Turkey and quickly became widely popular. Tulip mania prevailed in Holland in 17th century and Holland people traded bulbs literally for a fortune.

It is commercially cultivated in Holland, Italy, Spain and France. It is very popular in the gardens of U.S.A., Canada, Europe and other places as early spring flower. Tulip production dominates in Holland where it is being grown in about 10,800 ha followed by Japan (300 ha), France (293 ha), Polland (200 ha), Germany (155 ha) and New Zealand (127 ha). In all Holland produces about 4.32 billion tulip bulbs out of which 2.3 billion of bulbs are exported and 1.3 billion tulip bulbs are plated for cut flower and bulb production. In India, its commercial cultivation is being commenced in temperate region of Kashmir (J&K) where 15,00,000 bulbs of different varieties have been imported from Holland to produce flowers and bulbs. It will be biggest garden of tulip and will make a spectacular display of tulip varieties and will be a great attraction for the tourists during spring *i.e.* April to May.

There are about 3000 cultivated varieties registered and many more are added every year. Tulips come in a huge variety of bright colour including white, yellow, pink, orange, red, black, purple, bicolour and more. Generally cultivars are grouped by flower form, height and time taken to bloom. Flower form reflects, singleness, doubleness or plain or fringed petals where height means short (less than 20 cm), medium (20-45 cm) and tall means more than 45 cm. These varieties are classified in following group as advocated by Royal General Bulb growers association in 1996.

Single Earlies. Cultivar are mostly short stemmed easily flower and highly suitable for formal planting. Important cvs. are Keisers kroon (bright red edged without yellow). Rising Sun (golden yellow); General Da Wet (soft orange); Pink Beauty (brilliant cherry-rose with white stripes).

Double Earlies. These cvs. are mainly short stemmed and early flowering. Important cvs. are Mr. Van der Hoef (bright clean yellow); vuurback and Scarlet Cai Nival (red); Electra (Deep rosy violet).

Single Late Tulips. These cvs are single flowered, mainly long stemmed and late flowering.

Darwin Hybrid Tulip. These cvs. are mid seasoned long stemmed (60-90 cms) and single flowered cup shape flowers. Important cvs are : Gloria Swanson (crimson); Scotch Lassie (deep lavender); Zwanen Burg (pure white); Bishop (dark purple); Niphetos (soft yellow); Father Hugos Rose (Rose).

Lily Flowered. These cvs. have pointed and curled back petals with variable stem length single and mid or late season. These cvs. are very graceful and high in demand as cut tulip. Important cvs. are : Captain Fryatt (reddish violet), White Duchess (white), Yankee Girl (Buffy salmon).

Cottage tulips. These evs have larger flower than Darwin. Important cvs. are Advance (red); Rosabella (Pink), Marjorie Bowen (Salmon) ; Golden Harvest (deep lemon yellow).

Breeders. These cvs. are of unusual art shades and smoky overlays of hazy lavender or purple and excellent as cut flower.

Late Doubles. These cvs. are late and heavy so need a stake during rains. Important cvs. are Eros—(rosy lavender); Mount Tacoma (white); Uncle Tom (blackish red).

Parrots. These cvs. are single flowered with fringed, curled and twisted petal and bold splashes of green on their outer petals and look like as made of feather. Important cvs are Red Champion; Violet Queen, Sunshine, Scarlet Therese, Fantasy (salmon pink).

Rembrandt. The cvs of this group have flower with broken flower colour, striped or marked with brown, bronze, black, red, pink or purple on red, white background.

Multi flowered tulips

Fringed tulips. These cvs have petals with crystalline fringes, single flower, mid to late season with variable length. Honey (White), Canary (Red with yellow fringes), Fabio (Red with yellow combinations of white, yellow, green and red inside), Curly Sue (Deep violet colour) etc.

Kaufmanniana tulips (***Tulipa kaufmanniana***). Cvs of this group are very early flowering and dwarf (20 cm) exterior of flower has bright carmine blush.

Fosteriana tulips (*Tulipa fosteriana*). These are early flowering, very broad leaves, green or gray green sometime mottled or striped. Stem medium to long and large elongated flower.

Greigii Tulips (*Tulipa greigii*). The cvs have usually mottled or stripped foliage, leaves usually spread out and bend down toward ground. Late flowering with variable shape of flower.

Tulip grows happily in temperate region as bulb requires low temperature to break the dormancy. Thus, it can be forced also to regulate flowering for longer duration. Cool temperature (20-25°C) and ample of sunshine is good for plant growth and flower production. It blooms in spring season so rains, frost or hailstorms during this period are not desirable.

Tulip being a bulbous crop, soil should be deep, friable, loose, and sandy in texture which is good for its growth and bulb production. It prefers neutral pH and hence soil with lower pH can be corrected by liming. The soil with higher pH can be corrected to certain limit by adding gypsum in the soil. The quantity of these will be determined by the severity of acidity or alkalinity.

Propagation. Tulip is propagated vegetatively by off sets of bulbs which are separated from mother bulbs and planted to grow into full-fledged bulbs. Seeds for propagation are used by to grow seedlings into flowering bulbs by breeders.

Planting. Tulip bulbs are planted in fall *i.e.*, November and December and produce flowers during spring. Two planting methods are used depending upon the purpose of growing *i.e.*, for landscaping or commercial. Best landscape effects are created by planting in groups of 5-13 or more at close distance of 15 cm or when planted in informal designs under deciduous trees in different colour combinations. Single row planting should be avoided.

For commercial planting, long beds of 100-200 m and 1 m wide are prepared about 15-20 cm above the ground and bulbs are planted at distance of 10 cm and 15 cm deep. Each bed is separated with about 30 cm wide furrows used for irrigation. Due to close planting initially one or two weedings are required. Later on it is not required. For forcing the tulips in the field, it is covered with polythene during January to March so that early flowering and early harvest of bulb is achieved. These bulbs are against forced. Mainly forcing of tulip is done by progressive growers having cold rooms and green house facilities.

Important Disease of Tulip

Bulb crown and Root rot (fungi—*Penicillium* sp., *Sclerotium rolfsii*; *Pythium* sp., *Fusarium* sp., *Rhizopus stolonifera, Rhizoctonia solani*, *Aspergillus* sp.). There are various funguses responsible for bulb crown and root rot and displays various symptoms depending upon the severity of infection. Symptoms like poor sprouting, saddening or yellowing of leaves, stunting or death of plants are observed. In severe case rotting of bulb is observed. Roots may be shimy, soft, dry, hard or powdery in texture. Colour may be dark or light grey, black or brown with mouldy growth on bulb or scales. To control the disease, use disease free planting material and follow proper field sanitation. Removal and destroying of infected plants help rot in controlling the disease. Treatment of bulbs with systemic fungicides like Bavistin (0.2%) has been found useful.

Botrytis Tulipae. Minute, yellowish spots, elongated in the direction of leaf veins and surrounded by a darker, water-soaked area appear on leaves. As they enlarge, the regions, become more depressed and their colour charges to a whitish grey or brown. Such regions appear or plant stem and flower. Small black structures the size of pin head often are found on the outer bulb scales. To control this, three to five years rotations are recommended. Chemical like vinclozolin (Ornalin), Chlero thalonil, Ipriodion or benomyl or bavistin can be used.

Virus (Tulip Breaking Mosaic). Viruses are very common in tulip which spreads by cutting of flowers or by the attack of several species of aphids. To control this disease, destroy the diseased plants when found and control the aphid population by spraying 0.1% Rogor (1 ml/L of water).

***Zephyranthes* sp. (Amaryllidaceae).** It is native of South America and is commonly called as Zephyr lily. It is highly suitable on the side of path, under the trees, for rockeries and for mass effect of flowers as ground covers. Plants grow 15-20 cm and leaves are narrow. There are three common species *i.e., Z. grandiflora* syn. *Z. rosea* is a native of Mexico and Guatemala. It has dark green flat leaves and produces pink flowers with the advent of showers *i.e.,* July-August or few flowers after showers in autumn and spring. *Z. candida* is native of Argentina and has needle like leaves of light green colour and produces white flowers in September-October. *Z. sulphurea*—It has flat long leaves of light green colour and produces yellow coloured flowers in late rainy season and reappear in February-March with the advent of rains. The bulb production is very rapid in all the three species which can be separated in March-April or October.

***

24

Shade Loving Plants

Such plants grow well in shade or partial shade and are not able to tolerate direct scorching sunshine. In the gardens such conditions are available under the trees or artificially constructed green houses, chick or lath houses or in the verrandah of the buildings. Shade loving plants are mainly grown in pots and are highly prized for their ornamental foliage and hence, are also called as 'Foliage plants' or 'Pot plants'. Many such plants produce attractive flowers, too. These are used for interior decoration. Pots are kept hidden in wooden crates, brass pots or with other material in such a way that it harmonizes with surroundings. There is a wide choice of such plants and their horticultural varieties which provide a good opportunity for selecting a right plant material to suit different situations.

The pot mixture used for such plants should be friable and light enough to drain out extra water. Organic matters such as leaf mould helps to retain moisture. Rich soil should be avoided to cause undue stimulation of growth in plants which is not desired. Similarly, over feeding is not advisable otherwise plants produce unattractive ragged growth. The weak fertilizer should be applied once in two years. The pot mixture for such plants should be comprised of 2 parts garden soil : 1 part sand : 2 parts leaf mould : 1 part farm yard manure. The soil should be sterilized by heating to kill harmful bacteria and fungi.

For better growth of plants adequate water should be applied at proper time. Excessive watering or less application of water hampers the normal growth of plants. When to water and how much to water are two important factors which should be considered carefully. Watering should be done when plant exhibits initial symptoms of water stress. It can be judged by condition of plant and soil colour. At the time of watering, about 10 per cent more water than the actual requirement should be applied. It will ensure wetting of whole root zone. Extra water will drain out from drain hole. Sprinkling water on foliage or sponging helps in better growth of plant. Plant growth also depends upon availability of fresh air and light. If plants are being kept inside the rooms, they should be transferred to airy environment before any symptoms of sickness are exhibited and thereafter shifted inside again. Inside the rooms, the plants should be kept near or away from the light source depending upon the requirement of light by the plant. Repotting of plants should be done when plant becomes pot bound. It should be done either in rainy season or February-March. At the time of repotting, extra root should be cut away, and refilled with fresh potting mixture. Dry and yellow leaves should be removed from time to time. Feed the pot plants with liquid manure containing bone meals and spray systemic fungicides and insecticides to avoid the attack of diseases and pests.

Description of Important Shade Loving Plants

***Aglaonema* sp.** (Araceae)

The group includes plants having erect habit with long, slender pointed leaves with depressed lateral veins and conjugated appearance. *A. angustifolium* has dark green and shining leaves. *A. costatum* has white spotted leaves. *A. silver King* and *A. pseudobracteateum* are variegated, ornamental and are much prized.

***Alocasia* sp.** (Araceae)

These are lovely plants with sagittate-peltate leaves, and bushy growth. There are several species but common types sold in the market are :

A. sanderiana 'Van Houtte'. The leaves have variegated and prominent veins.

A. macrorhiza variegata. The leaves are variegated and very attractive.

***Anthurium* sp.** (Araceae)

It is native of West Indies. This genus comprises more than 200 species which are valued for ornamental leaves and colourful spathes which are rigid and waxy. Plants are rhizomatous and can be propagated through division. Important species and varieties are :

A. andreanum. Leaves are long lobed, heart shaped, green in colour and are very long. Spathe is waxy and red. Other attractive sports are album, giganteum, rubrum.

A. ornatum. Leaves are ovate-cordate and bright green, spathe is cup shaped, white turning purplish rose.

A. veitchii. Leaves are cordate, long about 90-100 cm, pendant of rich green colour.

A. cordatum. Leaves are broadly sagittate with wide sinus of olive green. Spathe green but reddish from inside.

A. magnificum. Leaves are large, heart shaped, velvety, olive green with prominent white veins. Spathe and spadix are green.

A. crystallinum. Leaves are large, beautiful, heart shaped and green with contrasting white veins. Spathe is green and linear.

A. podophyllum. Leaves are large, leathery, lobed with finger like segments.

A. lindenianum. Leaves are fleshy, sagittate, leathery, glossy green, borne on stiff wiry long petioles.

Aphelandra auriantiaca (Acanthaceae)

It is native of Mexico. Plants are erect and leaves ovate, smooth, green grey in vein area. Flowers are borne in bracted spikes of brilliant scarlet-red flowers and orange in throat. *A. squarrosa* is another species with handsome waxy, dark green ovate leaves with silver white veins. Flowers of yellow colour are borne on long spikes with orange bracts.

***Araucaria* sp.** (See trees)

Asparagus (Liliaceae)

It is native of South Africa. It is another common plant grown in pots for ornamental foliage. Asparagus are tuberous rooted. Important species are :

A. densiflorus cv. *Sprengeri.* It has profuse branches which are soft to touch.

A. densiflorus 'Myers'. It is showy plant dense with stiffly erect, plume like branches, dense needle like 'foliage' is rich green.

A. densiflorus pyramidelis. Branches are erect and zigzag on grey stem. Leaves are thread like in dense cluster of bright green colour.

A. setaceus syn. *A. plumosus* is commonly called as 'Fern Asparagus'. Plants have thin wiry stem and fern like rich green fronds.

A. scandens deflexus. It is a climber with green stem. Branches are zigzag in nature. Leaves are light green in colour.

Aspidistra elatior (Liliaceae)

It is native of China. It is small plant highly suitable for pots. Leaves are tough, leathery, oblong and blackish green in colour measuring about 60-75 cm long.

A. elatior variegata is another decorative species having alternately striped and banded green and white in varied widths.

Begonia (See bulbous plants)

Chlorophytum comosum variegata (Liliaceae)

It is commonly called as 'Ribbon plant' and is native of tropical Africa. It has large rosettes of arching, green, linear leaves having white margins. Long recemes appear from the centre with first flower, then develop tufts of leaves with aerial roots.

C. comosum vittatum is commonly called as 'Spider plant'. It has narrow linear, recurving leaves with white centre. Successive plantlets develop from the long flowering recemes which are pendant in nature.

Justicia brandegeeana Syn. *Beloperone guttata* (Acanthaceae)

It is native of Mexico and is commonly called as 'Shrimp plant'. It has wiry stem and leaves are ovate and hairy. Bracts are reddish brown and overlapping to give showy effect. Flowers are white.

B. guttata 'Yellow Queen'—It has yellow bracts instead of reddish brown.

Caladium bicolor Syn. *Caladium hortulanum* (Araceae)

It is native of tropical America and is a tuberous herb. Leaves are large peltate—heart shaped and mostly beautifully marked in many colours and patterns. There are numerous varieties, the important of which are Ace of Heart, Bleeding Heart, Cinderella, Fire Chief, June Bride, key Stone, Pink Beauty, Red Ace, Red Polka, Red Frill, Tom Thumb.

Caiathea sp. (Marantaceae)

It is native of tropical America. The plants resemble Maranta. Plants are small and are prized for their attractive foliage. Important species are *C. zebrina, C. magnifica, C. medio-picta, C. insignis, C. metallica, C. picturata,* etc.

Plectranthus scutellarioides Syn. *Coleus blumei* (Labiatae)

It is another beautiful foliage plant which makes a garden in itself. These are prized for their richly coloured leaves of different sizes in different patterns. The plants can be easily propagated by seeds or terminal cuttings. There are several varieties which are grouped as follows :

Light centered. Varieties with cream, yellow, light pink or any other pale shade in the centre of leaf but having a wider green margin.

Bright Centered. Leaves are bright red with coppery centre and have wider green margin; *Fringed*—The margins of leaves are fringed or deeply indented, with variable colours, self speckle and central blotching is there with leaf fringed; *Mammoth*—Varieties with large leaves (15-25 cm long and 8-12 cm wide) variable colours and often reflexed foliage; *Irregular leaved*—Leaves are small having a longer petiole and twisted; *Basket*—Plants have prostrate growth and leaves are very small green with brown or red colours; *Narrow leaved*—Leaves are shorter about 2.5 cm wide, irregular and deeply indented. *Speckled*—Varieties with a green ground or pale shade of any other colour speckled, splashed or blotched with a darker shade, usually brown or red. Important varieties are : Rainbow, Red, Sunset, The Chief, Pyreness Gold, Majesty, Frilled Fantasy, Christmas Cheer, etc.

Cordyline sp. (Liliaceae)

It is native of New Zealand, India, Malayasia to Polynesia and commonly called as Red dracaena. It has many varieties differing in shades, intensity of colour, shapes and sizes of leaves. The plants remain colourful in semi-shade. Direct exposure to sun causes scorching whereas dense shade results into dull colour. Easily propagated through cuttings and air layering. Latter method helps in faster growth of plants. Important species are :

C. terminalis. Plants are slender growing about 3 m high, leaves are leathery, sword shaped, and are of green coppery colour with red shade. Other attractive varieties are : *C. terminalis*—bicolour, *C. terminalis*—firebrand and *C. terminalis*—tricolour.

Crotons—Codiaeum variegatum (Euphorbiaceae)

It is native of South India, Ceylon, Malaya and Sunda islands. It is beautiful tropical shrub which is valued for its colourful ornamental foliage. Leaves are thick, leathery and glabrous. Shape of leaves vary from ovate to linear, entire or lobed, flat or twisted. Young and emerging leaves are green to yellow and later on colour changes according to varietal characteristics. Propagated easily through air layering. Seeds give rise to new forms. There are numerous varieties. The important ones are Beautiful Lady, Day Spring, Duke of Windsor, Golden Ring, Imperials, Indian Headdress, Jungle Queen, King George, Mona Lisa, Sunrise, Show Girl, Undulatum, Mosaic, etc.

Dieffenbachia sp. (Araceae)

It is native of Costa Rica, Columbia and Equador. Plants are dwarf, sturdy and thick stemmed. Leaves are large oblong pointed, glabrous and vary from green to variegated. Important species are :

D. amoena. Leaves are dark green and marked with cream white bends and blotches along lateral veins.

D. amoena. Tropic Snow is another prized variety for its ornamental foliage. Leaves are leathery, glossy, deep green, highly variegated from centre with cream and nile green. 'Gigantea' is another popular variety. Leaves are broad ovate, smooth fresh green with irregular creamy white blotches.

Other important varieties are Mars, Jupiter and Venus.

Dracaena sp. (Liliaceae)

It is native of tropical America. Plants are highly valued for their ornamental foliage. Leaves vary from green to variegated and are lanceolate in shape which are 30-40 cm long. Plants are grown in pots and are kept in shade and semi-shady situations. Important species are :

D. godseffiana. Plants have spreading, wiry stems and thin leathery and elliptic leaves. Leaves are glossy deep green with irregular yellow spots.

D. sanderiana. Plants are erect with neat growth. Leaves are lanceolate twisted, green and with white margins.

D. fragrans. It is common species which grows upto 4-5 metres. Branches are erect. Leaves are large, oblanceolate and recurved.

D. deremensis bausii, D. deremensis warneckei are other colourful species.

Excoecaria cochinchinensis* Syn. *Exoecaria bicolour

It is a dwarf bushy plant. Leaves are elliptic in shape. Upper side of leaves is shining olive green and lower side is rich red colour which makes them very elegant. Veins are prominent.

Ferns

There are more than 1,000 species of ferns found growing in humid hilly regions but only few ferns can be grown in the garden in the plains of north India. For raising large number of ferns, special structures like chick houses are needed.

These ferns, due to their elegant foliage and diversity are preferred in the gardens which are grown in pots, hanging baskets, or under the trees. For successful raising of ferns, fertile and porus soil, adequate humid conditions and maintenance of optimum temperature in extreme climates are important-factors. Important ferns which are grown successfully in plains are : *Asplenium nidus* (Bird's net fern), *Adiantum peruvianum* (Maiden hair fern), *Adiantum capillus-veneris* (Venus hair fern, hair fern), *Blechnum occidentale*, *Nephrolepis exaltata*, etc.

Ficus elastica (Moraceae)

See Chapter on trees.

***Fittonia* sp.** (Acanthaceae)

It is native of Peru. Plants are small, growing low and foliage is ornamental which is known for their beauty of network of veins.

F. argyroneura. It is a creeping herb with flat, papery, oval leaves green with white veins.

F. versehaffeltii pearcei. Leaves are thin-papery with rose-pink veins.

Graptophyllum pictum (Acanthaceae)

It is native of New Guinea. It is a small shrub with oval pointed leaves, purplish green marked with yellow centre vein and pink stem.

Hedera canariensis (Araliceae)

It is native of Morocco and is commonly called as 'Algerian Ivy'. Leaves are broadly ovate and shallowly lobed. *H. canariensis variegata* commonly known as 'Ghost tree' is very effective with variegated leaves.

***Heliconia* sp.** (Musaceae)

It is native of tropical America and West Indies. The plants are similar to banana and grow for handsome foliage. They grow well in semi shade. Common species are :

H. metallica. It is a slender perennial with long red stalks. Leaves are oblong velvety and lush green with white midrib. *H. humilis, H. psittacorum* and *H. bihai* are other good species.

***Kalanchoe* sp.** (Crassulaceae)

It is commonly called as bryophyllum and is native of tropical Asia and Africa. Plants have thick leaves and serrated or toothed margin. Vegetative buds are present in the axil of north of leaf. Propagation is very easy through leaf cuttings. Common species are :

K. pinnata. It is native of India and other tropical regions. Plant is erect and has oval, green tinged red leaves.

K. tubiflora. It is slender, erect, with pinkish brown stem with many almost cylindrical, pinkish leaves.

K. velutina. It has creeping branches. Lower leaves are ovate and crenate, upper leaves narrower. Flowers are yellow to salmon in colour.

K. mortagei syn. *K. crenate.* It is tall branching succulent. Leaves are opposite, lanceolate, fleshy, waxy green with red doubly crenate margins. Flowers are red.

K. marmorata. It is a short plant. Leaves are fleshy, broad obovate, pinkish to purplish green blotched purple on both sides.

K. tomentosa. It is commonly called as panda plant and is valued for its beautiful look. Plants are erect with fleshy spoon shaped leave; and marked brown with light brown stripes.

***Maranta* sp.** (Marantaceae)

It is native of tropical America and is grown for its ornamental foliage. The common species are :

M. bicolor. Plant is slow growing and producing oval leaves. These leaves have chocolate spots on upper side of leaves and red blotches on lower side. Plants produce tuber and propagation is done by dividing in rainy season.

M. leuconeura perchoveana. Plants are slow growing. Leaves are oval of about 15 cm length. Leaf surface is pale green with a row of chocolate spots on both sides of leaves along with mid rib.

Other species are : *M. arundinacea* and *M. arundinacea variegata.*

***Monstera* sp.** (Araceae)

It is commonly called as 'carimen' and is native of tropical America. There are numerous species but commonly growing species is *M. deliciosa.* It is a tree climber and produce aerial roots. Stem is woody and has nodes very closely. Leaves are large, green about 80 cm long, pinnately cut and perforated with oblong holes. Cone like fruit is edible. *M. deliciosa variegata* has white spotted leaves.

***Peperomea* sp.** (Piperaceae)

It is native of tropical America and is commonly called as 'Pepper face'. It is very good for dish garden. Leaves are obovate or spatulate, waxy, green, fleshy and light green underside.

P. obtusifolia variegata has beautifully obovate, waxy leaves, light green variegated with broad area of creamy white.

Other good species are : *P. japonica, P. peltifolia, P. verschafeltti, P. glabella, P. glabella variegata, P. bicolour,* etc.

Pilea microphylla* Syn. *P. musocsa (Urticaceae)

It is native of West Indies and is commonly called as 'Artillery plant' Plant is small densely branched with sub-erect fleshy stems. Leaves are green, small and oblong, green It makes excellent pot plant.

P. cadierei. It is commonly called as 'Aluminium plant'. It is a fast growing plant Leaves are thin, opposite, obovate, and margins are toothed. Leaves have attractively shining silver spots giving them a painted look.

***Philodendron* sp.** (Araceae)

It is a larger group comprising of about 300 species and their hybrids. It is native of tropical America. There is a great variation in leaf shape, size, colour and variable spadex. It is widely used for interior decoration and also grown as pot plant. The beautiful species are :

P. domesticum. Leaves are lush green, arrow shaped, pale veins raised and assending *P. domesticum variegatum*-a mutant with irregularly variegated leaves and splashed with cream white and yellow spots.

P. bipinnatifidum. It makes good growth and has stiff leaves measuring about 90 cm, waxy green, bipinnate having 10-12 segments each side.

P. ilsemannii. Leaves are leathery sagittate, almost entire, white or cream with grey dark green spots.

P. imbe. Leaves are long sagittate with veins at right angle and somewhat red beneath are petioles in red.

P. imbe variegatum. Leaves are waxy, saggittate and irregularly variegated. One side of leaf may be dark green with white spots.

P. oxycardium. It is tall climbing vine and leaves are dark green and cordate. Leaves are velvety when young and glossy at maturity.

P. oxycardium variegatum. Leaves are heart shaped spotted with ivory white colour

Other important species are *P. florida, P. florida variegata, P. wendlendii, P. wandlandii variegatum,* etc.

***Pleomele* sp.** (Liliaceae)

It is very similar to dracaena and is still grouped along with dracaena.

P. reflexa variegata. "Song of India". It is very common as pot plant. It is self branching, slow growing but grows about 3 m high, densely covered with leathery leaves which are margined by two bands of golden yellow or creamy shades.

P. reflexa is also common which has deep glossy green leaves, waxy and reflexed.

Tradescantia spathacea* Syn. *Roheo discolor (Commelinaceae)

It is native of Mexico and is commonly called as 'Moses in the Cradle'. It is a dwarf plant with rosette of waxy, lens shaped leaves which are dark green on upper side and glossy purple beneath. In the leaf axils, little white boat shaped flowers are produced.

R. spathacea 'vittate' is a variegated form with leaves striped length wise on the surface which is pale yellow and tinted red.

Ruscus aculeatus (Liliaceae)

It is commonly called as 'Butcher's Broom'. It is an erect and bushy plant. Leaves are ovate, stiff and spiky. Easily propagated through leaf which has growing bud in centre.

Sanchezia nobilis* Syn. *Sanchezia nobilis* var. *glaucophylla (Acanthaceae)

It is native of Ecuador. Leaves are large lanceolate, leathery, glossy green with contrasting prominent yellow veins. Easily propagated through terminal cuttings during rainy season.

Epipremnum aureum* Syn. *Scindapsus aureus* Syn. *Pothos pendula (Araceae)

It is native of Solemen islands and is commonly called as 'Pothos' or 'Money Plant' or 'Devils ivy' and is quite commonly grown plant. It can climb up with the help of aerial rootlets. Leaves are ovate, waxy, green with yellow variegation in some varieties. Young leaves are small whereas old leaves are large.

S. aureus 'Marble Queen' commonly known as Silver Pothos, is a mutant with white variegation and tinting of leaves.

Other attractive species are *S. pictus* and *S. pictus argyraeus*. The plants can be easily propagated through stem cuttings.

***Syngonium* sp.** (Araceae)

It is native of tropical America and is a climber for shady situations. Plants are highly valued for their attractive foliage. It produces aerial roots and is easily propagated through cuttings. Important species are :

S. podophyllum. Leaves are thin, sagittate, 5-7 lobed in later stages.

S. podophyllum albolineatum. Very ornamental with silver white centre and veins. Mature leaves are green.

S. wendlandii. Leaves are trilobed having white veins in the young leaves whereas mature leaves are green.

***Tradescantia* sp.** (Commelinaceae)

It is native of tropical America and is very good plant for pot as well as for hanging baskets. Branches are brittle. Leaves are oval and plants grow luxuriantly in shade. Easily propagated through stem cuttings. Important species are :

T. albiflora laekenensis. It has creeping stems, and leaves are delicate, small, ovate, pale green with white.

T. flaminensis. It is another beautiful species. Leaves are bluish and purplish beneath.

T. sillamontana (white velvet). Leaves are ovate, fleshy, deep green covered with fluffy white wool, underside and the stem is purplish, and brittle.

T. blossfeldiana variegata. It is another attractive sport with the fleshy leaves striped cream to yellow and pale green, midrib purple, underneath deep purplish and white.

Tradescantia zebrina* var. *zebrina* Syn. *Zebrina pendula (Commelinaceae)

It is native of Mexico and is commonly called as 'Wandering Jew'. It is very similar to Tradescantia. Plants are fleshy and have creeping stem. Leaves are small, fleshy, deep green to purple with two bands but clear purple beneath. It can be easily propagated by cuttings.

T. zebrina quadricolor. Leaves are small, purplish green, broadly banded glistening white with pink and carmine red and silver stripes.

T. zebrine discolour. It has trailing stem. Leaves are large, glossy coppery and edged with metallic purple with two narrow silver bands and is purple beneath.

25 Cactus and Succulents

These are group of plants which have developed a special capacity to store water in thick fleshy leaves or stems. They thrive best in sunny situations and are light loving. They need little care except when actively growing. These plants are native of drier parts of the world *i.e.* slopes of Andes mountains in South America, the arid zones of Bolivia, Mexico, Argentina, Brazil, Peru, other parts of North America, Canada and South Africa. All the cactus are succulents on account of storing water but all the succulents are not cacti. There is a clear cut distinction between both. Cactus is characterized by the presence of areoles, which often look like wooly cushions carrying spines, hairs or glochids and the flowers arise from or near the areoles. The spines in cactus are modified leaves which provide shade against scorching sun and help in conservation of moisture besides protecting against birds and beasts.

Cactus and succulents are very popular amongst gardeners, amateurs, and hobbysts which are used to adorn sunny situations of gardens, houses, window sills and rock gardens. Popular names have been given to them like 'Mother in low's chair' (*Echinocactus grusonii*), 'Oldman' (*Cephalocereus senilis*), 'Silver torch' (*Cleistocactus strausii*), 'Moon cactus' (*Selenicereus*), 'Old Lady' (*Mammillaria hahniana*), Mr. Red cap (*Gymnocalycium mihanovichii* var. *Friederickii*), Bishop's hood (*Astrophytum myriostigma*), 'Arab's turban' (*Crassula hemisphaerica*), 'Sea onion' (*Bowea volubilis*), 'Old man's tooth' (*Lithops* sp.), 'African milk barrel' (*Euphorbia horrida*), 'Beads' (*Sedum stahlii*), Airplane plant (*Crassula falcata*), 'Elephant foot tree' (*Beaucarnea recurvata*), etc.

All the cactus plants belong to family 'Cactaceae' which is further divided into following three tribes *viz.*

Tribe I : *Pereskieae*

This is the primitive form and near to original forest trees. Plants are leafy bushes, look like citrus trees and have stalked flowers in clusters resembling wild roses. Spines are present but not glochids or barbed bristles *e.g. Pereskia grandiflora* commonly known as rose cactus and *Pereskia aculeata* are members of this group.

Tribe II : *Opuntieae*

Plants are characterized by the glochids or sharp, easily detached bundles of barbed bristles in the aerioles. Stem or branches are fleshy, loosely joined either flat or pad like or cylindrical having spines but without ribs. Flowers are without petioles and tubes and are generally of wheel shape *e.g. Opuntia* sp.

Tribe III : *Cereae*

This is the largest group and members are characterized by their ribbed, fleshy stems of continuous growth in columns or globes. Glochids are absent from areoles. Flowers are tubular and are very showy except Riphalis section. They are attached directly on the plant without stems.

This tribe being the largest one, has further been divided into 8 subtribes. They are :

1. *Cereanae*. Members are known as column or torch cacti. Branches are numerous and ribbed. Funnel shaped flowers appear on the top *e.g.* cereus, cephalocereus, cleistocactus, Trichocereus.

2. *Hylocereanae*. These have vine like habit with aerial roots clinging to trees and rocks. Mostly are night blooming cacti with large flowers. Important genera are : *Selenicereus, Hylocereus, Weberocereus, Cryptocereus,* etc.

3. *Echinocereanae*. These are low, growing in clusters or in single and heavily ribbed. Flowers are bell shaped, showy and appear on the side of the stem. Important genera are : *Echinocereus, Echinopsis, Rebutia, Chamaecereus,* etc.

4. *Echinocactaenae*. These also grow low and are small and sometimes forming clusters, ribbed and spiny. Flowers are most beautiful *e.g. Astrophytum, Echinocactus, Ferocactus, Notocactus, Gymnocalycium,* etc.

5. *Cactanae*. These are commonly called as Melon cactus or Turk's cap. These can be easily identified by the presence of wooly or bristly, cap like structure on the top of the globular body. Flowers arise from the tops *e.g. Melocactus*.

6. *Coryphanthanae*. It is popularly known as Pincushion cacti. Plants are globe shaped. Stems are small forming large clusters with tubercles or nipples in rows. Important genera are : Mammillaria and Coryphantha.

7. *Epiphyelanae*. These are mostly epiphytic and tropical. Branches are flat and leaf like of usually without spines. Flowers are showy *e.g.* Zygocactus and Epiphyllum.

8. *Rhipsalidanae*. It is commonly called as chain cacti. These are freely branching with slender, cylindrical or flat branches. Stems are pendant forming dense masses on trees of tropical forests. They bear tiny flowers and berries. Important genera are *Rhipsalis* and *Rhipsalidopis*.

Growing of Cacti and Succulents (Plates 13 & 14)

Climate. Cacti are adaptable to wide range of climatic conditions. The extremes of temperature *i.e.* above 35°C and exposure to direct scorching sun results into yellowing and finally rotting. Plants should be shaded to protect them. Very low temperature also inhibits the growth of cacti. The plants also have a resting period. During this period cacti need little care and watering at longer intervals. Active growth starts again during spring season and continues till autumn.

Soil. Cacti thrive well in porus and rich calcareous soil. Therefore, for making a pot mixture care should be taken to include all necessary gradients which are beneficial for their growth. The general ideal pot mixture for cacti consists of 2 parts garden soil : 1 part sand : 1 part leaf mould : 1 part well rotten manure : 1 part old mortar or lime stone gravel and ½ part charcoal. Small amount of bone meal or basic slag should also be added. Plants become succulent and susceptible to rotting by excessive addition of chemical fertilizers and thus, frequent application should be avoided. Undecomposed organic matter causes rotting of roots.

Propagation. Cacti can be multiplied sexually as well as asexually. Seeds are collected from ripened fruit and are sown from March to September in pots, seed pans, or boxes. The mixture for sowing the seeds, consists of 1 part garden soil : 1 part sand : 1 part leaf mould. Seeds are sown and gently covered with sand and then with glass plate. Seeds germinate but seedlings grow slowly and it takes long time to make a specimen plant from seeds.

Vegetative methods are very popular to multiply the cacti. Division of off set or clumps or root suckers is an easy method which is generally followed in *Mammillaria*, *Rebutia*, *Trichocereus*, *Cleistocactus*, *Echinocereus*. Another easy method employed is by cuttings which is conveniently used. Cuttings of convenient length are made and kept for drying 7-30 days before planting. Then these cuttings are planted in sand and new growth shows that rooting has taken place.

Grafting is a useful method and grafted cacti are generally preferred. Weakly growing plants are grafted over strongly growing rootstocks. The ideal season of grafting is spring or autumn. Genera like *Cereus*, *Trichocereus*, *Harrisia*, *Opuntia*, *Cleistocactus*, etc. are generally employed as rootstock. Plain cuts are provided on stock and scion. Scion is placed over rootstock and pressure is provided by placing some flat wood in cross ways or by tying with a rubber band string for a month. Then the pressure is released. Wedge grafting is also very successful method for propagation.

Watering. Cacti and succulents do not need liberal watering. Judicious application of water is required for successful growing. In winter months *i.e.* during rest period watering is to be done very sparingly. Little frequent watering is done during summer months when cacti grow. Every time the soil is to be drenched completely so that water comes out of the drain hole and over-watering should be avoided.

Description of Important Cacti

Pereskia aculeata

It is native of West Indies and coast of South America and is commonly known as 'Lemon vine'. It is a woody plant, and have needle like brown spines. Leaves are waxy green and elliptic. Flowers are creamish in colour and possess lemon like scent. *P. grandiflora* known as 'Rose cactus' is another common species. It is a tall shrub growing about 5 m with abundantly spiny trunk. Leaves are rich green, fleshy and waxy. Flowers are produced in terminal clusters of rose colour and are very similar to the wild rose.

Opuntia sp.

It is a commonly grown cactus for fencing. There are large number of species and some are ornamental. Common species used for fencing are : *O. cylindrica, O. caribaea, O. tetracantha* and *O. subulata.* Other common species are : *O. leptocaulis, O. linbericata, O. ficus-indica, O. mammillata, O. microdasys, O. streptacantha.*

Opuntia monocantha variegata is a highly ornamental having patches of different colours like pink, red, and yellow.

Disocactus flagelliformis Syn. *Aporocactus flagelliformis*

It is popularly called as 'Rat's tail cactus'. Stems are long about 1.5-2 m, slender and hanging. Stems. are closely ribbed and densely covered with small brownish spines. Flowers are tubular which are produced on long hanging stems and last for several days. This is ideally suitable for hanging baskets. Other important species are : *A. mallisonii, A. conzatti,* and *A. martianus.*

Astrophytum myriostigma

It is commonly known as 'Bishop's cap'. The usual number of ribs is 5 but it ranges from 3-8. The dark surface is covered with silvery scales. Plants are spineless but have prominent areoles. Large numbers of small, sweetly scented yellow with reddish throat are produced. It is ideally suited for small pots. There are several varieties like 'Column arc', 'Nudum', 'Quadricostata', 'Tulensis', etc. Other popular species are :

A. asterias. It is popularly known as 'Sea Urchin' cactus. It has 8 spineless ribs and is of flattened hemispherical shape. The epidermis is greyish green which is covered with white spots.

A. capricorne is difficult to grow. It is a cylindrical and tall growing about 30 cm. It has 9 ribs with long interwining spines.

Cephalocereus senilis

It is popularly called as 'Old man cactus' on account of presence of long grey hair which cover the entire column. Length of column varies from 15 cm to 10 m. Ribs are closely attached. Important species are : *C. flumiensis, C. cometes, C. leucostele, C. royenii*, etc.

Cereus sp.

The word cereus is derived from Latin word *cera* which means wax or torch. It is commonly called as 'Night blooming cacti' on account of nocturnal habit of flowering. Young plants are straight columnar in shape but with the age they grow vigorously and are useful root-stocks for propagation purpose. Flowers are white or pink. There are about 26 species and common species are :

C. hexagonus. It grows in a column upto 9 m high and 15 cm thick. It has blue-green to grey stems and branch freely.

C. argentinensis. It is from Argentina. It is another vigorously growing and freely branching species.

Other important species are : *C. jamacaru, C. peruvianus* and *C. repandus.*

Echinopsis chamaecereus Syn. *Chamaecereus silvestrii*

It has short, finger like freely off-setting stems. Plants are prostrate, pale green, and covered with short white spines. Flowers are large and of scarlet colour.

Coryphantha sp.

These are small, globular or cylindrical cacti, divided into tubercles. They have attractive spine formations and look attractive. Flowers are large and are produced at the top. Important species are *C. echinus, C. cornifera, C. clava, C. pallida, C. vivipara*, etc.

Echinopsis candicans Syn. *Trichocereus candicans*

It is commonly called as 'Barrel' or 'Hedge hog' cacti. These are large, globular and have sharp continuous ribs. The flowers are long, funnel shaped, usually white or rose coloured, scented and open in the evening. Important species are *E. campylacanthah, E. calochlora, E. bridgesii, E. tubiflora, E. eyriesii*, etc.

Echinocereus sp.

These form clumps by branching from the basis of shoots at ground level. In some species, plants grow in a large plant body which do not usually branch. Echinocereus are divided into three groups according to plant growth.

1. Plants with weak spines—*e.g. E. subinermis*—is a branching plant. The stems are 8-9 cm thick and have 5-8 ribs with very short spines. Other species are : *E. pulchellus* and *E. knippelianus.*

2. Prostrate, mat forming plants *e.g. E. blanckii, E. pentalophus, E. procumbens,* etc.
3. Erect plants, branching from base *e.g. E. pectinatus, E. oklahomensis, E. delaetii, E. engelmanii, E. fendleri,* etc.

Ferocactus sp.

The general name has been derived from English word 'Ferocious', on account of its lool due to presence of fierce spines. But spines are beautifully coloured. Plants are slow growing. Flowers of red or yellow colour are borne on the top of the plant. Important species are : *F. acanthodes, F. covillei, F. johnstonianus, F. latispinus, F. melocactiformis,* etc.

Gymnocalycium sp.

It is commonly called as 'Chin' cactus. These are globular plants and have few ribs which are divided into tubercles. There is a cleft below each tubercle, giving a 'chin' like effect. Length of spines and shape vary greatly with the species. Some have short, thin spines, while others possess long stout, and curved ones. Flowers are quite large of about 5 cm across and last for several days.

The important species are : *G. denudatum, G. baldianum, G. bruchii, G. dimsii, G. gibbosum, G. mihanovichii.*

Haageocereus sp.

The genus consists of many ribbed and columnar cacti which are closely covered with brightly coloured spines. The sweetly scented flowers are nocturnal. *H. decumbens, H. multicolorispinus* and *H. acranthus* produce attractive flowers.

Hamatocactus sp.

It is a small cactus. Plants are cylindrical, young and spines are deep red but fade to white upon ageing. Important species are :

H. hamatacanthus. Flowers are large and yellow with a red throat.

H. setispenus. It is a free flowering and producing yellow flowers with red throat.

H. uncinatus. Flowers are comparatively smaller and of a reddish brown colour.

Harrisia sp.

These are tall, slender, become sprawling and need support. They have several broad ribs and spines. Flowers are large, usually white and nocturnal. Pods are yellow or orange which do not split. Important species are *H. eriophora, H. gracilis* and *H. simpsonii.*

Lemaireocereus

These are tall, columns and branching from just above the ground. Flowers are white, small and open during day. Important species are : *L. beneckii, L. chichipa, L. dumortieri, L. marginatus, L. thunberi* and *L. weberi.*

Lobivia sp.

These are commonly called 'cob cacti' and have globular or cylindrical stem. Plants are medium sized having ribs usually divided into tubercles. Spines are stout usually directed upwards. Bristly hairs are also present which sometimes entirely cover the plant body. Flowers are multicoloured in red and yellow which open in day. *L. chrysantha, L. haageana, L. shaperi, L. pentlandii* are some of important species.

Mammilleria

It is popularly known as 'Bird's Nest Cacti'. The word mammilleria is derived from Latin word *mammilla* which means nipples or teats. They form clusters of globular green stems which are covered with silky hair. Spines are variable in size. The small bell shaped flowers are borne in circles around the top of the plant. Flower colour is usually cream or pale yellow. Many species also beautify the plant with colourful seed pod for months. The genus is comprised of about 300 species. *M. goldii, M. candida, M. bocasana, M. zeilmann lana, M. celsiana, M. gracilis, M. hahniana, M. chionocephala, M. wildii, M. discolor cristata* are important species.

Notocactus sp.

This is large group of vigorous and hardy plants. Plants flower when they are young. Large flowers are produced on the top of the plant and are yellow in colour with purple red stigma. Important species are *N. haselbergii, N. leninghausii* and *N. elachisanthus*.

Rebutia sp.

These are small, free clustering cacti. They are short lived (3-4 years), and are easily propagated through off-sets which root easily. Funnel shaped flowers are produced around the sides of the plants arising from near the base. Flower colour varies from red to orange, pink yellow, mauve and white. Most commonly grown species are *R. minuscula, R. senilis* and *R. xanthocarpa*.

Trichocereus candicanis

Plants are glossy green and have 9-11 ribs. These ribs bear large, round, white woolly areoles, with spreading golden yellow spines of variable lengths. Branching from the base is profuse and plants become many stemmed. Flowers are large, glistening white and have lily like scent. Other species are *T. theologonus, T. chilensis*, etc.

Description of Important Succulents

Agave sp. (Amaryllidaceae)

There are about 300 species and are native of America. Majority of the species are ornamental but *A. sisalana* and *A. fourcroydes* are cultivated for extracting fibre. *A. americana* is known as 'Century plant'. *A. americana marginata* is another important species. Leaf margins are wavy, toothed and broad having golden yellow bands at the margin of leaves. Other important species are : *A. filifera* and *A. victoriae reginae*.

***Aloe* sp.** (Liliaceae)

These are native of Africa, Madagascar and Atlantic islands. Leaves are produced in rosette which are thick and fleshy. Inflorescence arise from axils of leaves. Flowers are mainly orange, red or yellow. *A. variegata* is the most commonly cultivated species. This has transverse wavy bands of white markings on the leaves. It is stemless and produces dense group of rosette. Other important species are *A. fosteri*, *A. brevifolia, A. vera,* etc.

Mesembryanthemum cordifolium* Syn. *Aptenia cordifolia (Aizoaceae)

Plants are prostrate and stems are covered with glistening papillae. Leaves are heart shaped, short stalked, opposite, fleshy and covered with papillae. Small flowers are short stalked and purplish red in colour. Ideally suited for hanging baskets, rockery and window gardens. Flowering occurs during hot months and July to September.

***Cotyledon* sp.** (Crassulaceae)

These are perennial and shrubby in nature. They have either opposite and persistent leaves with thick and fleshy stems or leaves are alternative and summer deciduous. Easily propagated by seeds, stem cuttings and some by leaf cuttings. Important species are :

C. orbiculata. Plant has thick stems with opposite, unstalked, ovate-obovate and fleshy leaves. Leaves have grey waxy covering. Flowers appear in summer and are of yellowish-red colour.

Other species are *C. undulata* and *C. jucobensiana.*

***Euphorbia* sp.** (Euphorbiaceae)

There are about 800 species, widely distributed throughout the world. They look like cacti but actually these are different. Euphorbias are mainly identified by the presence of milk and flowers which are called cyathia and whose colour is due to colourful bracts. Important species are :

E. candelabrum. This grows about 10 m and becomes like a tree. The branches are four ribbed and bear leaves near the tips. The crests of the ribs are wavy, toothed and bear pairs of spines.

E. neriifolia. Stem is five angled and becomes cylindrical when mature. Fleshy long leaves are produced on youngest portion of shoot.

E. splendens. Commonly called as 'Mother in law's Tongue'. It is a spiny shrub about a metre or taller. Branches are irregular and brown coloured. Leaves are variable and present over all the stem. Cyathia are scarlet coloured.

E. pendula. It has many branched, long pendant shoots. It is suitable for hanging baskets.

E. aphylla. This is a succulent shrub which produces forkly, cylindrical branches, *E. tirucalli* is similar in shape but it grows tall to about 6-8 m.

Other important species are : *E. angularis, E. meloformis, E. pseudocactus, E. triangularis, E. ramipressa.*

Furcraea foetida* Syn. *Furcraea gigantea

It is similar to agave with tufted leaves, margin, spiny, dentate or leathery Important species are :

F. selloa var. marginata. Leaves are variegated and margins yellow.

F. gigantea. Leaves are large oblong, oblanceolate, flat, glossy green with margin entire.

F. gigantea medopicta. Leaves are creamy white, yellow with longitudinal centre.

***Haworthia* sp.** (Liliaceae)

These are small delightful plants and are highly suitable for semi-shady situations, living rooms, and other situations. Leaves are usually in rosette, densely arranged and overlapping each other. These are easily propagated by separating rooted rosettes. Important species are :

H. fasciata. Leaves are sessile and triangular lanceolate. Upper surface is flat but lower is markedly covered with white tubercles arranged in distinct transverse bands.

H. cymbiformis and *H. limifolia* are also important from ornamental point of view.

***Lithops* sp.** (Aizoaceae)

These are popularly called as 'Flowering stones' on account of their similarity to surrounding gravels, stones or pebbles. The generic word is derived from Greek words *i.e. lithos* (stone) and *ops* (likes). Plants grow singly or in clumps. Plant bodies consist of single pair of brownish-grey leaves with and reticulate surfaces. Sometimes surfaced is dotted. Flowers are short stalked, white or yellow in colour. Very easily propagated by seeds and cuttings. Important species are *L. bella, L. divergens, L. salicola*, etc.

Beaucarnea recurvata* Syn. *Nolina recurvata (Liliaceae)

It is a small tree with swollen trunk at the base and tapering towards apex. Leaves are long, narrow and recurved in tuft.

Portulacaria afra (Portulaceae)

It is succulent shrub and very common for hanging baskets. Stem is horizontally spreading. Leaves are thick, glabrous, opposite, obovate and shining green.

***Sanseveria* sp.** (Liliaceae)

These are native of South Africa and East Indies. Leaves are stiff, erect, sword like and flat in some species and cylindrical in others. They are arranged in rosette which arise from stoloniferous rhizomes. Flowers are white and inconspicuous. Propagation is done by division. Commonly grown species are :

S. marginata. Leaves are thick, green and having yellow strip on the margin.

S. cylindrica. Leaves are round, arching and rigid measuring 120-150 cm. These are usually furrowed and several grouped to a shoot.

S. trifasciata. Leaves are dark green have whitish wavy bands over the entire surfaces except wide green marginal band. *S. trifasciata* var. *laurentii* is another handsome species.

S. hahnii. Leaves are broad, resolute and are attractively marked.

***Sedum* sp.** (Crassulaceae)

There are about 300 species of this genus which are native of mountainous region of Central Africa, Madagascar, Peru, Bolivia and Philippines. The word sedum is derived from word 'Sedo' which means to sit and it is in reference to the method by which some species are attached to the rocks. Common species are :

S. compactum. Low growing herbaceous plants with obovate to obtuse leaves on short rosette.

S. morganianum. Leaves ate thick, fleshy and numerous.

S. nudum. Leaves are alternate, obovate, oblong, thick and blunt at the tip.

***Yucca* sp.** (Liliaceae)

It is not succulent but is a xerophytic plant. Leaves are long, lanceolate and green. Important species are :

Y. gloriosa. Commonly known as 'Spanish dagger' which grows to about 240 cm in height. Stem is thick, short and topped by dense rosette of sword shaped leaves which are flat, glaucous, grey-green and rough.

Other notable species are *Y. elephantipes* and *Y. schotii.*

***

26

Palms

Palms are group of the plants which have great ornamental value and are planted in the garden for special effects as most of the species have unbranched single stem. Tufted leaves at the end of stem creates most picturesque effect. These are mostly planted as specimen, or group planting or as avenue plantation in big palaces. It is common sight to find certain palms like *chamaedorea elegans*, *Chrysalido carpus lutescens*, *Livistonia chinensis*, *Caryota urens*, etc. in pots being grown for home landscape or for beautifying the office buildings.

Many of the palms are native of tropics and comprises of about 212 genera and 2780 species and belonging to family Palmae. It is common saying that feet of palms should in the water and head should be in the fire. The palms widely vary in form, size and habit of growth. All palms are endogenes, do not increase in thickness and do not show any distinction into bark, wood and pith. Palm like *Attalea cohune* has very short stem and leaves originate directly form ground whereas palms are as tall as 15-20 meter. These palms are divided into two main categories according to shape of leaves *i.e.* feather shape and fan shape. In feather shaped palm (date palm, coconut palm) main vein runs out of the side of long mid rib, the leaf being divided into long and narrow segments whereas in fan shaped leaves (Chinese palm) main veins radiate outsides in all directions resulting into semi circular to circular leaves. Flowers are produced on spikes which hang down gracefully in many species. Palm inflorescences are highly diverse, from large panicles with many orders of many branching (calamus) to simply spikes. The flowers may be bisexual or unisexual borne either on the same plant (monoecious) or on separate plants (dioecious) whereas some species have bisexual and unisexual flowers on the same plant (polygamous).

Palms produce seeds in abundance and germinate slowly and makes slow growth which makes many years to become plant of desirable shape and size. Seeds are covered with thick fiber in many species which makes germination slow. They can be sown at any time but spring is the best time for sowing. Seeds should be sown in well drained fine soil and covered sufficiently. Seedlings are potted separately in pots enough big to accommodate to root. As the palm plants throw long roots and suffer much if they are not removed early and repotted in bigger pots. Many palms throw away root suckers and palms are increased by dividing these suckers, though separation is a tough job. After separation suckers should be planted in ground or pots as desired.

Growing of palms in pots does not require frequent repotting and prefers to be in the pot bound and thrive in under sized pots. Repotting is done to its next size and intact roots system is placed in the pot with fresh soil and farm yard manure mixture. Top dressing once or twice a year with oil cake or with fertilizer should be done. To keep plants healthy regular watering specially over head watering, syringing the leaves is highly beneficial.

Certain palms have high economic value and are being cultivated commercially in the world like coconut (*Cocos nucifera*), Areca nut (*Areca catechu*), toddy palm (*Borassus flabelliformis*), sugar palm (*Arenga saccharifera*) and date palm (*Phoenix dactylifera*) etc. Oil is also extracted from the oil palm (*Elaeis guinensis*) and species of orbigyna. Palms also produce a variety of fibers which include coir which comes from the husk of the coconut.

There are very few diseases and pests which attack the palms. Scales are sometimes seen. To control these, stem, and leaves should be kept clean by washing with the water. Sometimes beetles (Rhinoceros) attack young shoots and best killed by harpooning them with a thick wire through into holes or by putting malathion (0.1%) in hole and plugging it with mud.

Most common palms which are being grown in house or as ornamental plants in general landscape are described here :

Livistonia chinensis

It is commonly called as chinese palm and commonly grown in pots. It is fan leaved palm, slow growing but very ornamental. It is propagated easily through seeds.

Dypsis lutescens Syn. *Chrysalidocarpus lutescens*

It is feather leaved palm, slow growing highly suitable for pot culture as well as for group planting in the lawn or at the enterance of house or buildings. It forms large clumps and grows about 6-8 meter high. It has large arching leaves which makes it more ornamental. It is propagated through seeds or division of suckers.

Bismarckia nobilis—Bismarck Plam (Palmae)

It is native to the island of Madagascar which is situated of the east Africa. It has stout stem with huge crown of blue green foliage. This palm may reach an ultimate height of 17-18 m. with a spread of 6 m and, hence, it needs a large space for its planting and ultimate impressive look. Leaves are palmate of bright light blue colour, waxy and are up to 3 m. across. Leaf stem is about 2 m in length and posses small tooth.

For impressive growth, it requires full sun light but is tolerant to some shade. It happily grows in warmer climate but is occasionally damaged with freezing temp. After its establishment it is a drought resistant.

Propagation is done by seeds which germinate easily in 6-8 weeks.

This palm is best planted where it can serve as a focal point. It serves as living sculpture when planted against a back ground of dark foliage. Bismarck palm is rapidly gaining popularity as it is a spectacular species and due to drought tolerance and free from disease problems and nutritional deficiencies.

Butia capitata

This single trunked palm is native to Argentina, Brazil and Uruguay. It is one of the most cold hardy palms tolerating up to –10°C. It grows slowly up to a maximum height of 6 m. Blue grey leaves arch downwards towards the stocky stem making a beautiful round canopy. Arching leaves differentiate is from other feather leaves palms. *Butia* looks at its best as a specimen plant but can be used for avenue plantation also, even under power lines due to small size and slow growth. it can tolerate harsh urban conditions like air pollution, poor drainage, compacted soil, and drought.

Caryota urens

It is commonly called as fish tail palm on account of resemblance of small divisions or leaflets similar to tail of fish. It is quick growing large, feather leaved palm with large broad leaves which are finally cut up. Tall palm growing upto 6-15 meter high with very large leaves. It is mostly liked for its drooping spadices.

Chamaerops humilis

This slow growing palm is one of the two palm species native to Europe and is hardier than most palms. It has a shrub-like clumping growth habit, with several stems growing from a single base and grey-green, fan-shaped leaves, borne thickly in a bushy head. The petioles have sharp needle like spines. It can be used as specimen and near a patio or foundation or as border plant. By removing suckers from the base of the main trunk, it may also be trained as a single trunked plant. It is drought and wind resistant, slightly salt-tolerant and grows best in moist rich soil. It can be used as indoor plant but grows very slowly in the shade.

Roystonea regia Syn. *Oreodoxa regia*

It is called as royal palm or bottle palm. It's beauty is in rigid swollen stem usually barrel shape of which half top is smooth green and basal half is of ash colour with having ring marks of old leaves. It bears large terminal pinnate leaves with long sheath stalks forming a cylinder around the summit. It is highly suitable as a specimen in the garden or for garden avenues. It is propagated through seeds.

Howea belmoreana* Syn. *Kentia belmoreana

It is very handsome, hardy palm with feather leaves. Stem and leaf stalks are smooth and spineless. Mostly planted in pots for decoration of plant house or gardens.

Chamaedorea elegans

It is small palm with feather shaped leaves. It is slow growing. Leaves are most elegant. It is most suited to pot culture.

Rhapis excelsa

It is dwarf palm with fan shaped leaves deeply cut into segments. It is highly suitable for bush planting in shady place of garden or for pot culture. It is propagated by root suckers which are thrown very profusely.

Washingtonia filifera

It is a small palm with fan shaped leaves which are plaited and have fiber like apandages making it more attractive. However, leaf stalk has stout spines. It is highly suitable for pot culture.

Phoenix roebelenii

It is a highly ornamental species. Plants are dwarf. Leaves are light feathery, arching pinnate which makes it more graceful. It may produce branches from the base. It is propagated through suckers.

***Phoenix canariensis* (Canary island date palm)** Palmae

It is native to Canary Island. It is a very impressive looking palm owing to canopy of dense leaves (30-40). It grows very slow and reaches the height of 3-4 m in 15 years. Stem is thick and massive. Leaves are pinnate feather like fronds arching at top about 5-6 m long and leaf fronds spread about 6-12 m. Leaf petiole is toothed. Flower are borne on long fronds and are dioecious small and off white color which appear in late spring or early summer. Seeds are bold and germinate moderately.

For better performance it needs sunny situation and requires infrequent deep irrigation otherwise grown up plants are quite drought tolerant.

In landscaping it is highly suitable for road side planting and for oasis effects and xeriscaping.

Canary island palms are massive, tough and durable. It has a leaf scar pattern on the trunk which adds to-its impressive look.

Syagrus romanzoffiana* Syn. *Arecastrum romanzoffianum

It is commonly called as Queen palm. It is fairly rapid growing palm and grows about 8-10 metre in height. The presence of 10-15 long, arching, feather like glossy green leaves near the crown makes it more handsome and attractive. It is easily propagated through seeds.

Wodyetia bifurcata (Foxtail Palm) : Palmae (Plate 12)

Amongst all palms, Foxtail palm displays most spectacular foliage. The generic name of this palm is given in the honour of Wodyeti an Australia Aboriginal bushman who was the last male in his line of Melville Range Aborgines. The species name bifurcata comes from the Latin, meaning twice forked which is due to the unique shape of leaflets and fruits of this palm.

It is native of northeastern Australia specifically north eastern Queensland. The climate in this region of Australia is tropical with prolonged dry season. It is consider threatened species in Australia but after introduction in US it has become very popular ornamental palm in the southern parts of Florida, California and Texas. In India it has been introduced very recently and is extensively used in various landscape designs.

It is valuable for its robust trunk and its unique bushy leaves. In garden Fox tail is used as an accent, specimen or group planting of three or five for stunning mass effect. Fox tail can also be effectively planted in rows along street and drive ways. It can also be used for large pots and patio decorations.

It is a thorn less and has a slender closely ringed bottle shape columnar trunk. Single straight stem grows upto 10 m. A mature fox tail palm has 8-10 leaves each with characteristic resemblance of tail of fox and a crown of foliage which is 5-6 m, across. The pale green arching fronds have leaflets that radiate out at all angles from the leaf stem thus appearing like tail of a fox. The foxtail palm bears white blossoms of both sexes at the base of its crown and a single palm is able to produce fertile seeds. It bears colourful clusters of red to orange red fruits, each containing just one seed. Seeds germinate in 2-3 months after sowing and some may take upto 12 months. After germination, seedlings grow moderately under good conditions. The foxtail Palm grow well in full sun but it also grows well in partial shade. It is hardy and easy to grows in variable temperature but it must be protected against freezing condition. It tolerates wide variety of well drained soil including alkaline stone soils. The mature foxtail palm has a deep root system that makes it quite drought tolerant. However, regular watering helps in better growth.

Foxtail palms tolerate light frost and rarely attract pests and diseases and resistant to lethal yellowing.

Cycads

Though these plants do not belong to palm group but resembling like palms. Thus are described here :

Cycads are about 200 million years old on the earth *i.e.* they have been found growing from the Triassic age which coincides with the age of reptiles and Dinosaurs. Cycads are very slow growing and highly suitable for ground planting or for larger pots and tubs. Stem is stout and short, marked with old leaves scars. Leaves are tough, unbranched and pinnate. Propagation is mainly done by large bulb like buds which appear on the stems and grow freely when planted in well drained soil under humid and shady conditions. Male and female flowers are produced on separate plants and female flowers need hand pollination for seed setting. Seed should be collected on ripening and sown with in a month. Seedling grows very slowly. Important species grown in the garden are as below :

Cycas revoluta

It is very common plant grown in the pots in the houses or grown in big lawns as a specimen. The plants are very slow in growth. Stem is short and cylindrical and is usually unbranched and terminated at the top by a crown of deeply cut pinnate leaves of variable length ranging from 1-2 meter. Plants throw suckers which should be separated. Bulbs are also used for the multiplication. Seed is seldomly used for propagation.

Cycas circinalis

It is another species commonly grown in the house and gardens. Leaves are comparatively bigger and plants are taller than *Cycas revoluta*.

Dioon spinulosum

This attractive, easily grown cycad is native of Mexico. It can be grown as a replacement or alternative for the *Cycas* species. Stiff leaves are pinnately arranged, about 5-7 ft in length and extending radially out from the trunk. The leaflets are small, flat, have small marginal thorns and are tapered to a sharp point. Dioon may be grown in partial shade, filtered sun or full sun and prefers moist to wet soil with good drainage and regular watering for optimal growth. It can grow in soils having few nutrients, in limestone-rich soils and on moderate slopes. Mature and established plants can tolerate temperatures as low as –3.9°C for short time duration, without significant foliage damage. It also makes a wonderful pot plant in brightly lit interiors, on a porch or balcony. It can also be grown as specimen plant or as understory plant beneath a large trees or any structure that allows good passage of light.

Zamia furfuracea

It is also called Cardboard Palm and is native to Mexico. It has 3 to 4 feet leaves that emerge from a central point forming a rosette. When grown in bright sunlight the plant becomes a 3 feet high clump of tightly overlapping leaves that will slowly grow to about 6 feet in diameter. Leaves are pinnately arranged and have thick, leathery, oval leaflets. They are slightly fuzzy and feel a like a cardboard when rubbed. Trunk is thick and fleshy which serves as a water reservoir in periods of drought. It is cold hardy and grows well in a neutral, well-drained sandy soil. Cardboard palm makes a very good specimen plant, can be used in mixed foundation plantings or in perennial beds. It also makes a good house plant which can survive occasional neglect and harsh indoor environments.

Zamia pumila Syn. *Zamia integrifolia*

It is a small, tough, woody cycad native to southeast United States, the Bahamas and Cuba. Leaflets are thick, often revolute, with prickly petioles. The trunk remains low and grows to 3-25 cm high, but is often subterranean. It forms a multi-branched cluster over time, with a large, tuberous root system, which is actually an extension of the above-ground stems. This cycad is relatively cold hardy and leaves can drop in severe cold conditions. It prefers lightly shaded, well drained sandy soils.

27

Ground Covers

Ground covers are the group of plants which form a low canopy of foliage or spreading type in nature and cover the ground very densely so that weeds can not grow or establish under them. These group of plants are well suited to formal or informal style of gardening. They beautify the gardens with their colourful foliage and can be trained in various shapes. These are perennial in nature and easy to propagate. By using ground covers, the maintenance cost of garden is reduced drastically and hence, these are getting very popular for low cost landscaping.

Various situations in the gardens like flat or raised grounds, sloppy land, underneath trees, shady situations and along side roads or paths can be effectively beautified by the selecting ground covers judiciously. On flat ground, rangoli type effects can be created by planting ground covers to create the picturesque effects. The monotony of larger gardens can be easily broken by these colourful ground covers. Front areas of a house or gardens can be decorated permanently and can be made trouble free area. If more than one ground covers are to be used at one place, uniformity in growth must be ensured.

Selection of ground covers

The choice of ground cover depends on requirement of a particular situation, desired effects and personal choice. The selection can be done from sufficiently low growing perennial herbs to low growing shrubs. The best ground cover is that which grows fast, vigorous and will provide a good cover in very short time.

Ground covers are divided in to two groups according to their habit of growth and colour of foliage.

According to habit of growth

***(i)* Upright and spreading type.** These grow upright about 15-30 cm and spreading type of growth which cover the larger area in short period *e.g. Duranta* goldiana, *Alternanthera ficoidea, Iresine lindinii, Cuphea hyssopifolia, Opheophogon japonicus*, etc.

(*ii*) Trailing type. These type, trail on the ground and cover larger area if not kept under control *e.g. Ipomea charteruse, Ipomea tricolour, Wadelia trilobata, Zebrina pendula, Tradescentia flamiensis, Assyasitasia coromandeliana*, etc.

According to colour of foliage

(*i*) Green colour foliage. These plants have green colour foliage *e.g. Wadelia trilobata, Alternanthera ficodea, Cuphea hyssopifolia, Opheophogon japonicus*, etc.

(*ii*) Red colour foliage. These plants have red, crimson or combinations and thus make good contrast with green or other coloured foliage plants *e.g. Iresine lindenii, Oxalis, Ipomea tricolour, Alternanthera versicolour, Tradescantia flamiensis, Zebrina pendula* etc.

(*iii*) Golden colour or other colour. These plants have golden, yellowish green or greenish yellow or white colour foliage *e.g. Duranta goldiana, Ipomea charteruse, Alternanthera variegata*, etc.

Propagation. Propagation plays an important role as large number of plants are required to cover a larger area and the purchase of such a large number of ground cover can be a costly affair. Luckily, these ground covers are easily propagated either by division or stem cuttings or terminal cuttings. To facilitate better and profuse rooting, the basal end of cutting should be treated with seradix-l. This treatment results in profuse rooting and better establishment of cuttings. These cuttings are made during either in Feb.-March or July-August. However, under mist propagation or under modified structures, these can be propagated round the year.

Preparation of field and planting of the ground covers. These ground covers grow very well in wide variety of soil, however, soil should be well fertilized, well drained and neutral in pH. Field should be well prepared by digging or ploughing and by mixing five kg well rotten farm yard manure, 10 g N, P_2O_5 and K_2O per sq. m (40 g calcium ammonium nitrate, 62 g single super phosphate and 16 g of Muriate of potash per sq m). After mixing, according to layout plan, root cutting of desired type are planted during rainy season *i.e.* July-September or February-April in the cool hours of the day. It is desirable that before planting a light watering should be done which helps in better establishment of plants. These plants are planted at about 10-15 cm apart. After planting immediately watering should be done.

Maintenance of ground covers

For better growth of ground covers timely irrigation, weeding and hoeing is required. It has been observed that ground covers need frequent watering. During rainy season no watering is required unless rainfall fails. During winter season irrigation at 10-15 days interval is required whereas from Feb. to June irrigation at interval of 4-5 days is to be done. Initially, weeding and hoeing is required but it is not required in later stages when ground covers have grown sufficiently.

Trimming of ground covers (Plate 15). To maintain the plants in desired shapes, a regular trimming is required with hedge shears. Upright ground covers need trimming more frequently than trailing type. Normally plants are trimmed about 15-20 cm above the ground. But it can be further increased according to the plan. Certain ground covers like Oxalis or *Opheopogon japonicus* do not require any trimming.

Insects, pests and diseases. Generally, ground covers are free from the attack of insects, pests and diseases. However, certain ground covers are attacked by particular type of insects. It has been observed that *Ipomea charteruse* is commonly attacked by leaf boring insects and several short holes are made on the leaves during rainy season which *mar* the beauty of the plants if it is not controlled. It can be controlled by spraying Rogor 30 EC @ 0.1%.

The brief description of commonly grown ground covers is given below :

***Duranta erecta* cv. Golden Dewdrop (Verbenaceae) (Plate 16)**

It is recently introduced ground cover in Indian gardens and has gained popularity in very short period. New emerging leaves are of golden colour which makes it very attractive. Plant can be maintained as low as 10 cm by constant trimming that it can attain the height of 60-75 cm. It is easily propagated by terminal cuttings.

***Iresine lindenii* (Amaranthaceae)**

It is commonly called as *Lal patti* due to red-crimson colour foliage. It is very common ground cover and grows upto 15-30 cm and can be trimmed as low as desired. It is also easily propagated by terminal cuttings.

***Alternanthera pulchella Syn. Alternanthera versicolor* (Amaranthaceae)**

It is most commonly used ground cover due to its attractive foliage colour and makes fast growth. Plants grow about 15-20 cm. Leaves are of reddish scarlet, orange, purple and red colour. It is very easily propagated by cuttings. *Alternanthera ficoidea*—the colour of foliage is green and it makes good contrast. *Alternanthera variegata*—the colour of foliage is green and white or green and yellow.

***Sphagneticola trilobata* Syn. *Wedelia trilobata* (Compositae)**

It is trailing type, grows very fast and cover the larger area in very short duration. It produces small star like yellow coloured flowers which makes it more attractive. Every node produces roots so it is propagated very easily by cuttings and terminal cutting. It stands trimming and can be kept in desired shapes.

***Ipomoea tricolor* (Convolvulaceae)**

It is trailing type and produces coloured leaves reflecting purplish, yellowish and greenish effects. It is comparatively slow growing but easily propagated by cuttings.

Ipomea charteruse (Convolvulaceae)

It is also trailing type and produces golden yellow coloured leaves. It is tropical in nature and flourishes very well in tropical climate whereas under sub tropical climate its foliage gets killed during winter months. During rainy season it is commonly attacked by leaf borers which makes small holes in the leaves and mar the beauty of golden foliage.

Cuphea hyssopifolia (Lythraceae)

It is a small shrub hardly growing about 30-45 cm with a small leaves and produces purplish flowers. *Cuphea alba* produces white colour flowers. Plants need high moisture conditions for better growth. It is easily propagated by stem cuttings or by dividing the plants.

Tradescantia zebrina* var. *zebrina* Syn. *Zebrina pendula (Commelinaceae)

Plants have trailing stem and easily breaking at nodes from where it strikes roots easily and hence, it is propagated very easily by stem cuttings. Leaves are fleshy and have greyish streak on the upper surface. It grows fast in semi shady situations and makes good ground cover. Zebrina pendula var quadricolour has purplish green, broadly banded, glistening white with pink and carmine and silver strips. *Zebrina pendula* discolour has large glossy coppery leaves edged with metallic purple with two narrow silver bands and is purple beneath.

Tradescantia fluminensis (Commelinaceae)

It is very good, hardy colourful plants which is used for colourful effects as a ground covers. Leaves are sessile, ovate, large of purplish red colour and makes a very good contrast with green of other ground cover. It is very easily propagated by stem cuttings and grows very luxuriantly in semi shady situations without any difficulty.

Strobilanthes scaber (Acanthaceae)

It is dwarf plant with trailing habit and is easy to grow. Plants have shining dark green leaves. The flowers are yellow colour produced in great abundance in clusters during marks. It is propagated by cuttings. It makes good ground cover in shady situations.

Asystasia gangetica* Syn. *Asystasia coromandeliana (Acanthaceae)

It is a herbaceous perennial plant which grows very fast under partial shady situations. Plants have trailing habit and bear numerous lavendar coloured flowers with a yellow tube. It can with stand adverse weather condition. It is easily propagated by cuttings.

Lantana montevidensis* Syn. *Lantana sellowiana (Verbenanceae)

It is hardy plant with trailing habit and hardy grows about 30 cm. Leaves are dark green and crinkled. Flowers of pink mauve colour are produced in abundance in the form globular head. It is easily propagated by seed or cuttings.

***Oxalis* sp.** (Oxalidaceae)

It is easy to grow and hardly grows about 15 cm. Leaves are trifoliate and of which crimson colour which makes very good contrast with green colour ground covers. It is very good ground cover for partial shady situations. It is easily propagated by separation of bulbs from mother bulb or by seed.

Mesembryanthemum corodifolium Syn. Aptenia cordifolia (Aizoaceae)

Plants are prostrate and stems are covered with glistening papillae. Leaves are cordate, short stalked fleshy and covered by papillae. Small flowers are short stalked and purplish red in colour. Ideally suited for rockery and thus requires less watering. Flowering occurs during May to September. It is easily propagated by stem cuttings.

Ophiopogon japonicus (Graminae)

A small grass with little broader green leaves. It makes a good ground cover for partial shady situations. It is multiplied by dividing the plants.

Browallia speciosa (Solanaceae)

It is commonly called as Bush violet or Amethyst violet. It is a slow growing, upright flowering ground cover. It needs pinching of branches to make plants bushy. It produces small violet flowers. Propagation is done through seeds as well as through cuttings. It is sensitive to frost injury.

Juniperus horizontalis (Conifereae)

It has trailing stem and branches grow horizontally above the ground. It grows about 30 cm high and its branches are highly matted and hence, it makes very good ground cover. It is preferred at prime location in the garden or in front of house.

28

Interior Landscaping

Urban area are bursting to the seams due to natural growth as well migration from rural areas. Land is getting costlier and future of cities lies in vertical growth. Getting down from a flat in high rise building, winding through the roads laden with fumes and spending the day in a multistory office and back home makes the routine of city dwellers. The indoor air in cities is generally considered even more polluted than outdoor air due to lack of proper ventilation and natural light penetration. The air in homes, in schools, and in offices can put people at risk for health problems. Some pollutants can be chemicals, gases, and living organisms like fungi and pests. There are several sources of indoor air pollution which cause problems such as sore eyes, burning in the nose and throat, headaches, or fatigue. Other pollutants cause or worsen allergies, respiratory illnesses (such as asthma), heart disease, cancer, and other serious long-term conditions. Buildings with polluted interiors causing such symptoms are often termed as 'Sick Buildings'.

Some of the sources of indoor air pollution are :

1. Combustion pollutants which are gases like carbon monoxide and Nitrogen dioxide or particles emanating from improperly vented or unvented fuel burning sources like gas stove, driers or fireplaces. They can cause headache, dizziness, weakness, nausea, eye, nose and throat irritation, shortness of breath, and an increased risk of respiratory infection.
2. Volatile organic compounds (VOC's) like formaldehyde, benzene, trichloroethylene, toluene which are found in paints, paint strippers, carpets, cleaning supplies, varnishes and waxes, pesticides, building materials like plywood and furnishings, adhesives, office equipment, moth repellents, air fresheners, and dry-cleaned clothing. These chemicals evaporate into the air when these products are used or even during storage. These compounds irritate the eyes,

nose and throat, and cause headache, nausea, and damage to the liver, kidneys, and central nervous system. Some of them can cause cancer.

3. Asthma triggers are commonly found in homes, schools, and offices and include fungi, dust mites, second hand smoke, and pet dander. They cause symptoms including coughing, chest tightness, wheezing, and breathing problems. An asthma attack occurs when symptoms keep getting worse or are suddenly very severe.
4. Fungi produce spores that float in the air, land on damp surfaces, and grow. Inhaling or touching molds can cause hay fever-type symptoms such as sneezing, runny nose, red eyes and skin rashes.

Optimum humidity level for indoor air is between 30 to 50%. However, indoor air usually has much lower humidity which increases the potential damaging effect of pollutants by drying the mucus membrane of the respiratory system and making it more susceptible to pathogens and pollutants. Dry indoor air also increases the amount of bioactive dust *i.e.* dust laden with mites and fungal spores which triggers allergies. Along with the methods like controlling the source of pollution, improving ventilation and changing filters of air conditioners heating units, plants can act as biological filters for these pollutants.

Lack of land in house and office obliterates the opportunity to develop and enjoy a proper garden and the only option left is to grow plants in containers and bring nature into the house. These plants constitute the interior landscape and are also the key to healthy indoor air.

IMPROVEMENT IN AIR QUALITY BY INDOOR PLANTS

National Aeronautics and Space Administration (NASA) have demonstrated way back in 1986 that common houseplants have the potential for improving indoor air quality and reducing sick-building syndrome by removing pollutants, including formaldehyde, benzene, carbon monoxide and nitrogen dioxide from the contaminated air.

Volatile organic compounds from the air are absorbed by plant leaves and some of them are destroyed within the plant system through biological processes. Roots also play active role in mitigation. Some VOC's absorbed by leaves are translocated to root zone, where microbes break them down and a substantial amount is taken up by roots directly from the air they use for respiration.

Plant foliage takes in carbon dioxide during photosynthesis and produce oxygen as a by-product. Photosynthesis stops at night and plants respire, giving off carbon dioxide, just like humans do. However, plants reduce CO_2 level by about 10% in offices in air-conditioned building and by about 25% in the naturally ventilated building. Plants also reduce Carbon monoxide concentrations by 8-14% in buildings with or without air-conditioning as compared to unplanted ones. Plants themselves also release some VOC's called Biogenic Volatile Organic Compounds (BVOC's) which appear to play active role in controlling microbes and mold spores in surrounding air.

Microbes present in growing media are responsible for majority of the pollutant removal achieved by the plant system. Many soil microorganisms are capable of degrading toxic chemicals. This is confirmed by the fact that even when the plant(s) are removed from the media, the quantity of VOC in surrounding air continue to decrease. Similarly, the plants held in the dark also continue to remove VOC despite of the fact that stomata are generally closed. Plants provide photosynthates in form of exudates for the microbes which stimulates their development in the media.

Efficiency of plant systems to remove pollutants is improved when air was forced through the media. Hydroponic system have been modified to develop botanical indoor air biofilters (BIAB) which are used in conjunction with ventilation system of buildings for supplying clean air to the occupants. BIAB's are composed of plants, porous media and hydroponic nutrient solution delivery system to sustain plants and associated microbial community. They are installed in the form of vertical green walls. Water flows gravimetrically from top and is recirculated. Fans located behind the plants draw air through the media and exhaust air is circulated back into the building. Biofilters in form of green walls are promising technology which can substantially reduce costs of supplying indoor air through physical methods since they rely on biological processes which can go on for longer periods without much maintenance and replacement of parts like conventional filters.

Indoor plants release water vapour via transpiration through the leaves and into the air along with some evaporation from media, raising humidity of surrounding air. This creates convection currents resulting in pulling down of polluted air into the soil around the roots, where microbes break them down into a source of food and energy. Increasing relative humidity has a soothing effect on people and dust particles laden with allergens settle down. Foliage plants around the periphery of a room can reduce dust accumulation on horizontal surfaces by as much as 20%. Reducing particulate matter in a room is also beneficial for computers and other electronic equipment by reducing the amount of dust collecting on circuit boards and clogging air vents.

Pollutant removal potential varies from plant to plant and all the indoor plants have not been tested for this potential. However, the following plants have been proven to remove pollutants :

Areca palm (*Dypsis lutescens*), Bamboo palm (*Chamadorea elegans*), Boston fern (*Nephrolepis exaltata*), many species of Dracaena, Dendrobium orchid, Dumbcane (*Dieffenbachia varieties*), Dwarf date palm (*Phoenix roebelini*), English ivy (*Hedera helix*), *Ficus bennendijkii* 'Alii', *Chrysanthemum morifolium*, Gerbera daisy (*Gerbera jamesonii*), Golden pothos (*Epipremnum aureum*), Fern (*Nephrolepis obliterrata*), Lady palm (*Rhapis excelsa* or *R. humilis*), Peace lily (*Spathiphyllum* varieties), Philodendron species, Rubber plant (*Ficus elastic*), Schefflera, Spider plant (*Chlorophytum comosum*), *Ficus benjamina*.

Indoor plants do remove pollutants to some extent but cannot be relied upon as sole air purifying method. They can be used to augment the efficiency of conventional air cleaning devices. However, indoor plants have so many other benefits and use that they are indispensable in creating healthy living environment. House plants have been associated with general feeling of well-being. When foliage plants are in the room, people feel more attentive. Plants in the offices help to reduce stress felt by employees and give a feeling of being content or happy. Green office interiors reduce absenteeism and sick days of the workers thus enhancing office productivity. Workers in office with foliage plants report fewer symptoms like coughing, hoarse throat, and fatigue as compared to the ones without plants. Plants in classrooms boost the learning potential of students and enhance grades. However, for plants to be instrumental in providing good indoor air quality, the buildings must be designed with plants in mind, not as an afterthought on completion of the building. With the future of cities lying in vertical growth, pot plants will be the only means for all time connection of nature.

SELECTION OF PLANTS FOR INTERIOR LANDSCAPING

Plants for the interorscape must be selected keeping in view the pros and cons of specific location for them. Generally, people go to a nursery and select the plants which appeal to them. Such plants selected solely on the basis of personal liking and then fitted into the available spaces without considering the conditions required for their survival and growth, ultimately leads to many failures.

Before finalizing the plants, all available spaces in the building intended for placing plants should be carefully analysed for availability of light, ventilation and humidity along with the volume of space available.

Light

Natural light is available only near the windows and intensity and duration of light varies with distance from the window and orientation. Tint of window panes or the source of artificial light can also alter the quality of available light. All these factors affect the growth of plants and thus dictate the selection of plants for a location. Light requirement of plants largely depends on their natural habitat. Many of the indoor plants are of tropical origin and grow as understory plant in rainforests where they get filtered, diffused light. In their natural homes, they grow in uniform light and temperature throughout the year and do not undergo any seasonal change. Such plants can grow in low to medium light intensities, mild temperature and need high moisture content in air as well as growing media.

Interior spaces can be divided into high, medium and low light areas as given below :

High Light. Locations within four feet of large south, east, and west facing windows.

Medium Light. Locations in a range of four to eight feet from south and east windows and west windows that do not receive direct sun.

Low Light. Areas more than eight feet from windows as in the center of a room or an inside wall. Northern facing windows often fall into this category, even when close to the window. The areas that receive only artificial light generally have low light situations. Plants for low, medium, and high light locations are listed in Table 28.1.

TABLE 28.1

Low Light	*Medium Light*	*High Light*
Aglaonema commutatun	*Asparagus densiflorus*	*Aloe barbadensis*
Aglaonema modestum	*Asparagus setaceus*	*Alternanthera ficoidea*
Aspidistra elatior	*Brassaia actinophylla**	*Araucaria heterophylla*
Chamaedorea elegans	*Brassaia arboricola**	*Beaucarnea recurvata*
Epipremnum aureum	*Chlorophytum comosum* 'Variegatun'	*Cissus antarctica***
Monstera deliciosa	*Dieffenbachia amoena*	*Crassula argentea*
Sansevieria trifasciata	*Dieffenbachia maculata*	*Fatshedera lizei***
	*Dracaena fragrans**	*Hibiscus rosa-sinensis*
	Dracaena godseffiana	*Hoya carnosa***
	*Dracaena marginata**	*Podocarpus gracilior*
	*Dracaena sanderana**	*Rhoeo spathacea*
	Ficus benjamina	*Sedum morganianum*
	Ficus elastica 'Decora'	
	Ficus lyrata	
	Ficus retusa	
	Gynura aurantiaca	
	Hedera helix	
	Maranta leuconeura	
	Nephrolepis exatata	
	Pandanus veitchii	
	*Peperomia caperata**	
	Peperomia obtusifolia	
	*Philodendron bipennifolium**	
	Philodendron selloum	
	Pilea cadierei	
	Polyscias balfouriana	
	Saintpaulia sp., and hybrids	
	Spathiphyllum sp.	
	*Syngonium podophyllum**	
	Tradescantia fluminensis	
	Zebrina pendula	

*May also be conditioned to grow in low light.

** May also be conditioned to grow in medium light.

Further the duration of light varies with latitude and is drastically reduced in winter months as we go north. However, south facing windows are likely to get higher intensity due to inclination of sun in sub-tropical areas. Some other less obvious factors can also affect indoor sunlight levels. These include : the color of walls and floors, types of window coverings, roof overhangs, outside awnings, nearby buildings and trees that filter or block incoming light.

Plants grown under optimum light conditions are vigorous, compact and bushy with true vibrant colour, normal sized leaves and sturdy stems. Plants grown at a reduced light intensity have smaller leaves and dull colour. They often grow leggy and necessitate frequent pruning for compact form. Many flowering plants fail to bloom under such conditions. If light conditions are unavoidably poor, keep plants drier than those in bright light and reduce fertilizers. Such plants may survive for several months to a year, while gradually deteriorating in appearance and vigour.

When light levels are higher than required by the plant, leaves show an overall yellowing resulting from the destruction of leaf pigments. Ultimately, large brown spots of dead tissue develop, a condition called leaf scorch or leaf burn.

Artificial light can be used to supplement low natural sunlight or to light up dark areas and create good growing environment for plants. Cool white fluorescent lights alone or in combination with warm fluorescent lights are generally most suitable and economical. Typically, a fixture holding two 40 watt tubes in positioned approximately 12 inches above the plants. Most plants need 12 to 16 hours of artificial light per day for good growth.

Temperature

Rising income levels have led to increased control of living environment by artificial cooling and heating with human comfort in mind rather than the growth requirements of indoor plants. Fortunately, the most comfortable temperatures for humans fall within the optimum range for most foliage and flowering plants. Daytime temperatures of 20 to 26°C and a night time range of 16 to 20°C is satisfactory for most of the plant species grown indoors. Many flowering plants bloom for a longer period at the lower end of these day-night temperature ranges. Temperatures in different rooms of same building can vary with orientation. A minimum/maximum thermometer should be used for routine monitoring of such differences and considered when placing the plants. Sudden temperature changes can be harmful to interior plants. If a plant is positioned near a window, it must be protected from heat and intense sunlight during the day and from cold conditions at night. Shades and curtains can be used or the plant can be moved away from the window to avoid temperature and light extremes. Plants should never be placed near heaters, fireplaces or on television sets.

Humidity

Most of the indoor plants are native to moist tropical regions of the world. The most common problem faced by these plants is dry air which results in the leaves of these plants turning yellow or showing brown tips and margins. Some fail to bloom while in many others, flower buds shrivel and drop suddenly. Moderate humidity range of 30 to 50% is optimum for growth of a variety of flowering and foliage plants. However, cactus and other succulents do well in a humidity range of 5 to 15%. Hygrometer should be used to determine the relative humidity of different locations throughout the year. One of the simplest ways to increase the humidity around plants is to place them in groups where evapo-transpiration creates humid microclimate. Placing trays holding an inch or two of wet pebbles under plant containers also increases humidity. Water should be added to a point just below the surface of the pebble layer. Humidity increases around the plant as this water evaporates. Air conditioned rooms have drier indoor air during the summer season and mist vaporizers or room humidifers can be used to add moisture. In homes, sensitive plants can be grown in naturally humid rooms such as the kitchen, bathroom or laundry room.

DESIGNING WITH INDOOR PLANTS

Design process and principles for interior landscaping are same as explained in chapter on Landscape gardening. Indoor plants can be used for the following purpose :

Aesthetic use. Foliage plants can serve as background for art pieces or sculptures and enhance their prominence. Well maintained plant topiaries are like sculptures themselves and serve as focal points. Bonsais, dish gardens and terrariums can also be used as ascent points. Aesthetically arranged plants can turn a bare wall into an interesting feature. Pots can be placed in iron stands studded to walls for this purpose if space is a limiting factor. Plants also help to soften the sharp architectural lines and merge large patios or balconies with the surroundings. Pot plants can effectively frame views inside the building or control sight line and directing it to a particular feature. Plants can also be used to partially hide objects of interest and gradually revealing vistas.

Architectural use. Indoor plants placed at strategic locations provide privacy from neighbours and help to screen unpleasant views. Row of pots can divide a large space into small parts in a pleasant manner, like spaces in between tables in a restaurant or separating waiting area in an office. Large number of pots arranged in atriums can provide altogether new vistas.

Engineering use. Indoor plants can control movement of people in large spaces like shopping malls. Acoustic control can also be achieved to some extent by placing plants with dense foliage which can muffle noise. Pot plants near large glass windows, particularly on western side can help in glare reduction and increase comfort level. Plants in balconies can be used to direct air currents inside the adjoining rooms.

SELECTION OF CONTAINERS

Selection of container for a plant is as important as selecting the plant itself. At the simplest level, the purpose of a container is to hold the right amount of growing medium for the plant. Primary requirement is that the container should match the size of plant. Generally, a container should be one third as tall as the plant measured from soil level to the highest leaf. Plants that are too small for their containers look out of proportion and grow poorly since the soil stays overly moist for a long time. Plants that are too large for their containers also look out of proportion. They become root-bound and often topple over, since their pots don't have enough weight to hold them up.

Wide choice is available as far as material of container is concerned and each one has its pros and cons as listed below :

Clay or terracotta pots. Available in a variety of shapes and sizes, terra-cotta pots can be used in any location and their earthy colour will enhance the beauty of almost any plant without dominating it. Made of porous clay, terra cotta pots can 'breathe' which keeps potting soils cool and removes excess moisture away from plant roots, keeping them healthy. The major problem with clay pots is that they are relatively fragile and dry out quite rapidly, especially in sunny locations. However, glazed clay pots can retain water for a longer duration. Since overwatering is a major problem with most beginners, these pots are ideal for them. Initial cost of these pots is very low.

Cement pots. Cement pots come is many shapes and up to much larger size than clay pots. They are more durable but less porous and weigh much more than terracotta. They are more suitable for outdoor use and large plants.

Plastic pots. Plastic containers are durable, retain moisture well, but are expensive than clay or cement pots. The most desirable feature of plastic pots is their low weight, which makes them ideal for moving around with less chances of damage. Black plastic pots are cheaper but are less durable and should not be used in sunny locations as they absorb heat and will get very hot, which can damage tender roots. Light colour containers reflect the heat and keep the roots cool. Although they have a long life, but once broken, they form non-biodegradable waste and add to environmental hazards.

Wooden containers. They are most natural containers for gardening, retain water well and are relatively lightweight. However, they are prone to rotting, seasonal expansion and shrinking and termite attack. Creative wooden planters can be built easily with a few nails from scrap wood. Wooden containers are best used as outer planters in double potting system with a plastic tray placed inside for collecting excess water.

Stone pots. Pots made of natural stones like marble and sand stone look stunningly beautiful. They are very durable but heavy and are best suited for outdoor use where they are not likely to be moved for years. Stone planters can be very costly depending on design and finishing. They age very beautifully and change colour over time which gives a rustic look.

Ceramic and metal containers. Ceramic and metal containers come in intricate designs and are usually very costly. Whereas ceramic ones are very fragile, metal containers are very durable. Porosity is poor in both the types and extra care has to be taken for proper drainage. These types of containers are also best suited as outer cover in double potting.

Creativity in selection of containers has no limits and plants can be seen growing in anything which can be made to hold media, ranging from old shoes to tea pots, old watering cans to bath tubs.

SPECIAL GARDENING

Terrarium

Terrariums are miniature gardens developed in an enclosed or partially enclosed glass container. The glass walls allow light to reach the plants and give protection from dust, drafts, and temperature changes. Water vapours from transpiration get condensed on contact with the cooler glass walls which runs down the sides of the container and re-enter the growing media. In case of enclosed terrariums, water keeps on circulating and supplemental watering is rarely needed.

Terrariums were first used in early nineteenth century by a British Physician with interest in Botany, Dr. Nathaniel B. Ward and were initially named 'Wardian case'. They were glass-topped boxes, resembling a miniature greenhouse used to transport plants collected in other parts of the world back to Britain. Within the glass enclosure, plants could be grown in near natural environment along with protection from the harsh conditions on ship during long voyage. Plants inside glass enclosures were aesthetically pleasing and easy to maintain which led to quick adaption for interior decoration.

Terrarium can be developed in any clear glass container. Cloudy or tinted glass reduces the amount of light and adversely affects plant growth. Size of the container can be selected according to availability of space and the number of plants to be used. Container with a removable lid helps to control moisture. If the container is without lid, a piece of glass or plastic can be used as cover.

Containers used for terrariums do not have drainage holes and provision must be made for removal of excess water from the media. A layer of moss under the media can serve as a drainage layer in small containers. For large terrariums, small pieces of clay pots, charcoal, sand or fine gravel should be used as bottom layer.

Media layer should be porous to allow good drainage and should be 1 inch to 3 inch thick depending on size of the container. Potting mixtures for terrarium can be made by mixing equal parts of garden soil, sand and leaf mould or sieved FYM. High level of fertility is not desirable; else the plants will quickly outgrow the container. One teaspoon of 12-32-16 fertilizer can be mixed with two litres of media.

A terrarium can be used to grow a single specimen plant, or a miniature garden can be created with combination of plants and accessories. Slow growing plants which require high humidity are ideal for growing in terrariums. Arrangement of plants depends on display location and shape of the container. For an all side open view, tall vertical plants should be placed in centre and surrounded by short, prostrate plants. For single side view, tall, vertical plants should form the backdrop and gradual reduction in plant height towards the front. The effect of height can also be enhanced by grading the base layer and small rocks, pebbles, shells, figurine etc. can be used to create miniature garden effect in larger containers.

Plants should be gently removed from their pots without damaging the root system and planted at same depth as original. The plants should not be crowded against walls of container or against each other. The foliage should be adjusted slowly to natural form. Small amount of water can be used to remove media from leaves and walls without making media soggy. Media should be kept moist but not wet in any case.

Terrarium should be placed in bright light but direct sun exposure should be avoided. Fogging of containers during initial phase can normally occur and partial opening of lid can mitigate this situation by venting excess humidity. When the terrarium's water cycle is optimised, water starts condensing in form of light film and trickles from the walls back to the media. After this stage is reached, terrarium rarely needs water. The miniature gardens usually remain proportional for about a year with minor trimming of aggressive plants. Plants suitable for terrarium are listed below according to their height :

Tall Plants (6″ to 12″). *Pilea cadierei*, *Syngonium podophyllum*, *Pilea microphylla*, *Asparagus* spp., *Begonia* spp., *Dracaena sanderiana*, *Iresine herbstii*, *Miniature Aspidistra elatior*, *Dwarf Aglaonema* spp., *Codiaeum variegatum*, *Euonymus japonicas*, *Adiantum* spp., *Pellaea* spp., *Fittonia verschaffeltii*, *Calathea makoyana*, *Maranta leuconeura*.

Short, Rounded Plants (< 6″). Begonia Red, *Ajuga reptans*, *Asplenium nidus*, Miniature *Pelargonium* spp., *Alternanthera* spp., Dwarf *Ophiopogon japonicas*, Miniature *Peperomia* spp., *Hypoestes phyllostachya*, Dwarf *Sansevieria trifasciata*, *Oxalis* spp., *Drosera* spp., *Saintpaulia ionantha*, *Viola odorata*.

Short, Prostrate Plants (1″ to 3″ tall, spreading). *Ficus pumila*, *Dwarf Hedera helix*, *Hemigraphis alternate*, *Selaginella* spp., dwarf *Epipremnum* spp., *Callisia elegans*, *Tradescantia* spp.

Dish Garden

As the name suggests, this is a 'garden' planted in a shallow dish of glass, ceramic, or metal. Preparation of dish gardens is almost similar to terrariums. Since, the dish gardens need frequent watering, a drainage hole should be drilled if possible, otherwise a gravel layer at the bottom can serve the purpose coupled with shallow watering.

Whereas, moisture loving plants work better in terrariums, cacti and succulents make excellent dish gardens. The dish selected for such garden should be provided with a drainage hole and media recommended for succulents should be used. Small desert plants like aloes, agaves, euphorbias and echeverias can be arranged in interesting landscape design and sand can be used for finishing. Small rocks and pieces of wood can be added to create interest. Over watering is detrimental to desert plants, thus watering should be done only when the media becomes dry. Most of the cacti and succulent need bright light and can be placed in sunny locations. Desert dish gardens should be fertilized very carefully as excess nutrients can damage the plants and cause them to rapidly outgrow the container.

A miniature water garden can also be created in a dish which works as hydroponics system. Soil should be completely removed from the roots by washing it off and plant should be cleaned of dead or damaged parts. Here, the dish does not require drainage hole. One inch layer of washed aquarium stones should be placed at the bottom and subsequent layers of charcoal and course sand should be added. Plants should be put in place and gravel should be added to cover the roots. Water containing weak fertilizer solution should be added thereafter, to fills about one-half of the gravel level. This level should always be maintained and the solution should be changed about every six weeks. Plants like *Aglaonema* sp., *Syngonium podophyllum*, *Coleus blumei*, *Codiaeum variegatum*, *Dieffenbachia* sp., *Schefflera arboricola*, *Hedere helix*, *Cordyline terminalis*, *Chamaedorea elegans*, *Philodendron* sp., *Epipremnum aureum*, *Spathiphyllum* sp. and *Tradescantia* sp. can be grown in such containers. Small aquatic plants like *Pistia* and *Lemna* species can also be used to create miniature water gardens.

Hanging Baskets

Top of the windows and balconies generally have the highest levels of light found in most buildings. Hanging baskets provide an opportunity to utilize this light to grow a wide range of plants. Almost all of the flowering pot plants can be grown in these locations.

Any type of container that can hold at least one litre of growing media by volume can be used for hanging. Containers with less volume often dry out rapidly. Metal frames, wires or cords can be used to hang the container. The metal frame itself can be used as hanging basket by lining it with moss. Containers must be provided with drainage holes at least three-eighths of an inch in diameter, spaced 3 inches apart near the bottom. A large variety of baskets made of plastic is readily available at nurseries and garden centres. Hanging baskets are also available with self-adhering saucers. Such hanging baskets can be used anywhere while preventing water drippings on furnishings.

Growing media containing 2 parts of sandy loam soil, one part coarse sphagnum moss or cocopeat and one part coarse aggregate (washed cinders or medium gravel) can be used in hanging baskets. Media should be kept wet but must not become soggy due to overwatering.

Plants like Ivy, Begonia, Geranium, Fuchsia, Sedum, Tradescantia, Callisia, Ferns, Pothos, Orchids, Asparagus sp. Lantana, Petunia, Verbena, Portulaca, Sweet Allysum, *Impatients walleriana*, Chrysanthemum, Verbena, Pansy, Nasturtium etc. can be easily grown in hanging baskets.

Window Boxes/Windowsill gardens

Indoor pot plants or dish gardens can be placed in specially made or fit to size boxes in windows. This gives an opportunity to grow a variety of plants which is usually not possible inside the rooms and creating beautiful transition to the exteriors. Window boxes and hanging baskets in multi-storey flats can lend a green touch to the entire building creating hanging gardens. Window boxes can be made of light wood, steel or fibre with adequate provision for firm anchoring to the walls. Sun facing window gardens need frequent watering due to additional heat generated by concrete. Almost all the annuals can be grown in windowsills to add rainbow of colours to the building.

29

Lawn

Lawn is a natural green carpet. It is an important feature of a landscape and a garden without a lawn is not considered complete. Lush green lawn provides a great satisfaction to owner and becomes a centre of the garden for major activities. Lawn provides a place for taking rest after tiring job of the day, holding parties, social functions, passive and active recreation. A patch of green grass in home gardens add beauty and softness against walls and other concrete structures. It harmonises with the surroundings and provides a natural green colour. It is not difficult to make and maintain a good lawn if little care is taken. Therefore, it is worthwhile to take a little pain during making and maintenance to get the desired effect.

Selection of Grass

There are mainly two types of grasses which are used for planting lawn. The selection can be done according to choice, area and availability of water.

***(i)* Bermuda grass** (*Cynodon dactylon*). It is commonly called as doob or Hariali. This is very commonly used for planting lawn due to its fast growth hardiness, less water requirement and response to frequent mowing. This makes an excellent turf. It is highly suitable for large areas and playgrounds on account of its tolerance to wear and tear and has good recuperation habit. Foliage gets killed by frost and regrows from rhizomes when soil warms up.

***(ii)* Korean grass** (*Zoysia japonica*). It is native of Japan, Korea and Philippine islands. This kind of grass is recently introduced in India and has become very popular in short span of time due to its velvety growth and more cold tolerance. It makes cushion like turf. This is highly suitable for smaller areas and home lawns. Water requirement of this grass is comparatively more. It is tough grass and is difficult to mow. Lawn should be clipped at 1.5-2.5 cm height. It grows slowly and frequent mowing is not required. Due to its slow establishment, its usage for sport grounds is limited.

There are other species like *Z. matrella* (Manilla grass) and *Z. tenuifolia* (Korean velvet grass) which can be tried.

Preparation of Soil

Different types of soils have been found suitable for making lawns. The soil should be fertile and well drained with good water holding capacity, for an ideal lawn making. An ideal pH of soil is 5.5-7.0 but it grows well even at a high pH of 8.2-8.4. During summer months *i.e.* April-June, soil should be dug upto 45 cm depth by spade. The clods of earth are exposed to scorching sunrays for killing weeds and other harmful insects and micro-organisms. Few days before the rains start, break the clods and 40-50 cartloads/acre of well rotten farmyard manure should be incorporated. During mixing of farmyard manure, pick out stones, roots of weeds and undesirable matters. Rolling should also be done lightly. During rains soil will settle down and weeds will germinate which should be removed.

Before planting the lawn, ground should be levelled giving a gentle slope for drainage of extra water. Generally 2-3 cm slope is necessary per 50 m towards the natural direction of drainage. There should be a provision of an outlet to drain extra water.

Planting of Grass

Following are different methods of planting lawns :

Dibbling of roots. Though this is little slow method for making lawn yet it is most common and the cheapest method, hence preferred. In this method, grass pieces with roots taken from area free from weeds, are dibbled at a distance of 10-15 cm both ways. After planting water is applied liberally if rainfall is not there.

Seeding. This method has not become so popular but can be followed for making lawn. Before sowing the seeds soil should be tilled fine followed by light rolling. Ten to 12 kg seed of grass is required for an acre. Seeds should be uniformly and evenly sown on a windless day. Bulk material like sand or ash should be used for even distribution of seeds. The ground should he lightly rolled again. Watering should be done with a watering can fitted with a spray nozzle. Flood irrigation should be avoided before germination of seeds otherwise seeds get collected at the slopy places.

Turfing. In this method, the pieces of uniform thickness of turf along with soil are made from the place where grass is compact and free from weeds. Then these pieces are spread on the ground evenly and beaten with a stick covered with a fine soil. Liberal watering is done. In very short period grass starts growing.

Bricking. To replace the few unhealthy patches in well maintained lawn, this method is used. In this method, piece of lawn alongwith soil in the shape of bricks are taken out and are planted and watered immediately.

Planting on polythene. In this method, lawn is grown on thick polythene (800 gauge thickness). First polythene is cut into suitable pieces of desired dimension. Mixture of soil (3-4 cm thick) is spread over the polythene sheets and grass is grown over. It can be rolled and taken to place and rolled out where temporary effect is to be created.

Management of Lawn

Mowing and rolling. Newly planted lawn is not mowed till it has established well. In the beginning the grass should be trimmed with the help of scythe. Light rolling should be done on dry ground to suppress the upright growth to anchor the grass firmly in the soil and to keep the level of ground. Interval of mowing of an established lawn depends upon season. The height of lawn can be maintained by adjusting the height of shears of lawn mower but it should not be more than 5-6 cm. In rainy and winter season mowing is required at an interval of 7-10 days. During spring season mowing is required at the Interval of 15 days whereas during summer it is done at monthly intervals. Different types of mower like hand driven, bullock driven or tractor driven are available and can be selected depending upon the size of lawn.

Irrigation. Water requirement of lawn depends upon season, type of soil and the grass used. Grasses are surface feeder and, hence adequate watering should be done frequently. During autumn and winter months, watering is done at an interval of 10 days whereas during spring and summer months it should be done at an interval of 7 days. Generally flood irrigation is followed but sprinkler system is better and should be used wherever possible. In sandy loam soil, frequent watering is required than in heavy soils. Korean grass needs more frequent watering than Calcutta grass.

Weeding. Regular mowing controls the population of weeds by removal of upper portion of the weeds and starvation of roots. But still there may be large population of weeds which grows fast and needs removal. Hand weeding is possible in small areas and it is the only solution to check motha (*Cyperus rotundus*) which is a problematic weed. It perpetuates by nuts (rhizomes) deeply placed underground. Weedicide for dicot weeds can be used *e.g.* 0.5% solution of 2,4-Dichlorophenoxy acetic acid.

Fertilization. The regular application of fertilizers keeps the grass to grow luxuriantly and maintains the lush green colour of the lawn. There are many arbitrary recommendations about fertilizer applications. It has been observed that the application of 50-60 g/sq m of mixture (2 parts Calcium Ammonium Nitrate : 1 part Super Phosphate : 1 part Potassium Sulphate) during rainy season and February-March maintains the growth of grass. Spraying of urea (30 g/10 L of water) is also useful for the growth of grass.

Scraping of Lawn. It is done to renovate the old lawn when it becomes old and grass becomes compact. After 3-4 years during summer months *i.e.* during June, lawn should be scraped completely with the help of Khurpa and raking should be done both ways. Before the start of rains top dressing of mixture consisting of garden soil, sand and sieved leaf mould (1 : 2 : 1) should be applied to cover the upper 3-5 cm. Soil for top dressing should be homogeneous, clear, free of pests and seeds of weeds and should be similar in texture to that of the turf soil. It helps in improving the renovation of stolon or rhizomes, fertility level and levelling the ground. Bonemeal @ 1 kg/100 sq m should also be applied with the onset of rains or application of irrigation water. Thus grass will grow luxuriantly and will make a compact and healthy turf.

Diseases and Other Problems

Fairy ring. It is caused by soil borne fungi like *Marasmius ordeades, Psalliota campestris, Leipiota morgani,* etc. These fungi cause a circular ring of thin coloured or dead grass. Rings may not be complete and given the appearance of an arc or horse shoe. Grass turn brown and rings widen in the circles from few cms to metres. This can be controlled by drenching soil with bordeaux mixture (4 : 4 : 50) (4 kg Copper sulphate : 4 kg hydrated lime : 50 L of water) or by spraying other Copper fungicides like Blitox.

Pale or yellow lawn. Yellowing of lawn may be due to several reasons and hence, proper care must be taken in checking the problem. This may be due to water logging conditions, due to poor drainage in heavy rainfall area or lack of water. Deficiency of nitrogen is also responsible for yellowing of grasses which could be prevented by frequent application of nitrogenous fertilizers. Hardening of soils is also responsible for this malady which creates oxygen deficiency in roots. Loosening of soil with the help of khurpa or rake will reduce yellowing.

***

30

Landscape Gardening

Man might have felt the necessity of cool and shady place during summer and open and sunny situation during winter for taking rest after tiring jobs of hunting. These must have been situated near water bodies where hunting was also convenient. This gave a concept of selection of site for taking rest and hunting. Later on, planting of such sites with trees might have started at an early stage. This was the beginning of landscape gardening. Frescoes of flowers of old times reveal the association of human with flowering plants which they used in making gardens and planting on roadsides. Man has been attracted by natural objects like water bodies, mountains, rocks, moon, stars, sun rise, sun set, colourful plants, animals, birds, fishes etc. He likes beauty and dislikes ugliness.

Landscape gardening is a very fascinating and interesting subject. It is an aesthetic branch of Horticulture which deals with planting of ornamental plants in such a way that it creates a picturesque effect. There are several definitions and expressions to define this subject. Landscape gardening can be defined as the decoration of a tract of land with plants and other garden materials so as to produce a picturesque and naturalistic effect in a limited space. According to Chambers' dictionary, the definition of landscape is the appearance of that portion of land which the eye can view at once and landscape gardening is the art of laying grounds so as to produce the effect of a picturesque landscape. So landscape may or may not include plants. According to Bailey, landscape gardening is the application of garden forms, methods and materials for the improvement of the landscape and the landscape in this connection is any area large or small on which it is desirable to develop a view or design.

Landscape gardening can also be defined as the beautification of a tract of land having a house or other object of interest on it. It is done with a view to create a natural scene by the planting of lawn, trees and shrubs. Further, landscape gardening is both an art and science of the establishment of a ground in such a way that it gives an effect of a natural landscape. Another simple definition can be "Landscape gardening is the imitation of nature in the garden". It can also be defined as "improving of total living environment for the people". In light of all the foregoing definition landscape gardening can be defined more appropriately as making a garden or to tame an area which shall be landscape or picture. The expression of landscape may be gay, bold, retired, quiet, etc. This expression will conform to the place and the purpose. It should be a picture and not a collection of interesting objects.

Since the landscape gardening is the making pictures on the ground with plant and other material, landscape designer should be proficient in art, ornamental gardening, ecology and physiology. He should be an architect and engineer to appreciate the relationship between plant form, colours and buildings.

Natural Elements of Landscape

We live on a vast planet which is characterized by different types of landscape depending upon prevailing geographical and agroclimatic conditions. There are mountains, hills, glens, valleys, seas, rivers, forests, plains, deserts, lakes, swamps, streams, etc. which comprise major part of natural landscape. At certain points, there is harmony between natural elements like ground forms, vegetation and even animal life. The landscape of such place is beautiful and conveys the feeling or mood of the landscape character like exhilaration, sadness, eeriness or awe. There are many qualities of natural landscape beauty like the picturesque; the ethereal, the serene, the delicate, the idyllic, the graceful, the majestic, the bold, etc.

Man has copied the natural elements for improving landscape around him and converted certain areas in the form of garden for his pleasure.

Principles of Landscaping

It is very difficult to lay down principles of landscaping, applicable everywhere. However, general principles of landscaping are as follows :

(*i*) The ideal landscape garden is like ideal landscape painting which expresses some single thought or feelings. Its expression may be gay, bold, retired, quiet, etc.

(*ii*) Beauty and utility should be harmoniously combined.

(*iii*) Area should be divided into different parts and plan should be conceived for each area. Overall plan should be such that the observer catches the entire effect and purpose of the plan without stopping to analyse its parts.

(*iv*) Let the garden and building merge into each other. There should not be stopping abruptly particularly in front of building. The view of garden from the windows and doors should be very attractive. Planting around the building, climbers against wall and on the porch, decoration of verandah and rooms with attractive foliage, flowering plants, hanging baskets serve to unique the building with garden. Every part of the compound should be planned in such a way that it gives surprising effect to visitor.

(*v*) Simplicity of design should be aimed at which helps in the execution of the plan.

(*vi*) An ideal landscape should have open space.

(*vii*) Over crowding of plants and object should be avoided.

Factors Affecting the Landscape Design

There are several factors which affect the making of suitable design for particular site. These factors are :

Human choice. Man's ultimate desire is to make his living pleasurable and surroundings comfortable. His dominance in making designs and selection of plant material is very well evident. Therefore, different styles of gardening have come into existence.

The site. This is an important factor and according to site, suitable design is made. In formal style gardening, the site is selected according to plan. Topography of the site also affects the design.

Views. Distant views of mountains, hills, woods, valley, etc. are preferred from the place of garden.

Heritage. One inherits the knowledge of botany and aesthetic sense and uses accordingly. Our rich heritage teaches us to use flowers and fragrant trees to improve the surroundings.

Climate. The climate of particular place affects the selection of plant material accordingly. Ideally suited plant material according to climate should be selected.

Soil. According to characteristics of soil types, suitable plants should be selected.

Styles of Gardening

Man's eternal desire is to make his living like that of paradise. This imaginary paradise has been created on the earth in the form of gardens. Therefore, the style of gardening has been influenced by political and geographical conditions. These styles have been changed from time to time with the new ideas and necessities. Broadly, the styles of gardening are grouped into three categories, *i.e.* (*i*) Formal style, (*ii*) Informal style and (*iii*) Free style of gardening.

Formal Style. Main features of this style of gardening are : First plan is made on the paper and then land is selected accordingly. Plan is symmetrical. These types of gardens are of geometric design *i.e.* squarish or rectangular. Therefore, the roads are cut at right angle. It has some sort of enclosure. Flower beds are also of geometric shapes. The arrangements of tree and shrubs are necessarily geometrical and kept in shape by trimming and training. Other features like fountains, water pools, cascades, etc. are used for further attraction. The examples of such style of gardening are Persian gardens and Mogul gardens.

Informal Style. This style reflects naturalistic effect of total view and represents natural beauty. This style is just contrast of above formal style. In this, plans is asymmetrical and according to the land available for making garden. Roads, paths are made curvaceous and bending. Water bodies are made of irregular shapes. Hillocks are made to create natural mountainous scenery. Flower beds are made of irregular shapes suiting to surroundings. Plants are allowed to grow in natural form and instead of trimming, annual pruning is done. Japanese gardens are the best example of this style of gardening.

Free Style of Gardening. This style combines the good points of both formal and informal style of gardening. Rose garden of Ludhiana is an example of this style of gardening.

Art Principles

Landscaping is making of pictures with plant material and, hence, its principles are same as those of art. Therefore, the understanding of such principles will help in using them properly. They are as follows :

Rhythm. Repetition of same object at equidistance is called rhythm. It can be created through the shapes, progression of sizes or a continuous line movement. Rhythm creates movement to the eye. In gardens, generally trees of single species of equal height and shape are planted to create this effect. In Mogul gardens, fountains and water canals have also been extensively used to create such effect. Now-a-days other objects like lights are also used to create the effect of rhythm.

Balance. It is very important to maintain the balance on both sides of the central line. The principle involved in making balance of see-saw game can help in understanding this. Equal weights can be balanced only when they are equidistant from the centre. If weights are unequal, the heavier must move towards centre for making balance. The balance may be formal, informal or symmetrical types. Imbalance will look lopsided and will distract the attention. In making the balance with the plants, their form, colour, texture etc. are kept in view.

One side planted with trees will not make balance if other side is planted with shrubs or climbers. Similarly, one side planted with tall upright trees will not make balance with other side if planted with dwarf or weeping trees. Making balance with colour is difficult untill and unless used properly. Red, orange and yellow colours are grouped into heavy and warm colours. On the contrast, white, green and blue colours are light and cool colours. Deeper shade is heavier than its lighter counterpart. Proper balance by selecting appropriate colours should be done to create the desired effect. Texture of leaves may be rough, coarse, fine or smooth, so proper balance is achieved between different types of texture. The consideration of bark colour and patterns should also be kept in mind to make the balance.

Accent or Emphasis. The accent or emphasis is created in the gardens to avoid the monotonous view. It is the method to stress the most important thing. This also serves as the centre of attraction. Landscape designer should know what to emphasise, where to emphasise, how to emphasise and how much to emphasise. Mostly unusual objects like tall fountain, tree, statue, etc. are used to create the effect of accent or emphasis. In English gardens, statues have been used extensively to create such effects.

Contrast. This principle is most useful in emphasising the best features of an object. It can be very easily understood by following contrast colour theory. Against green background, a fleck of scarlet colour, will make a contrast and will make scarlet colour prominent. In nature this is very common. Other contrast colour can also be used. Similarly, weeping growth habit against upright growth; dwarf against tall; rough texture against soft texture etc. are some of the examples which can be followed. It is also very important that one of the two contrasting objects must clearly dominate each other. In this way, one becomes feature whereas other acts as supporting background. The contrasting elements of equal power may create visual tensions.

Proportion. It is the relation of one thing to another in magnitude. When two or more objects are put together the proportions are established. In a landscape design, space provided for lawn, paths, herbaceous borders, shrubberry border, trees, buildings and other garden objects should be in a right proportion. It will create harmonious effect and look better. Such effects can be noticed in Persian and Mogul gardens. Out of proportion allotment of area in garden will distract the attention. Proportion helps in space organisation.

Harmony. It is an overall effect of various features, styles, and colour schemes of the total scene. The degree of harmony or unity of various elements of landscape is a measure induced in us and is called as beauty. Therefore, the beauty can be defined as the evident relationship of all parts of a thing observed. When different parts of landscape are correctly placed in right way, produces a harmonious effect. Such landscapes create

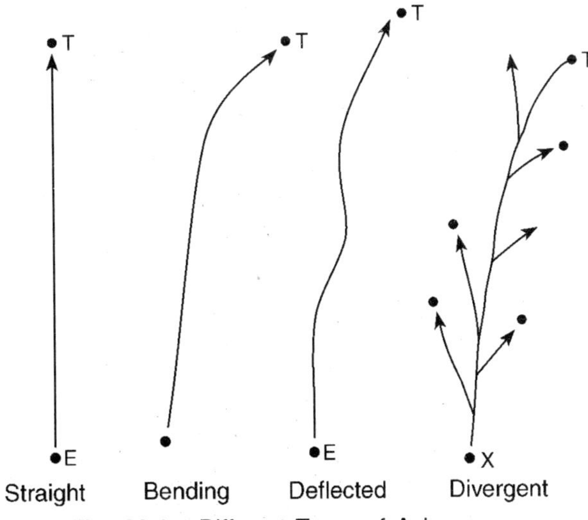

Fig. 30.1*a*. Different Types of Axis

picturesque effect and appeals to visitors. On the contrary, the absence of harmony or lack of unity is ugliness.

Some Important Terms used in Landscape Gardening

Axis. It is an imaginary line which divides garden into two parts. This also connects two or more points. It is presented in the form of a path, line of fountains or trees, etc. If this axis divides garden into two equal parts, it is called central axis. In formal style, axis is central whereas in informal style, it is oblique. This axis controls the movement in the garden from the entrance to the terminal. An axis in garden is directional, orderly or dominating. Axis shows certain characteristics in the garden, for example :

1. An axis maybe straight, bent or deflected but never divergent [Fig. 30.1(*a*)].
2. A powerful axis requires a powerful terminus *e.g.* Mogul gardens, (Fig. 30.1*b*).

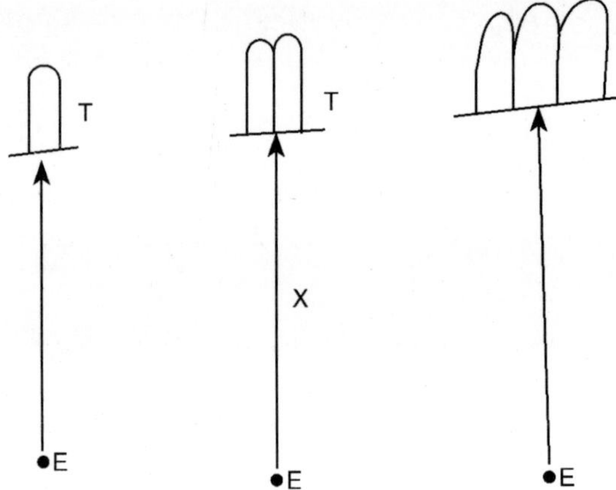

Fig. 30.1*b*. Requirement of Axis

When the axis terminates in a structure having entrance, one or three openings are better than two since they provide a receptive element rather than an obstruction.

3. The axis is a unifying element. (Fig. 30.1*c*).

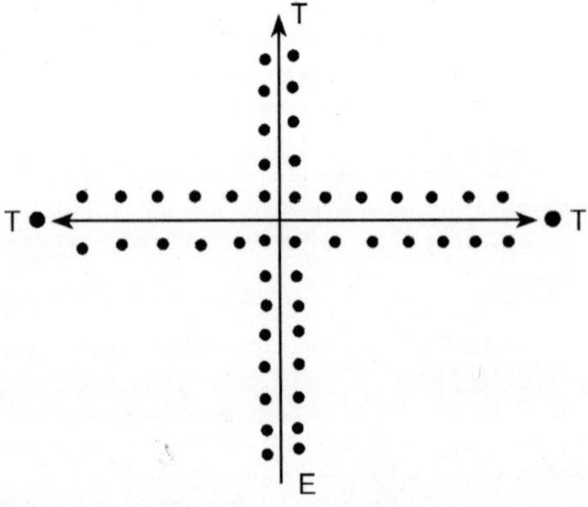

Fig. 30.1*c*. Axis of Unifying. E—Entrance, T—Terminus.

The symmetrical plan. In this plan, the different objects are in equilibrium about a central point or on either side of an axis. Symmetrical plan or formal plan is synonymous of beauty and is pleasant and handsome. This is because the symmetry is to be associated with plan clarity, rhythm, balance, unity, etc. Symmetrical plan being precise and disciplined, it requires precision in detail and maintenance and bold in concept.

Dynamic Symmetry. In such symmetrical plan, each pole generates its own magnetic field and between these two fields there is a field of dynamic tension. Such dynamic symmetrical plan is found in Taj garden.

Asymmetrical Plan. In such plans, there is absence of symmetry on both the sides of axis but balance, unity and harmony are maintained.

Circulation in landscaping. In landscape gardening it means a pathway from entrance to terminal point. Circulation varies with the style of gardening and topography. The more of circulation patterns, it has more points of views and attraction. Common circulation patterns are depicted in Fig. 30.2.

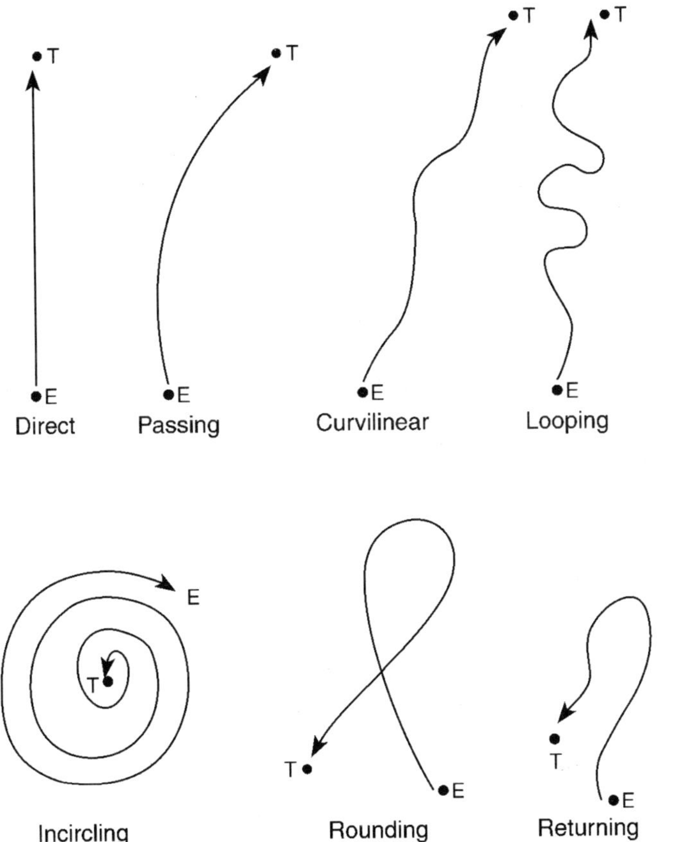

Fig. 30.2. Common Types of Circulation. E—Entrance, T—Terminus.

Vista. It is a three dimensional confined view of terminal building or dominant element of feature. It may be natural or man-made. Natural vistas are very common around the lofty mountains and snowy peaks. The best example of man-made vista is a view of main building of Taj Mahal from the entrance gate. Overall effect of vista may be of its characteristics. It may be calm or induce motion. There are three different parts of a vista *viz.* (1) it should be subject to close control, (2) it should have a viewing station to see object or objects, (3) it should have intermediate ground. These three should be satisfactorily united and thus result into an effect of totality.

31

Environmentals Horticulture

Urbanization, an unavoidable result of human social development, is a global phenomenon with an unprecedented pace. United Nation's Populations division estimates that between 2011 and 2050, the world population is expected to increase by 2.3 billion, passing from 7.0 billion to 9.3 billion. Majority of this growth is expected in urban areas where the projected growth is likely to be 2.6 billion, passing from 3.6 billion in 2011 to 6.2 billion 2050. Furthermore, the urban growth will be concentrated in the cities and towns of less developed regions particularly in Asia and Africa. Population growth is therefore becoming largely an urban phenomenon concentrated in the developing world.

Urbanization has brought numerous benefits to the society, such as concentrated populations and labor forces facilitating manufacturing of goods and providing services, extensive transport facilities expediting trade and economic development, cutting-edge communication and infomation technology, boosting education and scientific initiatives, health care and public facilities, job opportunities, and cultural diversity-all resulting in improved living standards. However, urbanization also has adverse bearing on the living environment, including encroachment on arable land and natural habitats, increased impervious surfaces, massive consumption of energy and natural resources resulting enormous production of waste and pollution, reduced biodiversity and estrangement of human beings from nature. The solution lies in sustainable development of cities with ecological approach.

Urban green spaces play a vital role in developing ecological cities, by improving the living environment of the urban area and infusing nature into the man-made concrete jungles. Urban green spaces offer numerous benefits which are listed below :

Environmental benefits

Temperature and energy use. Global surface temperatures have increased by about 0.3-0.6°C since late 19th century and Indian scenario is not different, as evident from Fig. 31.1, which shows general rising trend in temperature. Urban areas in particular behave as heat islands because of concrete structures, decreased wind, increased high-density surfaces and heat generated from human associated activities.

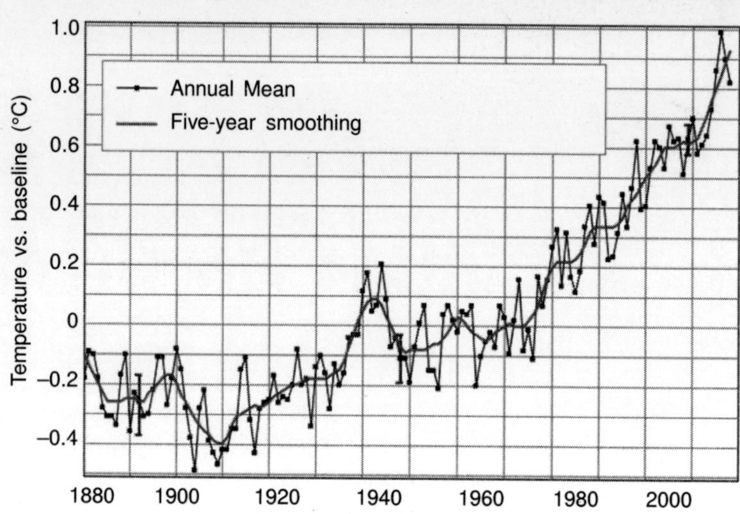

Fig. 31.1. Global Temperature Anomalies
(Source : NCDC, USA)

Plants can be successfully used to mitigate heat. Their leaves intercept, reflect, absorb and transmit solar radiations. Trees are called nature's air conditioners because when properly placed, they lower air temperature, shade buildings in the summer and block winter winds. Moisture from green spaces is emitted into the air, cooling it and helping to increase the comfort level in living and working spaces. Air under a tree's canopy can be as much as 5-10°C cooler compared to full sun. Daytime temperatures have been found to be about 2-3°C lower in a large urban park than the surrounding built up areas and this effect increases with increased vegetation cover. Cooling effect of green spaces is not restricted to their boundaries but can extend up to 300 m from the site.

Energy use and climate change. Trees can reduce building energy use and cooling costs. A single large tree can produce the cooling effect of 10 room size air conditioners operating 24 hours/day. Dew and frost are less common under trees because less radiant energy is released from the soil in those areas at night. Strategic placement of deciduous and evergreen species can reduce electricity bills of houses up to 20%. Networks of green space can provide transport routes for those in a local environment, reducing the need to travel by car and the level of pollutants emitted into the atmosphere. Parks and green spaces act as carbon sinks, removing carbon dioxide from the atmosphere and storing it away. Shaded parking lots and streets enhance fuel efficiency of vehicles by reducing air conditioning load. Reduced electricity consumption translates to reduced CO_2 emissions from power plants due to reduced demand. Green cover mitigates the effect of global warming on cities by reduction in surface temperature due to shading and evapotranspiration.

Air quality. Sulphur Dioxide, Oxides of Nitrogen, Suspended Particulate Matter, Lead, Ammonia and Carbon Monoxide are the major air pollutants, which cause respiratory and other health problems. Average PM_{10} level in most of the Indian cities is far higher than the WHO standard of 20 mg/m^3 rising to hazardous levels (more than 200 mg/m^3) in highly polluted cities. Plants help in controlling air pollution by acting as biological and physical nets. They cleanse the air by intercepting and slowing dust and other particulate materials causing them to fall out, and by absorbing pollutant gases. Plants increase humidity and thus help in settling of particulates. Increased wind turbulence by plants disperses pollutants. Fragrance of many flowers mask unpleasant odours thereby imparting freshness to air. Interiorscape plant can remove organic pollutants from indoor air. An efficient plant air purifier is one, which reduces the concentration of a given pollutant from harmful level to non-injurious level without itself suffering any ill effect. A list of pollution tolerant trees is given in table 31.1

Storm water management. Majority of the city surface is impermeable which necessitates the provision of artificial drainage systems to remove surface water. Heavy rainfalls put lot of pressure on storm water management system causing flooding particularly in low lying areas of city which has serious public health repercussions. The green spaces make highly permeable surfaces capable of absorbing substantial rainfall. By soaking up water and recharging natural groundwater supplies, green space reduces the volume and rate of run-off and immensely contributes to sustainable urban drainage systems. The rate of run-off for surfaces with plants and grass is estimated to be 10-20%, compared to 60-70% for hard landscaped urban areas. Plants intercept precipitation and slow it's decent to soil surface reducing runoff speed and volume as well as soil erosion. Pubescence on leaves help in water entrapment. Plants with horizontal branching and rough bark are most effective.

Noise abatement. Noise pollution from traffic, loudspeakers, construction and public work and industries affects human system. A noise level above 80 decibels creates tension and increase in blood pressure. Continuous exposure to noise levels of 115-120 db can result in permanent deafness. Along with the engineering methods, plants can be successfully used for noise reduction. Vegetation barriers partially deaden the sound from traffic, factories and loud neighbours. To be effective, the screen should be dense, tall and wide, planted close to the source of noise. Urban residential areas can be protected by a row of dense shrubs backed by row of tall trees, total 20 feet wide. Vegetation along the highways can screen vehicular noise from reaching the adjoining habitations. Plants also create "background" noise of rustling of leaves and wind through the branches that can help muffle other noises. Plants having thick and fleshy leaves with petioles, which allow the higher degree of flexibility and vibration are best suited for use as noise screens. Some of the trees suitable for reduction of noise pollution are given in table 31.1.

Carbon sequestration. Increase in CO_2 account for about 65% of the current direct positive radiative forcing due to greenhouse gases produced as a result of human

TABLE 31.1
Trees suitable for planting against various types of pollution

Air pollution	***Dust pollution***	***Noise pollution***
Tropics	**Tropics**	**Tropics**
Acacia auriculiformis	*Albizia lebbeck*	*Alstonia scholaris*
Albizia lebbeck	*Alstonia scholaris*	*Azadirachta indica*
Ailanthus excelsa	*Bombax ceiba*	*Butea monosperma*
Alstonia macrophylla	*Cassia fistula*	*Erythrina variegata*
A. scholaris	*Cordia dichotoma*	*Eucalyptus citriodora*
Azadirachta indica	*Dalbergia sissoo*	*Grevillea robusta*
Butea frondosa	*Erythrina variegata*	*Kigelia pinnata*
Cassia fistula	*Eucalyptus citriodora*	*Lagerstroemia speciosa*
C. marginata	*Eugenia cuspidata*	*Madhuca indica*
Lagerstroemia flosreginae	*Ficus benjamina*	*Mangifera indica*
L. thorelli	*Grevillea robusta*	*Morus alba*
Madhuca indica	*Heterophragma*	*Pterspermum acerifolium*
Mimusops elengi	*adenophyllum*	*Terminalia arjuna*
Parkinsonia aculeata	*Jacaranda acutifolia*	**Temperate areas**
Polyathia longifolia	*Kigelia pinnata*	*Alnus viridis*
Putranjiva roxburghii	*Lagerstroemia speciosa*	*A. crispa*
Terminalia arjuna	*Madhuca indica*	*Picia* spp.
T. mueleri	*Melia azadarach*	*Betula nana*
Thespesia populnea	*Millettia ovalifolia*	*Braya purpurascens*
Temperate areas	*Millingtonia hortensis*	*Dryas integrifolia*
Acer platanoides	*Morus alba*	*Salix planifolia*
A. saccharinum	*Pongamia pinnata*	
A. negundo	*Tamarindus indica*	
Betula pendula	*Terminalia arjuna*	
Forsythia suspense	*Toona ciliata*	
Ligustrum vulgare	**Temperate areas**	
Plantanus orientalis	*Acer negundo*	
Pyracantha coccinea	*Alnus incana*	
Quercus palustris	*Betula pendula*	
Q. rubra	*Carpinus betulus*	
Robinia pseudocacia	*Crataegus prunifolia*	
Ulnus spp.	*Lonicerea macckii*	
	Juniperus chinensis	
	Populus ferolinensis	
	Syrniga vulgaris	
	Viburnum lantana	

activities. The atmospheric CO_2 concentration has increased 30% since pre-industrial times, as a result of increasing emissions from fossil fuel combustion, land conversions, cement production and continues to increase by 0.4% per year. Indian is third largest CO_2 emitting country after China and USA and contributed 6% to global emissions in 2013. Plants remove CO_2 from the atmosphere during photosynthesis to form carbohydrates and return oxygen back as a byproduct. Plants therefore act as carbon sink and remain one of the cheapest, most effective means of drawing excess CO_2 from the atmosphere. One hectare of trees absorbs about 3.7 t of CO_2 from the air generating 2.5 t of life sustaining oxygen. (Table 31.2)

TABLE 31.2
CO_2 emissions from India

Year	*Million Metric Ton*	*Metric Tons per capita*
1960	120.58	0.27
1965	165.97	0.33
1970	195.14	0.35
1975	252.20	0.41
1980	348.58	0.50
1985	490.47	0.63
1990	690.58	0.79
1995	920.05	0.95
2000	1186.66	1.13
2005	1411.13	1.24
2010	2008.82	1.70

(**Source** : World Bank)

Water quality/erosion. Industrial effluents, sewage and domestic waste, excess nutrients and chemicals from agricultural systems are the major pollutants of the water bodies and underground water. Over exploitation of underground water for irrigation and domestic use has lead to lowering of water table in many states like Punjab and Haryana Plants prevent harmful pollutants contained in the soil from getting into water ways, reduce topsoil erosion, slow down water run-off and remove nutrients and sediments while increase ground water recharge. Reducing the flow of storm water reduces the amount of pollution that is washed into a drainage area. Vegetation along rivers, lakes and other water bodies decrease their siltation and reduces cutting of banks during floods.

Wind protection and air movement. Trees perpendicular to wind direction may reduce the wind speed up to distance of 2-5 times the height of the tallest tree on windward side and 30-40 times on the leeward side. Trees can be planted to funnel or baffle wind away from areas—both vertical and horizontal concentration of foliage can modify air movement patterns. Plants can also be used effectively to control snowdrift.

Glare and reflection control. Dense foliage absorbs about 70% of the sun's rays, reflect 17% and transmit about 13%. Trees and shrubs help control light scattering, light intensity and modifies predominant wavelength on a site. Trees block and reflect sunlight and artificial lights to minimize eyestrain.

Ecological benefits

Urban green spaces enhance biodiversity along with beautification of environment. Biological diversity has direct consumptive value in food, agriculture, medicine, and industry. It also has aesthetic and recreational value. Biodiversity maintains ecological balance and continues evolutionary processes. The indirect ecosystem services provided through biodiversity are photosynthesis, pollination, transpiration, chemical cycling, nutrient cycling, soil maintenance, climate regulation, air, water system management, waste treatment and pest control. Within urban areas the most valuable habitats are the woodlands, urban forests, ponds, rivers and riverbanks, parks and gardens and cemeteries, providing a variety of habitats for mammals, birds, insect, aquatic and plant life. Maintaining this diversity of green space is essential to meeting biodiversity conservation objectives as each space will offer different habitats for species. Since most of the urban green spaces are managed, introducing rare species is likely to be more successful and cities can be used as large conservatories particularly for plant species.

Economic benefits

Besides savings in the form of reduced energy consumption, plants provide timber, pulpwood, fruits, nuts, mulch, composting material, and firewood. Shading by trees extends life of paved surfaces and reduce maintenance cost. Greenery enhances community economic stability by attracting business and tourists, people linger and shop longer along tree lined streets and apartments/offices in wooded areas rent more quickly and have higher occupancy rates. Property values are higher in well-planned areas lined with trees. Urban green spaces also provide huge tangible value activities like team sports, bicycling, walking, picnicking and visiting a theme garden. All these activities in city parks are free of charge, but their value can be calculated by knowing the cost of a similar recreation experience in the private resorts or amusement parks *i.e.* willingness to pay.

Health Benefits

It has been scientifically proved that people's bodies respond physiologically to viewing and interacting with nature. One study showed that hospital patients with a view of trees from their windows had a more positive recovery course than those without. They had shorter hospital stays by 10%. Research suggests that views of natural environmental helped in quicker and more complete recovery from stress. More often a person visits urban open green spaces; the less often he or she will report stress-related illnesses. Even short duration (5 min) of activity in green space has a strong positive effect on mood and self-esteem, irrespective of activity intensity. Experience in natural environments certainly has soothing effect on mood; even nature sounds facilitate recovery from psychological stress. Green exercise as a health-promoting initiative for people experiencing mental ill health is equally as effective as existing programmes. Older individuals with a park within walking distance keep in better health than those without a park nearby. For people whose residential environment contains high levels of greenery, the likelihood of being more physically active is more than three times as high and the likelihood of being overweight and obese is about 40% less. Employees who could look out their office windows and see trees and nature remain happier at work.

Social benefits

The social ties between the residents of urban neighbourhoods are in decline due to changing work patterns, increased mobility and developments in communication. Neighbourhood open spaces and green spaces in particular, provide opportunities for social interactions that may help the residents to establish recognition and develop relationships. Public promenades, parks and squares, accessible to all classes, are serving as places to overcoming social differences. These are places to chat, relax and to enjoy the beauty of nature. However, for parks to realise their full potential in supporting social interactions and developing social ties, they need to be well-maintained and provide good recreational facilities. Children were found to lead more effective, self-disciplined lives in vicinity of green spaces.

In residential areas, barren, treeless spaces often become "no man's lands", which discourage resident interaction and invite crime. Green surroundings of buildings with properly placed and maintained trees have been shown to reduce crime.

Trees and other life forms supported by them have a positive value in scenic quality. Trees change skyline, add variety of form, colour and texture to otherwise dull and monotonous cities. They soften the harshness of concrete structures, complement architectural lines, enframe views and provide background settings to buildings. Dynamic behaviour of vegetation makes the surroundings interesting. Plants can be very effectively used to screen undesirable and disturbing sight lines.

People's participation

In a vast country like ours it is practically impossible for any government to undertake urban greening of its own. In a situation where potable water and electricity is not available so far to all the people, it is futile to think of beautification by public funding alone. People's participation is therefore very essential for realising the dream of beautiful, colourful cities.

People's participation in any program is driven by the benefits they get from it. Urban people will involve in bio-aesthetic planning primarily for their recreational needs and for integrating with nature. Increasing awareness about pollution has also forced the people to think positively about greening of cities. They should be further educated regarding other benefits like climatic amelioration, water quality improvement, decrease in energy consumption.

Municipal bodies should involve residents in planning of parks and open spaces so that their needs and preferences could be taken into account at initial stage. Development work should be carried out by government agencies and maintenance should be handed over to resident's associations of the area. This will inculcate a sense of responsibility in them and working with nature will make them understand natural processes in a better way. Many urban development agencies like Municipal Corporations, Improvement Trusts and Development Authorities of various cities have already adopted this approach. All the minicipal corporations in Punjab, many improvement trusts and almost all of the Development Authorities provide financial assistance to residents associations/park management committees to undertake routine maintenance on the basis of area of the park. Government of Delhi has established Delhi parks and Garden Society which provides financial assistance to Residents welfare associations to the tune of 30%-90% depending on category of the colony along with one time assistance for barded wire fencing and installation of tubewell. Municipal Corporation of Hyderabad also contributes 70% to the maintenance cost of parks managed by residents. Many cities like Mumbai and Kanpur encourage residents association, NGO's and other organizations to adopt parks for maintenance. Many Government agencies encourage people to develop their gardens beautifully, by organising garden competitions and flower shows. Kitchen garden societies and Agri-horticulture societies in some cities are providing technical know-how, plant material, and seeds on reasonable rates and services like lawn mower and sprayers to their members. Such societies should be encouraged in all parts of the country.

Educating the people is very essential to ensure their participation and the best method is to educate the children. Excursion tours to nearby parks and botanical gardens with few tips on environment could be a great fun for them. Organizing theme based science fairs and quiz competition can infuse love and respect for nature in them. Ministry of Environment and forests has introduced Urban Environmental Information System in Agra, Kanpur, Patna and Bhubaneshwar to provide information to the public in the form of local environmental report to be prepared by the municipalities on the status of socio-economic development and environment of the cities. Such efforts should be extended to whole of the country so that people could be made aware of the current state of their surroundings. NGO's, Agri-horticulture societies, Eco-clubs etc. should be involved in such campaigns for effective dispersal of information.

***

32

Bio-aesthetic Planning

On earth planet, during evolution, simpler forms of life transformed into complex systems and a variety of flora and fauna developed. To sustain life on earth, natural ecosystems developed in such a way that every form of life was dependent and complementary to each other. Dominance of flowering plants started during Cenozoic age (60-70 million years ago) and at the same time, dominance of mammals and insects started. Man appeared on the planet about 1,50,000 years ago and was a hunter-gatherer in the beginning. Hunting wild animals and gathering fruits and nuts were tiring jobs and man might have felt the necessity of cool, shady places during summer and warm and sunny situations during winter for taking rest.

Such places must have been situated near water bodies where hunting was also convenient, thus began the process of site selection and the experience gathered was later used in locating places for early settlements. Moving from stone to metal tools, from caves to primitive houses and from hunting-gathering to cultivation of plants was a long journey and well-developed civilizations occurred about 7,000-8,000 years ago. Various civilizations developed simultaneously in the regions of Mesopotamia, Indus valley, Egypt and China. With the advancement in agriculture and increase in population, large settlements developed, marking the beginning of urbanization. Man was no longer at total mercy of nature and instead started exerting control over it.

Simpler tools evolved into complex machines and thus began the era industrialization and mechanization of most of the human activities. Increased agricultural output coupled with advances in health care led to unprecedented increase in human population. More and more people moved to urban areas for comfortable life. Human control over nature further increased and started interfering with natural processes. Increasing human needs resulted in large-scale exploitation of natural resources, great reduction in forest cover and extinction of many species of flora and fauna. Increase in environmental degradation has aroused great concern amongst scientists and planners. Strategies for improving living conditions are being worked out and efforts are being made in this direction.

Although alienated from nature for gathering material comforts of life, inner human desire for reunion with nature is reflected through his love for natural objects like mountains, water bodies, rocks, colorful plants, animals, birds, fishes etc. Man is perhaps the only creature who has aesthetic sense and makes efforts to create orderliness and beauty. Paleolithic cave paintings of France and Spain, Greek sculptures of Apollo and Venus, Frescoes of Ajanta, Zen Buddhist gardens of Kyoto, Taj Mahal, Opera house of Sydney, all are the results of human quest for beauty. Natural forms, colors and textures have inspired most of the human art. Man's liking and understanding of colors could certainly be attributed to flowers and colorful feathers of birds. Fulfilling recreational needs through ornamental plants and wild animals can be traced back to earliest civilizations and this gave the concept of Bio-aesthetic planning.

BIO-AESTHETIC PLANNING

Professor Lancelot Hogben coined the term 'Bio-aesthetic planning' defined as planning of flora and fauna with the object of beautifying the country. With the development of wildlife conservation as a specialized subject, aesthetic aspect dominated the concept of Bio-aesthetic planning although increase in plant biodiversity also creates suitable habitat for many species of birds and small animals. Bio-aesthetic planning was thus recognised as the aesthetic branch of horticulture, which deals with planting of ornamental plants to create a picturesque effect. With the greater realization of environmental, ecological and socio-economic benefits of plants, scope of Bio-aesthetic planning has increased further. Potential of plants in combating pollution though realised long ago has not been utilised effectively so far. Bio-aesthetic planning can play important role in environmental amelioration of urban and industrial areas along with their beautification.

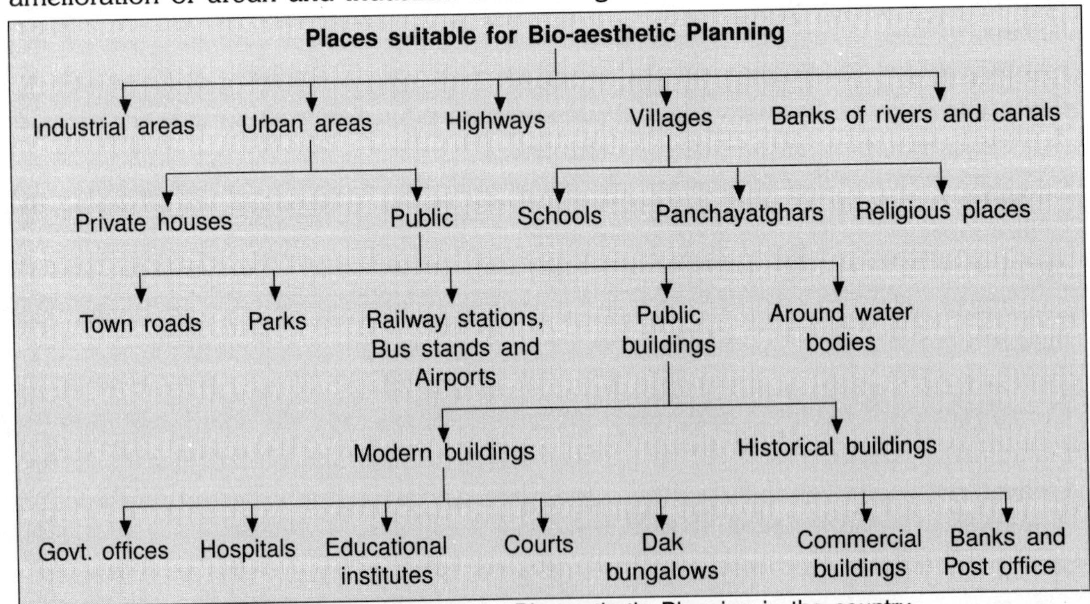

Fig. 32.1. Places suitable for Bio-aesthetic Planning in the country. (Modified from Randhawa, 1978)

Bio-aesthetic planning should be used for improving environment of all the areas deteriorated by developmental activities. The entire country should be brought under the purview of Bio-aesthetic planning. Priority should be given to public places, which belong to the community as a whole. A large number of persons, especially those who cannot afford gardens of their own, will thus be able to enjoy the splendour of nature. Different places suitable for Bio-aesthetic planning in the country are shown in Fig. 32.1. Public parks and squares, public roads, platforms of railway stations, bus stands, airports, compounds of courts, hospitals, universities, colleges and schools, ancient historical buildings, government office buildings, dak bungalows are the places in towns which are suitable for bio-aesthetic planning. Industrial areas, various categories of roads, banks of rivers and canals are neglected but potential sites for bio-aesthetic planning, which require urgent attention. Beautification of national parks, sanctuaries and biosphere reserves and wetlands will have immense ecological and environmental benefits apart from aesthetic value.

HISTORICAL PERSPECTIVE

All religions of the world teach humans to respect all other forms of life and protect them. Tree worship was possibly the earliest and the most prevalent form of religion. It was through the worship of trees that man attempted to approach and please god. Before man developed agriculture, he lived mostly on the edible roots, fruits and nuts of trees and the flesh of wild animals. Trees provided him shelter against inclement weather and also fruits and nuts as food. He used their wood for making implements and many other purposes. It was from wood alone that he obtained fire, which enabled him to cook his food and warm his cave dwellings. Apart from these, it is the beauty of their flowers, which excited his imagination. The narratives in the Bible centres round the planting of trees of life and knowledge in the Garden of Eden. Infact, every religion has descriptions of tree of life or world tree supporting all life forms and processes.

There are documented records of gardens as far as the ancient Babylonians, Sumerians and Egyptians over 7000 years ago. Egyptians were among the earliest civilizations to cultivate plants for their aesthetic values. Romans borrowed the art from Egyptian knowledge and refined the techniques. Temple gardens of India, large formal gardens of Italy, France and Natural English gardens of 18th century are landmarks in the history of beautification of environment. Major contribution of Englishmen was in the form of collection of beautiful plants and their introduction in different parts of the world. Concurrent with evolution of gardens in Europe but separated both geographically and attitudinally, were the gardens of Asia—first the Chinese, then Koreans and later Japanese. Oriental Asians perceived themselves as being part of the natural world, not separate and dominant over it. Zen Buddhist gardens of Kyoto were pinnacle of oriental aesthetic sense.

Trees of all species of the genus *Ficus* are protected in many parts of the world considering them as sacred. Local people seem to be often aware of their importance as affording food and shelter for a wide range of birds, bats and primates, and it is not difficult to imagine that such understanding was converted into widespread protection of *Ficus* trees at some point in the distant past.

In India tree worship was common in the third or fourth millennium B.C. Among the seals of Mohenjo-daro dating from that period is one depicting a stylised pipal tree with two heads of unicorns emerging from its stem. A seal from Harappa is engraved with the likeness of a weeping willow tree with hanging branches. It seems tree worship was widely prevalent in India before the rise of Buddhism. Mango was their popular fruit tree and shady trees like pipal, banyan, gular and pakur provided welcome shade against the scorching heat of the sun in the summer months.

Sal, Asoka, Palaksa and Pipal trees are associated with Buddha and are thus regarded sacred by Buddhists. Chinese traveller Hiuen Tsiang has described the garden of Lumbini in his travelogue, where Buddha was born under an Asoka tree. On conversion to Buddhism, Asoka (273-326 B.C.) abjured violence and became a messenger of good will and peace. For the first time in history a monarch encouraged arboriculture and adopted it as a state policy. He encouraged planting of trees in gardens and along roads in the form of avenues.

Trees like pipal, banyan, asoka, siris, sal mango, gular etc. represented in the sculptures of Bharhut, Sanchi, Mathura and many other places signify their importance in the lives of ancient Indians.

Sher Shah during his brief reign developed many roads, Grand Trunk Road being the most prominent one and promoted planting of shady trees along them. Planned gardens staged a comeback during moghul rule in India who were ardent lovers of plants and Babar, Jahangir, Shah Jahan and Faide Khan developed beautiful formal gardens. However, development by the moghuls was mostly limited to gardens around tombs, palaces and in forts, meant for their personal use. Except the avenues of Chinar along Jehlum River, they did not contribute much to general landscape of the country.

With the advent of British Empire came the era of large-scale introduction of plants from different parts of the world and floral treasure of India was greatly enriched. Many botanical gardens were established during that period, which proved to be important centres for collection, maintenance and evaluation of flora and education in botany.

In recent era, Dr. M.S. Randhawa was the flag bearer of Bio-aesthetic planning and strived hard for the cause of environmental amelioration by planting of flowering trees. He promoted the idea that rich colourfulness of nature in the form of ornamental trees should be conspicuously noticeable in the form of avenues along streets and roads in towns and even in-home gardens. He made great efforts in this direction and made arrangements for planting beautiful trees in the compounds of district courts, tehsil, schools and platforms of railway stations. He advocated garden suburbs around commercial city centres and created garden city of Chandigarh in collaboration with Le Corbusier of

France, which has inspired future generations of town planners and architects. Bio-aesthetic planning was included as subject under Ministry of Agriculture and Cooperation in List II of the Seventh Schedule to the Constitution of India, however, no concrete policy has come up in this regard, till date.

With a view of trees from their windows had a more positive recovery course than those without. They had shorter hospital stays by 10%. Less violence occurs in urban public housing where there are plants, due to nature-linked decrease in stress and irritability. Parks and open spaces provide opportunities for active and passive recreation to the people, enhancing interaction among them and building a sense of community.

Trees and other life forms supported by them have a positive value in scenic quality. Trees change skyline, add variety of form, colour and texture to otherwise cities will be dull and monotonous. They soften the harshness of concrete structures, complement architectural lines, enframe views and provide background settings to buildings. Dynamic behaviour of vegetation makes the surroundings interesting. Plants can be very effectively used to screen undesirable and disturbing sight lines.

INTEGRATING AESTHETICS AND NATURE CONSERVATION

India is one of the twelve mega-biodiversity countries of the world with 46,000 plant species and 81,000 animal species described so far from 70% of the total geographical area surveyed so far. To conserve flora and fauna from extinction many botanical gardens, national parks and wildlife sanctuaries have been created.

Wildlife Conservation

Approximately 1.56 lakh sq. km of area is protected under 89 national parks and 500 sanctuaries. Beauty of these protected areas lies primarily in their wilderness and that is what attracts nature lovers towards them. Wildlife has a significant positive impact on scenic quality evaluation. Mere expectation of seeing wildlife significantly increases landscape quality assessment. The wilderness should be given aesthetic treatment with proper planning, which can become an added attraction. The aesthetic and recreational approach places the main emphasis upon preserving the characteristic beauty of the landscape and upon providing ample access to and facilities for open-air recreation and for the enjoyment of beauty in those areas. The major features of the park are made easily accessible by providing roads, tracks and bridges and living accommodation in the form of hostels, tourist bungalows, hotels etc. Such utility features can be beautified with ornamental plants bearing attractive flowers and fruits, which will also attract many birds and small animals. Tourist infrastructure should be developed on the lines of African safaris to attract tourists.

Most of these areas are parts of the natural landscapes of that region declared 'protected' by law. Some of these were hunting reserves of former princely states. Although it is impossible to imitate nature but based on understanding of the natural processes one can develop a flexible system that accommodates the natural cycles, rhythms and unpredictability of time as a part of the aesthetics of the place. The landscape of Bharatpur (Keoladeo Ghana) Bird Sanctuary is a wonderful example of such process and aesthetics created through generation of a new wetland for recreational purpose. The once scrub land in the brackish area was developed by the Maharaja of Bharatpur based on the regions bowl shaped topography, potential to collect rainwater and attract wildlife. Manipulation of water has been done for nestling of resident birds as well as providing feeding grounds for the migratory birds. Emergent wetland species along with aquatic gardens are designed in some parts of the sanctuary.

There will be many such sites in our country, which if identified and developed properly can become great ecological assets—generating revenue through eco-tourism in addition to preserving and enhancing biodiversity. Care should be taken to use only the native species to avoid what is called species invasion like in the case of *Lantana*. Some of the important national parks and sanctuaries are :

- Kaziranga National Park (Assam)
- Manas Wildlife Sanctuary (Assam)
- Gir National Park (Gujrat)
- Great Himalayan National Park (H.P.)
- Dachgam National Park (J&K)
- Bandipur National Park (Karnataka)
- Peryar National Park (Kerela)
- Kanha National Park (M.P.)
- Keoladeo National Park (Rajasthan)
- Ranthambor National Park (Rajasthan)
- Corbet National Park (U.P.)
- Dudhwa National Park (U.P.)

Botanical Gardens

Many botanical gardens have been created for conservation of plant wealth of the country. These gardens serve as centres of collection, maintenance and evaluation of germplasm of important species apart from education in botany. Except the ones of national importance and those established by the British, majority have been planned for scientific purpose only and lack beauty. Botanical gardens and arboreta can be made attractive to general public if aesthetic aspect is integrated with scientific requirement. Proper laid out paths, some benches and a fountain can create the desired effect. Some of the prominent botanical gardens are :

- Lalbagh, Bangalore.
- Royal Botanic Garden, Kolkata.

- Royal Agri-Horticultural Society Garden, Kolkata
- Government Botanic Garden, Ootacamund.
- Lloyd Botanic Garden, Darjeeling.
- The National Botanic Garden, Lucknow.
- The Botanical Garden of Forest Research Institute, Dehradun

Landscaping of Roads

Roads running across all parts of the country are arteries of its economy. India has made tremendous development in this aspect of infrastructure with a total road length of 25.3 lakh kms of which 14.5 lakh kms are surfaced roads. The network will be further strengthened by projects like North-South and East-West highway corridors, Golden Quadrilateral and Pradhan Mantri Gram Sarak Yojna. It is apparent that as on date, the planning and designing of roads is normally done, almost purely, on the basis of economic and traffic flow considerations. However, the environmental impact of the road construction activities and the implications of air and noise pollution among others caused by the vehicular movement on the roads and the need to evolve an efficient approach to deal with such effect has received little attention. Bio-aesthetic planning can be a useful tool in mitigating the environmental impact of road construction and operation activities to a great extent. Problems like soil erosion, siltation of water bodies, loss of vegetation and wildlife, aesthetic degradation; air, water and noise pollution can be dealt with to a great extent by careful planning and maintenance of surroundings.

Plants should be selected with due regard to soil, rainfall, temperature and water level. The tree species should be either truly evergreen, nearly evergreen or be in leaves at least during height of summer. They must not become too wide to interfere with traffic flow and should be capable of developing clear bole up to a height of 2.5-3.5 m from ground level. They should be fast growing and wind firm and should not be thorny or drop too many leave. Trees should be deep rooted. Traditional roadside trees like Jamun and seedling mangoes should be preferred which have long life and are well adapted. A general list is presented in tables 31.1 to 31.3 and suitable tree can be selected according to requirement of site and agro climate.

Highways

Bio-aesthetic planning for highways should begin at design stage so that it could be a part of the master plan. Design objectives of highways should be to fit the road into the existing terrain and harmonise it with general landscape of the region. Existing features along the road alignment, like natural wooded areas, streams, ponds, rock outcrops, scenic vistas, historical buildings, monuments, gardens etc. can contribute significantly to general landscape and every effort should be made to preserve, incorporate and integrate these into the landscape. In the construction stage removal of vegetation and cutting of trees should be done judiciously and limited to minimum possible extent so that least habitat loss occurs. Plantation of ornamental and economically important species should be carried out to a sufficient distance on either side of the road. Functional planting applies to such problems as protection of slopes against erosion, screening of unsightly views, reducing headlight glare, providing shade in summer. Trees serve as biological reference points along border of the road and increase driver's perception of orientation. Arranged in varied forms and colours, they can reduce driver's fatigue and boredom. Besides pollution abatement, planting blends the road into the surrounding countryside and enhances the overall beauty. Vegetation also provides habitat for many species of birds, rodents etc.

TABLE 32.1
Flowering trees for Tropical and Subtropical areas

Scientific name	*Common name*	*Flower colour*	*Flowering time*
Acacia auriculiformis	Australian wattle	Yellow	Sep.-Nov.
Adansonia digitata	Monkey Bread tree	White	July
Albizzia lebbek	Siris	Pale yellow	Apr.-May
Albizzia procera	White siris	Yellow white	July-Aug.
Bauhinia purpurea	Kachnar	Purple	November
Bauhinia variegata	Kachnar	White	Feb.-March
Bombax malabaricum	Simbal	Red	Feb.-March
Butea monosperma	Dhak	Scarlet orange	Apr.-May
Callistemon lanceolatus	Bottle brush	Bright red	Mar. and Oct.
Cassia fistula	Amaltas	Deep yellow	Apr.-May
Cassia javanica	Java-ki-rani	Deep pink to rose pink	May-June
Cassia nodosa	Pink cassia	Pinkish white	May-June
Chorisia speciosa	Mexican silk cotton	Pink	Oct.
Cochlospermum gossypium	Yellow silk cotton tree	Golden yellow	Feb.-Mar.
Colvillea racemosa	Colville's glory	Orange	July-Oct.
Cordia sebestena	Lal lasora	Red	Jan.-Mar.
Crataeva religiosa	Barna	Pale yellow	Feb.-Mar.
Delonix regia	Gul mohur	Scarlet	Apr.-June
Dilenia indica	Chalta	White	July
Erythrina indica	Indian coral tree	Scarlet	Feb. Mar.
Gravillea robusta	Silver oak	Golden yellow	Mar.-May
Jacaranda acutifolia	Nili gulmohur	Blue	Mar.-May
Kigelia pinnata	Jhar phanoos	Dark purplish	April-May

Koelreuteria paniculata		Yellow	Sep.-Oct.
Lagerstroemia speciosa	Queen's flower	Mauve purple	Apr.-May July-Aug.
Madhuca indica	Mahua	Creamy	Feb.-Apr.
Magnolia stellata	Champa	White	Apr.-May
Melia azadarach	Drek	Lilac	Mar.-Apr.
Mesua ferrea	Nagkesar	White	April-May
Miletia ovalifolia	Moulmein rosewood	Mauve	April
Millingtonia hortensis	Akash neem	Silvery white	Jan.-Mar.
Moringa oleifera	Drumstick tree	Creamy white	Feb.-Apr.
Nauclea cadamba	Kadam	Yellow	June-Aug.
Peltophorum inerme	Copper pod	Golden yellow	Mar. May Sep.-Nov.
Pithecolobium saman	Rain tree	Pink	Mar.-Sep.
Plumeria acutifolia	Pagoda	White with yellow center	Mar.-April July-Oct.
Plumeria alba	Pagoda	White	Mar.-April July-Oct.
Polyalthia longifolia	Ashoka	Pale green	Feb.-April
Pongamia pinnata		Lilac	April-May
Pterospermum acerifolium	Kanak champa	Yellow	Apr.-June
Saraca asoca	Sita asoka	Orange red	Feb.-Mar.
Spathodia campanulata	Fountain tree	Orange crimson	Feb.-Mar.
Tabebuia speciosa	Mauve tabebuia	Pale mauve	June
Tecoma argentia		Yellow	Mar.-May
Tecomella undulata	Lahura	Orange yellow	March
Terminalia arjuna	Arjuna	Yellowish white	Mar.-June
Thespesia populnea	Bhendi tree	Yellow and crimson red	Oct.-Mar.

Roadside planting of trees may be in the form of avenues, groups or groves. At least 5-6 rows of trees should be planted so as to provide sufficient green cover. Certain restrictions imposed by engineering, traffic and safety requirements like future widening, setback distance, interference with utilities etc. should however be kept in view while planting. Location of trees and vertical clearance available across the roadway should be checked to ensure that motorists get clear vision of highway signs or signals at all times. Also, the foliage of trees should not come in the way of road illumination. Road dividers should be planted with hardy, short height shrubs to cut off headlight glare.

TABLE 32.2
Shade trees for tropical and Sub-tropical areas

Scientific Name	**Common Name**
Alstonia scholaris	Devil's tree
Azadirachta indica	Neem
Cederella toona	Tun
Eugenia jambolina	Jamun
Ficus benghalensis	Bargad
F. benjamina	Golden fig
F. infectoria	Pilkan/Pakar
F. religiosa	Pipal
Mangifera indica	Mango
Mimusops elengi	Molsari
Morus sp.	Mulberry
Swietenia macrophylla	Mahagony

TABLE 32.3
Trees for temperate areas

Scientific name	**Common name**
Acer spp.	Maples
Albizia julibrissin	Silk tree
Prunus serrulata	Cherry
Malus arnoldiana	Arnold crabapple
Cornus florida	Dogwood
Crataegus spp.	Hawthorn
Magnolia stellata	Magnolia
Ulmus spp.	Ulmus
Quercus spp.	Oak
Pyrus spp.	Pear
Platanus spp.	Plantrees
Populus spp.	Poplar

Tree planting along highways can be done according to the following two schemes

1. Front row of flowering trees and rear rows of utility trees. Front row should be planted with flowering trees of different species, each for 3-5 kms and the subsequent rows should be planted with species of economic importance such as timber, fruit and fuel wood. (Fig. 32.2)

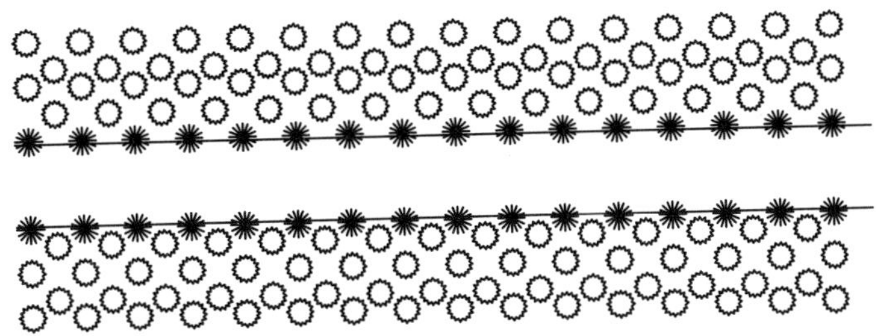

Fig. 32.2. Front row of flowering trees and rear rows of utility trees

2. Mass effect of flowering trees. Flowering trees should be planted as large groups for about 0.5 kms after every five kms of plantation of economically important species. Such groups should be arranged alternately on both sides of the road. This will break the monotony of singly planted species. (Fig. 32.3)

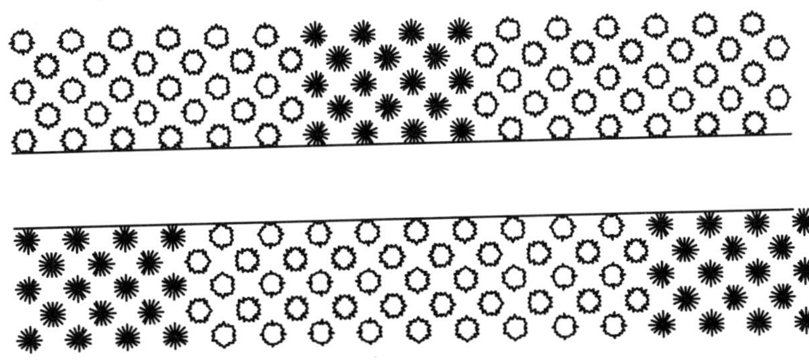

Fig. 32.3. Mass effect of flowering trees

Roadside amenities like rest areas, truck parks and off strip parking places should be created on all highways for safety, convenience and enjoyment of the travellers. Truck operators in particular have to undertake long journeys, which are often exhausting and boring. Roadside rest areas and truck parks should be planted with shady trees. Such areas should fit into the natural surroundings and become a part of general road landscape.

City Roads

Roads and streets are lifelines of the urban areas and city dwellers spend considerable part of their outdoor time on them. Without plants, these structures seem to pierce through the body of the city and create restlessness. To make them more enjoyable and safer, they should be lined with shady and flowering trees. An outer row of shady trees and inner row of flowering trees with cycle track/walkway between the two will be a good arrangement for wider roads. This will ensure pedestrian safety and reduction in number of accidents. Species of flowering trees should be sufficiently varied to avoid monotony. The central verge wherever available should be planted with small to medium sized shrubs to cut off headlight glare.

Trees selected for roadside plantation should not produce branches at low levels that obscure the view of vehicular traffic, cover the road junctions or the inner sides of curves of the road. Deciduous trees drop a large number of leaves on the road and footpath and render them unsuitable because of uneven surface and the slippery conditions they produce. Pavement should be kept green with turf grass as far as possible as this ensures proper aeration for trees. If tiling is to be done, porous tiles should be used. Lack of forethought is often seen in the form of tall trees interfering with overhead electricity wires, which are later headed back mercilessly. Due consideration should be given to overhead utilities while selecting plant species and tall shrubs should be preferred to trees in such situations.

Wherever single row is to be planted one of the following schemes could be adopted

1. One kind of flowering tree on both sides. In this scheme when the trees come to bloom, there is riot of colour and road looks beautiful for particular period. However, rest of the time road looks dull. Maintenance is easy in this type of arrangement. *E.g. Bauhinia variegata, Cassia nodosa, Lagerstroemia speciosa* etc. (Fig. 32.4)

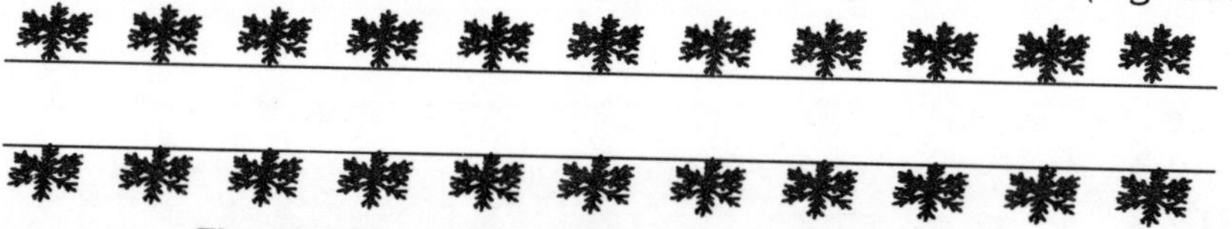

Fig. 32.4. One kind of flowering tree on both sides

2. Two kinds of flowering trees blooming at one time on both sides. Here, the two species are planted alternately on both sides of the road and different colours at same time avoid monotony. *E.g. Grevillea robusta* (Yellow), *Jacaranda acutifolia* (Blue), *Grevillea robusta* (Fig. 32.5).

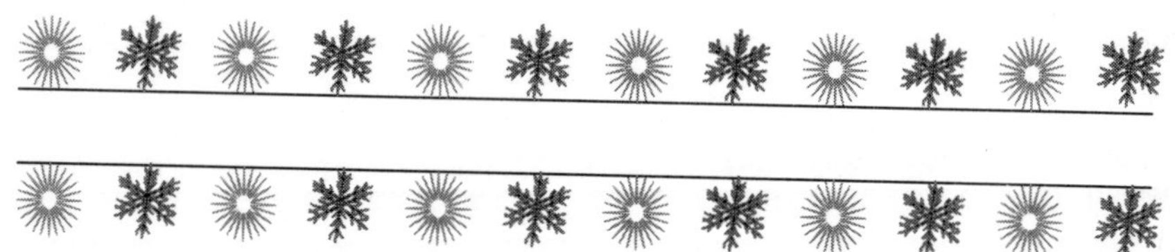

Fig. 32.5. Two kinds of flowering trees blooming at one time on both sides

3. Two kinds of flowering trees blooming at different time on both sides. In this case, one tree flowers in one part of the year and other in another part of the

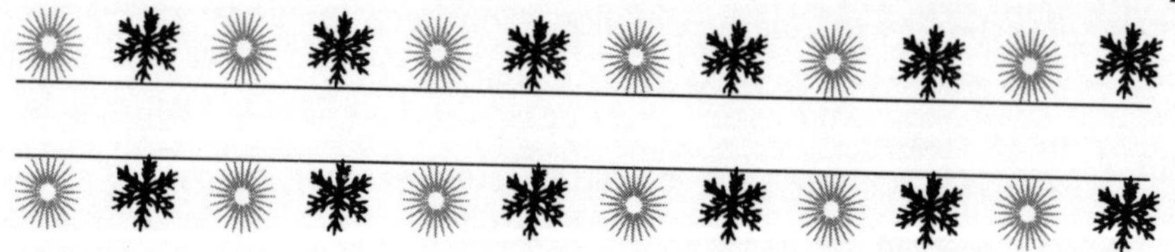

Fig. 32.6. Two kinds of flowering trees blooming at different time on both sides

ear. Roads remain colourful for longer period. *E.g. Cassia fistula* (Yellow in May-June), *acaranda acutifolia* (Blue in March-April), *Cassia fistula* (Fig. 32.6).

4. Shady trees only on both sides. This type of arrangement is suitable for oads with heavy pedestrian traffic like approach roads of Bus stops, railway stations and overnment offices. *Ficus infectoria, Mimuspos elengi, Alstonia scholaris etc.* (Fig. 32.7).

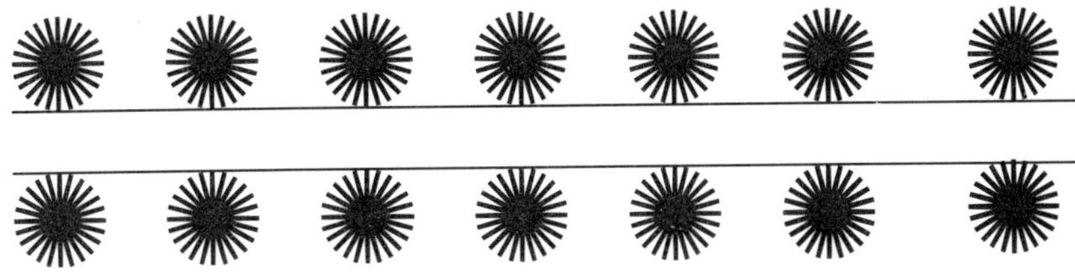

Fig. 32.7. Shady trees only on both sides

Rural Roads

Rural roads are the most neglected ones throughout the country as far as tree plantation is concerned. Here, preference should be given to plant multipurpose trees which besides providing shade and flowers, should also provide economic products such as fuel wood, timber fruits, fodder etc. Roads should be planted with small to medium sized trees preferably on the southern side so that fields are not shaded.

Urban Centre

Cities of today have generally expanded so much due to population pressure that nature, which used to be near at hand previously has receded far away to the countryside and citizens have little opportunity to be in contact with its various moods in different seasons. Chirping birds on beautiful trees in fresh morning breeze are like a fairy tale for most of city dwellers. The few public gardens and tree-lined avenues are the only places where the beauty of nature can be appreciated in the towns. Today's citizen lives in a concrete jungle, which cannot inspire or refresh him in any way. Strong structural lines of the urban areas accentuate the stress associated with city life, taking its toll on its inhabitants resulting in irritability, non-sociability and unwillingness to help others in need. Urban greening restores ecosystems that have been displaced or degraded by urban development and other human activities. The creation of healthy natural areas increases biodiversity while providing important habitat for urban wildlife. It also contributes to a city stock of "green infrastructure," which includes parks, avenues and green belts.

Proper town planning is a pre-requisite for bio-aesthetic treatment of our cities and towns. This can be seen only in new towns with wide roads and well laid out public parks, rather than in congested old towns with narrow crooked streets. Older parts of our towns like Delhi, Amritsar, Lucknow, Varanasi and Kolkata have ugly, ill ventilated houses joined together along narrow lanes. Deficient in open places, they offer little scope for bio-aesthetic planning. The few existing trees in such areas should be preserved as far as possible and wherever possible some open spaces should be created and converted into parks. Some open spaces are however available around religious places and historical buildings, which can be used to give a healing touch by planting flowering trees and shrubs.

New townships have greatest potential for proper bio-aesthetic planning. Chandigarh in Punjab provides good example, which can be followed in the development of new population centres throughout the country. Continuous increase in population is putting lot of pressure on limited land available in cities. House with a large garden of its own will soon be a luxury unaffordable, even for the upper class. The planner of today should be able to satisfy the urge of citizens to be in contact with nature without covering long distances. This is possible by developing green areas in the form of neighbourhood parks. Under ideal conditions, a garden should be near at hand, so as to be easily accessible by safe walkways from the dwelling places, so that the children can be taken there for playing. Such walkways can be fitted in, in the system of parks and gardens to enable children to reach their schools from home in complete safety and freedom from traffic hazards.

Gardens near commercial hubs can be very useful for resting and spending time while waiting for public transport. A continuous park system as against a number of small green patches will be much more useful and attractive. Land left by the way of open spaces around buildings for ensuring adequate light and ventilation should be utilized to the best advantage for microclimate improvement by way of plantation. Theme parks like *Bougainvillea* garden, Cacti garden, Seasonal garden, Fragrant garden, Rose garden etc. can be developed with an intention to create awareness among the citizens in respect of different categories of plant species and to encourage academic interest among the students. Large district parks of about 100 acres in size having water bodies, open spaces, wooded areas, open air theatres, eating joints and other recreational facilities should be developed near the cities. Such sites can serve as picnic spots and also serve the purpose of educating people about nature.

Many city parks, with little additional expense can be made more attractive, interesting and informative by suitably labelling the trees and shrubs and growing them with some consideration to botanical similarities. Such areas can serve dual purpose of botanical garden and public park. Intended use and available space should be the main criteria for selection of plants from amongst those suitable for the agro climate of the region. For planting at public places selection should be made from a variety of shade giving and flowering trees without going into their economic merits.

In case of individual buildings large trees help architectural composition and provide the required background and in fact look like structural members of the scheme. Shrubs and climbers also serve their purpose by the creation of various decorative patterns. Colourful flowering plants in window boxes add to the beauty of the building.

Compounds of Hospitals

Flowering trees, shrubs with fragrant flowers and shady trees should be planted in compounds of hospitals. Medicinal plants like neem should be preferred which also provide good shade. This will improve microclimate of the hospital complex and the beauty of flowers will make the otherwise dull environment, little interesting.

Educational Institutions

Maximum possible ornamental and economic species should be planted in educational institutions so as to educate the students about importance of plants in everyday life. Students should be involved in maintenance of the plants.

Courts, office buildings of the government and post offices should be planted with flowering and shady trees and shrubs so that visiting public may rest under the shade and enjoy their beauty while waiting. Railway stations, bus stand and airports, which are gateways to the cities, should be meticulously planted so as to accord welcome to the commuters and make the long waiting period, little interesting. Proprietors of commercial buildings should be encouraged and given all assistance in the planning and planting of ornamental plants.

Industrial Areas

Industrialization is one of the primary causes of environmental deterioration. Nevertheless, it has to persist and proliferate for continuous development of mankind. All that can be done is to make the industries more eco-friendly. Bio-aesthetic planning is thus very important for industrial areas.

Government of India has issued certain guidelines for industrial sitting according to which industries should be located at least 25 km from ecologically sensitive areas such as national parks, sanctuaries, lakes, biosphere reserves; archaeological and historic places and from major settlements (>3 lakh). Green belts should extend 500 m from the boundary of such areas. However, current scenario of the industrial areas demands strict enforcement of such guidelines for minimizing loss to the environment.

Bio-aesthetic plan for industrial areas should include creation of buffer zones around the areas with mixed plantation of economically important species to serve the dual purpose of revenue generation and environmental amelioration. A green belt of suitable trees having a width of 50 to 100 m between residential and industrial zones decrease dust level up to 52% and 40-60% of noise pollution. Five-hundred-meter belt can mitigate about 76% SO_2 and 67% NO_x from the atmosphere.

Premises of the industrial units should be planted with pollution tolerant flowering plants. This will help in improving working conditions for the industrial workers and increase their output. Deciduous species remove air pollutants in a better way by accumulating them in their leaves, which are all shed during dormant season. Thus, mixed plantation of evergreen and deciduous species is recommended. Suitable trees can be selected from table 1.

Historical Buildings

We have a rich legacy of historical buildings and heritage sites, which besides their archaeological importance are prime attraction for domestic as well as foreign tourists. Some of these have well planned gardens around them like The Taj while most of them have drab surroundings. With the intended promotion of heritage tourism in new tourism policy, the surroundings of these icons of our glorious past need to be revamped for presenting their true grandeur. Such sites should be beautified keeping in view their significance and type of vegetation historically associated with them should be preferred. Approaches to these sites should also be ornamented with flowering trees and shrubs as a welcome gesture.

Banks of Rivers

Rivers like Ganga, Yamuna, Sutlej, Vyas, Krishna, Kaveri, Godavari, Narmada, etc are regarded as sacred and many temples and ghats can be seen along their banks. These should be planted with sacred trees like kadam and aoska, which will greatly enhance their sanctity and also add colour and charm to them. Apart from beautification of banks, plants help in controlling erosion and silting of rivers and also help in reducing cutting of banks during rainy season. However, planting of riverbanks is not meticulously undertaken and on most of the area general agriculture and forced vegetables are grown. A national level body should be formed to undertake riverside plantation and serious efforts should be made for enhancing their beauty and maintaining ecological balance. Many wetlands of international importance have been identified along rivers *e.g.* Bhoj wetland, Bhopal (M.P.), Deepor Beel, Guwahati (Assam), Harike lake and Ropar headworks in Punjab, Pong Dam lake in Himachal Pradesh. Such areas have immense tourist potential which can be exploited by providing proper facilities and aesthetic improvement.

Banks of Canals

Banks of canals generally should be planted with moisture loving trees. Most of the tree planting alongside the canals has been done by Forest Department and economically important trees have been planted to generate regular income. Wherever they pass through or near a city or town or a highway crosses them, flowering plants should be preferred. Headworks of canals should be developed into pleasure resorts and water sports facilities should be created wherever possible. Canal side plantations also serve as important habitat for birds and small animals as they are not much disturbed by human activity.

Rural Areas

Our villages had a tradition of planting trees along the field bunds, panchayat lands and around wells. Although economic considerations always dominated over beauty but there used to be sufficient green cover anyway. Panchayats usually assembled under Pipal or Bargad trees, which used to be centres of village activities. Intensive agriculture has taken its toll and commercial plantations of Eucalyptus and Poplar have replaced long living fruit or timber trees at many places. Village ponds, which used to be prominent features of rural landscape, are fast vanishing or being converted into sewage and wastewater dump and every effort should be made not only to conserve them but also to make them attractive. They can be used as water harvest tanks especially in low rainfall areas where they can provide drinking and irrigation water during hot summer months.

Planting flowering trees on the approach roads, village schools, panchayatghars and religious places can beautify villages to a great extent. Common lands of the villages usually under the supervision of panchayats provide ample opportunity for beautification with a variety of flowering plants. A scheme was launched in Punjab for planting flowering trees and shrubs and fruit trees near the tube wells in villages and plants were provided to the farmers free of cost. Such schemes can greatly improve rural landscape and should be followed in whole of the country.

Role of Agricultural Universities

Of late, most of the agricultural universities in the country either have a separate Department of Floriculture and Landscaping or it is a part of Horticulture Department. Basic knowledge of ornamental horticulture is given to all the students at undergraduate level. Eight universities are offering post-graduate course and five are offering doctorate course in Floriculture and Landscaping. Students are given thorough training in science and art of ornamental horticulture with special emphasis on growing and maintenance of ornamental plants and their use in landscaping of parks, open spaces, buildings, roads etc. Services of such trained personnel should be utilized while planning residential, commercial, industrial and other areas suitable for Bio-aesthetic planning. In these universities systematic planting of ornamental plants has been done which serve the purpose of education and generating public awareness. These institutions have also played a key role in collection, multiplication and distribution of new plant material.

Rural people can be motivated towards bio-aesthetic planning by involving the panchayats. People here will be more interested if they are given exploitation rights like firewood and other minor products. Maintenance of roadside plantations in vicinity of the villages can be entrusted to the panchayats. This can open up new avenues of employment for rural youth.

Educating the people is very essential to ensure their participation and the best method is to educate the children. Excursion tours to nearby parks and botanical gardens with few tips on environment could be a great fun for them. Organizing theme-based science fairs and quiz competitions can infuse love and respect for nature in them.

Ministry of Environment and forests has introduced Urban Environmental Information System in Agra, Kanpur, Patna and Bhubaneshwar to provide information to the public in the form of local environmental report to be prepared by the municipalities on the status of socio-economic development and environment of the cities. Such efforts should be extended to whole of the country so that people could be made aware of the current state of their surroundings. NGO's, Agri-horticulture societies, Eco-clubs etc. should be involved in such campaigns for effective dispersal of information.

***

33

Gardening during Budha Period

Man is always attracted by natural objects like planets, mountains, sea, rivers, trees and other flora and fauna, from time immemorial. Man was always attracted toward, colorful objects. Out of many natural objects, trees attracted most. Since the trees are part of eco-system in which man lives and thus he has close association with them. Trees serve mankind in several ways like maintain balance between O_2 and CO_2, provides shade, fruits, timber, wood for fuel, medicines, etc. Trees also provide dense shade and living space and berries like fruits for birds. Long association has also been observed during Budha period and later on trees and other plants were utilized in making gardens. Therefore, it is necessary to understand the status of gardening during the period and association of these with Budha. There are several descriptions available about birth of Lord Buddha and association with trees. According to one mythology, Lord Budha was born as a royal prince in 624 BC, in place called Lumbini which was originally in north India but is now part of Nepal. His mother name was Queen Maya Devi and his father was Suddhodhana. One night Queen Maya dreamed that a white elephant descended from heaven and enter her womb. On that night she conceived a child who was pure and powerful being. When she gave a birth to a child instead of experiencing pain, the queen experienced a special pure vision in which she stood holding the branch of tree while gods Braham and Indra took the child painlessly from the side. There they proceed to heaven leaving infant after offering him ritual ablutions.

According to another mythology the events of birth of Lord Budha has been described by Thomas and according to him after queen Maha Maya bearing the Bodhi Sattva for ten months when her time came, desired to go to her relatives house and addressed to king Suddodhana, 'I wish, O King, to go to Deva daha, the city of my family'. The king approved and ordered that the road from Kapilvastu to Deva Daha to made smooth and adorned with vessel filled with plantain, banners and flags and seating her in golden palanquin borne by a thousand courtiers seat her with a great retinue. Between two cites, there is a grove of Sal trees (*Shorea robusta*) named the Lumbini groves. At that time from roots to the tips of branches it was one mass of red coloured flowers and from within the branches and flowers host of bees of five colours and various flocks of birds sported, singing sweetly.

When Queen saw it, and expressed desire to stop in the grove. The courtier brought the queen and entered the grove. She went to foot a Sal Tree and desired to seize a branch. The branches like the tip of a supple red bent down and came within the reach of hand. Stretching out her hand she seized the branch. She was shaken with throes of birth. So multitude set up a curtain for her and retired. Holding the branches and even while standing she was delivered.

According to another Legend, in the dream of Queen Maha Maya Bodhi Sattva descended in the form of elephant. It is further related that she seized not a Sal branch but a plaksha (*Butea monosperna*) at the moment of birth. These are two common trees in tarai of Nepal in which the birth place of Budha is situated. (Thomas, 1958, Life of Buddha, London pp. 33).

According to Chinese traveller Hiuen Sang who visited Lumbini in A.D. 630 and stayed upto 645 A.D. He has also described the birth place of Buddha. According to him to the north east of the arrow well about 80 or 90 li (1 li = 500 m), we came to Lumbini (Lavani) garden. Here is the bathing tank of the Sakya Muni, the water of which is bright and clear as a mirror and the surface is covered with a mixture of flowers (lotus and water lilies). To the north of this 24-25 paces, there a asoka flower tree (*Saraca asoca*) which is now decayed, this is the place where "Bodhi Sattva" was born, So Lord Budha birth has been associated with three trees *viz.* Sal, Asoka, Plaksha. There are other trees also which are associated with Lord Buddha. Buddha got enlighment under pipal tree (*Ficus religiosa*), at Bodh Gaya (Bihar). Buddha gave first sermon under the tree of mango at Sarnath (varnasi). Buddha died in Salvana- a grove of Sal (*Shorea robusta*) near Kushinagar (U.P).

The association of trees and other flowers with Lord Budha inspired the Budhist monk most and they were revered to them. Thus after the death of Budha, Buddhist priests had lot of leisure time and they preferred peaceful surrounding which are highly pertinent for tranquility of mind, worship and meditation. The monks planted flowering trees and flowers around their vihars (monasteries) and took great interest in gardening. Thus, in fact the evolution of gardening is intimately associated with the monasteries and temples. In India, Buddhist monks grew groves of flowering trees like nag kesar (*Mesua ferea*), kadamba (*Nucleon purpurea*), rukmini (*Ixora paviflora*) etc. whereas Chinese monks preserved rare trees like ginko (*Ginkoga biboba*).

In India, Budhism spread as a religion after King Asoka adopted it as state religion after Kalinga war. Emperor Asoka (273 to 232 BC) one of the ruler of India was most valour King who ruled over north India. His last major war was for the conquest of Kingdom of Kalinga (in Orissa) in 261 B.C. but long and brutal war resulted in toll of 1, 50,000 men and 1,50,000 men were wounded and thousands were captured. This horrific bloodshed shook him. Thus he gave up hunting and slaughtering of animals and adopted Budhism.

Ashoka actively promoted Budhism under his regime. Thus Budhism spread to Syria, Egypt, Central Asia, Burma, Ceyloen, Thai Land, Indonesia, China, Japan, etc. He sent his son prince Mahendra with a bodhi tree (*Ficus relegiosa*) to Ceylon in about 250BC, and planted at Anuradhpura. It is old historical tree in the world whose branches are supported by pillars.

Ashoka inscribed official orders on rocks and pillars at various places to spread the message of love and peace in effort to spread Budhism. During pilgrimage he visited several places like Vaishali, Lumbini, Kapilvastu, Shravanti, etc. along with his daughter. Where ever he went he caused pillars Stupas, to be erected in the memory of his visit. There is one such memorial pillar at Sarnath. On the top of a stone pillar of about fifty feet high, there are beautiful figures of four lions and it has been adopted as a official emblem by govt. of India. Ashoka chakra or known as Dharama chakra adorns the national flag of India. Besides, stupas and pillars, Ashoka built caves, dwellings, rest houses and Budhist Vihars which were duly planted with trees and flowering plants. Out of many, stupas of Bharhut, Sanchi and Mathura are very famous and depicts association with many flowering trees. In Bharhut sculpture, the Budha is shown nowhere but he is represented by symbols *viz* a lotus flowers symbolizes his birth, a pipal tree his enlightment, a wheel his first sermon and sutpa his nirvana. At Mathura the massive figure of Budha is shown with a back ground of pipal leaves. Cunnigham has identified seven Bodhi trees. The pipal (*Ficus religiosa*) is the bodhi tree of Sakya Mani. The other five trees are banyan (*Ficus benghalensis*), the bodhi tree of kasyapas; gular (*Ficus glomerata*) the bodhi tree of Kanaka muni; Siris, (*Albizzia lebbek*), the bodhi tree of Kraku echanda; Sal (*Shorea robusta*), the bodhi tree of vishwa bahu; (*Saraca asoca*)- the bodhi tree of vipasuvi. All these are shady trees and these trees have been given title gratitude as India is a hot country. Apart from these six trees, white lotus is symbol of sikhi.

The Tala tree of (*Borasșus flabelliformis*) is also depicted in one of the medallions. It is common tree of south and eastern India.

In ancient India tree worship was very common which is still being followed. Girls and young women are considered as human embodiments of maternal energy of nature. In awakening of asoca tress, by touching and kicking of the tress by their feets, they transfer their potency, and enable it to bring forth blossom and fruits. In the ancient city of Saravastia, Salabhanjika festival was celebrated with great zeal when the Sal trees flowered. The Sal trees in full blossom were being worshipped for offspring.

According to another jataka, trees have great spirits that is why they are being worshiped with perfumes, flowers and food. These spirits live in many kind of trees but the banyan, Sal and silk Cotton (*Bombax malabaricum*) were favorable. The trees are worshipped not for its own sake, but because it is the home of spirit. If the spirit leaves a tree, it withers and dies but the spirit is immortal.

All the tress have electro-magnetic energy which is passed in us when we touch them.

With the passage of time and changes in religioso, and socio-political conditions the impact of Budhist gardens eroded in India but they flourished and survived in Thailand, China, Korea and Japan. The Budhist gardens have been transformed into a new style of gardening in Japan Known Japanese styles of gardens in the hands of crafty gardeners and modified according to the local climate conditions and varied plants available.

34

Japanese Gardens

Japan is known for its gardens, tea ceremony, ikebana, bonsai and industrial goods. Japanese gardens are famous in the world for their unique style, natural and spiritual beauty and calmness. These gardens have been made by Zen-Buddhists for worship, meditation, tea ceremony and pleasure for members of royal families. Therefore; calmness and quietness is an essential feature. Japanese people are ardent lover of nature and therefore natural landscape elements of the country *i.e.* mountains, islands, rivers, lakes, streams, bridges, etc. have been created in the gardens in miniature. Hence Japanese gardens are also called as 'nature in miniature'.

The art of Indian style of gardening reached first in China, then Korea by Buddhist monks and finally it flourished and attained its peak in its own style in Japan by Buddhist monks and artistic people. The favourable climate, topography and liking of people are the factors responsible for making gardens of unique style which enjoys high esteem amongst garden lovers. In the beginning these gardens were developed around, Buddhist temples. Later on, these were spread to the imperial villas, castles, citadels, mansions, houses, etc.

HISTORY OF JAPANESE GARDENS

These gardens have passed through different developmental phases under the influence of religion, culture and likings of rulers of different era. Simple temple gardens of ancient period gradually paved the way for the development of artistic gardens of their unique style. Main historical periods can be summarised as follows :

Asuka Period (538-794 A.D.)

During this period Nara was the capital of Japan and Buddhism had started flourishing after its introduction in first century. Gigantic Buddhist temples were constructed. Among them Todaiji was the famous temple in which big statue of Lord Buddha is kept. In front of this garden, there is a big ground, planted with trees. There is a deer park also.

Other important temples at Nara are Kofukiji and Biodo inn at Uji. The influence of Chinese culture was prominent. Artificial hills or rockeries were constructed which were 100 feet high with running water as ornament which was brought from several miles to this place. More high artificial hills measuring about 200 feet where also made. The stones were selected for interesting shapes and kept near running water.

Heian Period (794-1185 A.D.)

The capital of Japan was shifted from Nara to Kyota. The gardens made during this period were highly decorative. Big shrines were made having large lakes, bridges, stepping stones. Round stepping stones in lake of Heianji temple is very famous for its style.

Kamakura Period (1186-1399 A.D.)

During this period stone, hills and lakes were essential feature. A lot of importance was attached to the selection of stones with religious and moral attributes. Various names were given to different islands like 'Hill island', 'Wood island', 'Cloud island', 'Misty island', 'Dry beach island', etc.

Another form of natural gardening developed during this period was dry landscape (Karesansui) based on symbolism. The most important example of such garden is Ryoanji Temple of Kyoto. True in its sense, gardens are devoid of traditional lake but raked sand and stones have been used for creating effect of the ocean. On the flat and rectangular plot fifteen stones of varying sizes have been placed in five separate groups of five, two, three, two and three. Design of lakes or rivers have represented through hollowed out beds and sand, pebbles and boulders. Inlet and outlet have carefully been shown. Artificial hillocks are called like 'Ocean Rock', 'Large River Rock', 'Offing Rock', 'Mountain Torrent Stone', etc.

Ashikaga Period (1400-1600 A.D.)

Yoshimitsu Ashikaga (1348-1408 A.D), the third Ashikaga Shogun, got the construction done of very important garden along with three storeyed golden pavilion called 'Kinkakuji' in the year 1397 at Kyoto. The three storey gold leaf covered Gold Pavilion is a breathtaking sight to see in any season. The first two storeys are based on traditional Japanese court architect. The top storey shows an influence of Chinese Zen school. There is a pond around it and the pavilion is reflected elegantly in the water. In spring, there is fresh green foliage of trees and during early summer pond is studded with lotus blossoms.

Another important garden of this era is Silver pavilion (Gin Kaku Ji) which was built 80 years later by Yoshimasa Ashikaga (1435-90 AD.) the eighth Ashikaga Shogun. Silver pavilion is not true to its name and is rustic naturally coloured.

Edo Period (1615-1867 A.D.)

During the Edo period, the making of small gardens dominated. The garden could be enjoyed from inside of a house. These gardens were similar to tea gardens.

Important garden of unique scenic beauty was built at Kyoto which is known as Katsura Imperial Villa. It was built in 1642 AD. by Kobori Enshu—a master of tea ceremony. The Shogun Hideyoshi was the ruler who provided patronage to this villa. It was contracted for merry making, playing games, drinking tea or *sake,* singing, dancing or reciting poetry, religious, social and political discourse by the members of imperial family and to enjoy in refreshing and inspirational natural setting. Therefore, the efforts have been done to create strong harmony between buildings, nature and man.

The important features of this garden are its different buildings which have been built in the Japanese style. There are different islands on which different buildings like Shokin-tei tea house, Enrindo, old Shoin, Middle Shoin, musical instrumental room, etc., are situated. These islands are connected with stone bridges and curvaceous paths. Strolling through trees, stone bridges, over islands and curvaceous paths, provides enjoyment of natural setting. There is a unique manicure garden too.

A six-year project was initiated in 1976 to renovate the major buildings of the villa which were taken apart, repaired and reassembled. This restoration has given a new look to the buildings which blend more harmoniously with the surroundings now.

Meiji Era (1876-1912 A.D.)

The capital of Japan was shifted from Kyoto to Tokyo in 1867 after the defeat of Shogun Yoshinobu in the hands of Emperor Meiji. There was a great industrial revolution during this period. Despite the influence of West on industrial sector, it could not influence the style of gardening. With the change in style or architecture, multi storey buildings have come in existence in big towns. This gave rise to the making of gardens in limited space around skyscrapers, mansions, roof tops, windows, etc. The techniques of growing flower have been changed and controlled conditions for growing of flowers have been introduced. With the change in political system and living conditions, tot lots, city parks, regional parks and national parks have been developed.

TYPES OF JAPANESE GARDENS

There are mainly three types of garden *viz*. (1) Hill gardens, (2) Tea Gardens, (3) Flat Gardens.

Hill Gardens

In Japanese language they are called as Tsukiyama Sansui which means hill and water because main features of such gardens are hills, streams and ponds along with other features. There are 3 to 5 hillocks. Hill-1 forms the central feature. It should be bigger in size and should have broad views. Hill-2 is employed as companion to No. 1, since it is adjacent and is somewhat lower and of secondary importance. Hill-3 is opposite side to No. 1. It represents lower hill. Hill-4 is small but elegantly introduced in the foreground of Hill-2. Hill-5 is placed in the remotest part of the garden visible from side of hill. Lake or lakelets are automatically formed. Different islands are made because hills, lakes and islands are complementary beauties. There are generally three islands *viz.* (1) Master island (2) Guest island which are connected with a bridge and (3) Central island.

Different stones are used to beautify surroundings. These stones represent different symbols like natural, mood, idea, spiritual and melodic symbols. There are five distinct shapes recognized for stones which are employed in the garden. They are : (1) *Statue Stone*—It is a tall vertical stone bulging out towards middle and finishing conically at the top (2) *Low Vertical Stone*—It is a short vertical stone which is rounded at the base and finishing in irregular blunted one. It resembles to bud of a magnolia, (3) *Flat Stone*—It is a low broad stone of irregular shape and has a flat top. It is little higher than ordinary stepping stone, (4) *Recumbent or Ox Stone*—It resembles trunk of a recumbent animal. It is a long curved and bent boulder, rising higher at one end than other (5) *Arching Stone*—It is medium in height, with broad flat top and bent over to one side forming an arch like structure.

The arrangement of these stones is done in composition comprising of two, triple or quintuple combination. Such compositions have different meanings when these are fixed at different places. For example, *Long life stone*—three stones placed together on the edge of a beach. *Good luck or death stone*—two stones used on the banks of lake. *Water gate stone*—making outlet of a lake.

In hill gardens the important rocks or stones used are about ten in number and fixed at specified places. They are : (1) *Stone*-1 is called as *Guardian stone*. It is a high standing stone which occupies the most central position of background. (2) *Stone*-2—1t is a companion of stone 1 and is placed opposite of the fall. (3) *Stone*-3—It is broad and flat and is called worshipping stone. It is placed in the foreground and in the centre of the island. (4) *Stone*-4—It is placed near the foreground and at one side of the garden. It is flat and broad based. It is called as interviewing stone or shoe removing stone. (5) *Stone*-5—It is placed on opposite side of the garden to stone 4. It is called as waiting stone. (6) *Stone*-6—It is called as Moon shadow stone which occupies an important position and placed between two hills. (7) *Stone*-7—It is called as cave stone and is standing stone which is similar to guardian stone and placed near central group of trees. (8) *Stone*-8—Seat of honour stone. (9) *Stone*-9—It is called pedestal stone and is the first stepping stone. (10) *Stone*-10—It is called as Idling stone which consists of a pair of stones. Stones are broad, low and round. *Stepping Stones*—These are the natural stones which are placed in zig-zag pathway. These are arranged in two, threes or sometimes four. While fixing these, regularity is avoided. The distance between two is kept about 10 cm and intermediate spaces are kept clean.

Trees. The selection of suitable trees and planting at fixed places play a major role in gardens. The trees have been given specific names because they are planted with certain objectives. These trees are : (1) *Principal tree-* It is a tree or group of trees planted in central part of background. These are surrounded by other trees which differ in foliage character. (2) *View perfecting tree*—It is planted in foreground on island. It adds beauty in lake scenery. (3) *Tree of Soltitude*—It is a tree or group of trees of thick foliage in the background on one side. (4) *Cascade Screening Tree*-It is a group of bushes or leafy tree (Bamboos) planted at the side of waterfall to hide of the portion of it. (5) *Tree of setting sun*—It is planted in west side to check the glare of setting sun. (6) *Distancing pine*—These are planted to give a forest type look.

Ornamental Water. Water is the life of garden and a necessary feature of Japanese gardens. It may be present in the form of big lakes or symbolically in the form of water basins. In hill gardens, water is mostly present in the form of lakes but it may be in the form of natural stream or river. The standing water may be in the shape of round, square, crescent shape, running water shape, etc.

In the tea gardens and flat gardens, water is present in the form of basins or wells. The size and shape of basins are of many types. It is proportionate to the surrounding buildings. Different shapes are 'Square' shape, 'Round Star' shape, 'Four God' shape, 'Bubble' shape, 'Genkai' shape, 'Stone Battle' shape. 'Mt Fuji' shape, etc. Near the basin wooden ladel is also kept to use the water for rinsing or drinking.

Rustic well of wood or stone wall may be present. The wall may be rough or smooth. These wells are complete with lever, rope, bucket, pulley, etc.

Garden lanterns. From very old time, the paved approaches or courts of Buddhist temples were decorated with lanterns made up of stone or Bronze. These lanterns were presented on the eve of festivals. The height of lanterns varies from 6-18′ and .are arranged singly or in combination with trees, shrubs or water basis. It is according to scale and character of building.

These lanterns are of mainly two types *i.e.* Standard and legged type. Standard lanterns have cylindrical standard whereas legged type may have three, four or six legs over which the main lantern is fixed.

There are hanging type lanterns too. In past these were used to be made up of brass but now-a-days lanterns made of papers are used frequently. These are hanged by a chain in verandah, house, or tea rooms.

Garden Pagoda. It is an ornament in Japanese garden which is purely decorative. It may be in the form of stone tower or pagoda. These pagodas may be standard type or legged type and have three, five or seven, nine or eleven separate roofs.

Garden bridges. There are many kinds of bridges in Japanese gardens which are used to connect different islands. In old days they were made up of wood and stone but now-a-days iron bridges are more common. The shape and size varies according to width. Mostly single stone or plank was used to make bridges. In many cases many pieces have been used artistically to make a bridge *e.g.* Yatsuhashi bridge was made in Iris field by uniting wooden planks in zig-zag manner. These planks were supported on wooden support in mud. Mostly garden bridges are arch shaped. Other types of bridges are 'Wooden Trestle' bridge, 'Peeping' bridge, 'Granite Slate' bridge, 'Curved bracket' bridge, 'Chinese Full Moon' bridge, etc.

Tea Gardens

A garden attached to tea house is called as Tea garden. The tea ceremony was started in China in 6th century by Buddhist monk Luwuh, who formulated a code of conduct of tea ceremony. The tea was served to monks who took it as medicine which helped in meditation.

The tea gardens are divided into three sections. The outer section is called as *Sotoroji.* There is waiting place called as *Machi ai* and *visitors* wait here before entering tea room. The inner section is called as *Uchiro ji.*

The guests or *samurais* when visited a tea garden for religious discourse or advice, used to wait at the waiting place and remove their shoes and swords. At waiting place there is a sword rack and stepping stones arranged in such a way that it facilitates the hanging of swords on the rack. The entrance of tea room consists of low door and guests are obliged to enter in bending postures which indicates humbleness and respect. The host used to enter from another door in the tea room.

The inner garden is made in such a way that it looks extremely simple and natural. Garden objects like stones, lanterns, rocks, water basins are selected which look as antiques. Trees, shrubs and grasses are often arranged in an unkept manner, so it looks wild. A rustic looking well is an important feature of such gardens. Pathway is made up of stepping stones in meandering circulation, connecting water basin, waiting shed, lavatory and gateway. Wall is covered with hedges which also look natural. The different stones in tea garden used are : *Kettle stone*—It is a flat stone suitable for placing the kettle. *Front stone*—It is placed in front of low basin stone. *Sword hanging stone*—These are double stepped stones and are used for hanging swords on sword hanging racks. First stone is used for standing. *Candle stick stone*—It is placed in one corner to be used for keeping lantern or placing candle when ceremonies take place after dusk. *Low basin stone*—It is hollow stone and is filled with water to wash the hands.

Flat Gardens

These are known as *Hira niwa* in Japanese language. Such gardens are devoid of hills, streams or ponds. Flat gardens are mostly planned for confined places or for laying out in front of buildings of secondary importance. Designs mostly adopted are either mount valley type or extensive moor and planting is done accordingly. For mount valley type surrounding should be thickly planted whereas for moor type it should be bare and open. Other garden ornaments like stones, well, water-basin, tree, etc. are used in such a way that it creates a scenic beauty. The surface of flat garden is covered with raked sand on beaten earth to give an effect of water current of ocean showing appropriate in let and out. In between, stones or boulders are arranged in an artistic manner *e.g.* Ryoanji temple, Kyoto. The rocks or pebbles are so arranged that they give an effect of diversion or rush of water.

Plant material used in japanese garden. For meditation green colour is more helpful whereas colourful plants distract mind during meditation and one does not feel peaceful. This is the reason, Japanese gardens are not planted with colourful trees. On the contrary, colourful plants are planted around castles to make these more attractive. Following is the list of plant material which is commonly planted in Japanese gardens.

Evergreen—*Pinus desiflora, Pinus thunbergii, Pinus parviflora, Pinus karaiensis, Abies barchyphylla, Abies firma, Podocarpus macrophylla, Quercus acuta, Quercus cuspidata, Nandina domestica.*

Deciduous trees like—Maple (*Acer parvifolium,* etc.). Birch (*Betula alba*), Elm (*Ulmus montana,* Ulmusparviflora), Willow (*Salix babylonica*), Poplar (*Populus tremula*), Mulberry (*Morus alba*) etc.

Amongst flowering trees and shrubs the commonly planted ones are : Cherry (*Prunus pseudocerasus*); *Camellia japonica, Rhododendron*, Azalea, Magnolia, *Wisteria sinensis, Wisteria alba*, Hydrangea, pomegranate, peach, plum, pear, Quince, etc. Other common flowering plants are Chrysanthemum, Carnations, Irises, Lotus, Peonies, Orchid, Freesia, etc.

Different kinds of bamboo like *Bambusa nana, B. aurea* are commonly planted.

35

Persian Gardens

Iran is the home of one of the world's oldest continuous major civilization with historical and Urban settlements dating back to 4000 B.C. The Medes unified Iran as a nation and became an empire in 625 B.C. The Achaemenid Empire (550-330 B.C) was the first Iranian empire to rule Middle east and central Asia.

Darius I (521-485 B.C) built a ceremonial capital at Persepolis and an administrative capital at Susa. The latter was a walled and moated city, a Paridaeza, a city in garden and a garden in city. When Cyrus younger showed the palace at Sardis to a Greek in 407 B.C. he called it Paridaeza, and explained. Precise geometric regularity was the outstanding characteristics of Persian garden. It was combined with mystical love of flowers, trees and above all water. Water was used in channels and trees were planted in rows.

Persia was conquered by Arabs in 7th century and deeply influenced by the Islam, but Islamic gardens were greatly influenced by Persian tradition of garden making. The Bulkwara palace in Samara is very much part of the Persian tradition and had a quadrant and pavilions. Similarly this style was widely copied in Moorish garden of Alhambra (Spain) to mogul gardens of India.

Persia was conquered by the Mongol including Genghis Khan (1219) and by Timur Timurlane (1381). They caused vast destruction but adopted the language and culture of Persia. Timur Timurlane transported Persian craftsman to his capital at Samar Kand now in Uzbekistan.

This was succeeded by Selencid Empire, Parithiens and Sessanid which governed Iran for almost 100 years. The end of Sessanid empire was a turning point in Iranian history. Islamization in Iran took place during 8 to 10th century and led to eventual decline of Zoroastrian religion in Persia and was absorbed by Islamic polity.

Iran had been monarchy ruled by Shah or Emperor from 1501 until 1979 almost without interruption but with the success of strong Iranian cultural revolution succeeded and then Iran become an Islamic Republic on 1st April 1979.

Climate

Part of Iran's terrain is comprised of hills and peaks which are covered with snow. The melting of snow becomes a source of water during spring and summer for irrigating the fields. Part of Iran is desert. Thus, Iran's climate ranges from arid to semi arid and subtropical along Caspician coast and northern forests. On the northern edge of the country (The Caspician-Coastal plain) temperature rarely falls below freezing during winter and area remains humid for the rest of the year. Summer temperatures on the other hand rarely exceed 29°C whereas rainfall is about 70 cm. in eastern part of plains and more them 170 cm in the western part. In the west, Zagros basin experience lower temperatures and severe winter below zero degree and experience heavy snow fall.

It central and eastern basins, the climate is arid with less than 20 cm rain. Average summer temperature exceed 38°C. The coastal plains of the Persian gulf and gulf of Oman in southern Iran have mild winter and very hot and humid summers. The annual rainfall of the area varies from 15 to 30 cm.

Iran enjoys sunny days and plenty of sun shine during most part of the year. During spring orchards of almond, peach and plum are in full bloom. The temperature of Tehran varies from –3-7°C during winter nights and 37°C in July.

Due to dry heat of Iran, shade is very important in the garden without which it could not be usable. Trees and trellises largely feature as biotic shade whereas pavilions and walls are also structurally prominent in blocking the sun. Similarly the presence of water in the garden is important. A form of underground tunnel below the water table called quanat is used to irrigate the garden and to improve the environments of the garden. Well like structures then connected to quanat enabling the drawing of water.

In past, Persian wheel, an animal drawn system to draw the water from well to the surface was used. Such wheel systems could also be used to assure surface water system. Such systems were earlier existed in chahar bagh style.

Persian Garden Style

The Persian gardens were born as a result of the harsh landscape of Iran. Persian gardens are synonym of Paradise. The Persian word paradise simply mean wall enclosure. It may have been driven from Avestan root *i.e.* pairi (around) and diz (shape). Avestan is an East Iranian Language. The Avstan word pairidieza is from old Persian word paridaida, paridaiza (wall garden) was translated into Greek-paradeisoz. Then adopted into Latin language as paradisus, and from there it entered into European language. *e.g.*

French–Paradis, German–Paradies and English–Paradise. Then the word entered into Semitic languages as well: Akkadian-pardesu; Hebrew-pardes and Arabic-Firdaws.

As the word expresses, such gardens would have been enclosed. The garden purpose was to provide a place for protected relaxation in a variety of manners *i.e.* spiritual and leisurly and is essentially paradise on earth.

The Persian word for enclosed space was pairidaeza—a term was adopted by Christian's mythology to describe the Garden of Eden or Paradise on earth.

Type of Persian Gardens

The Persian built lavish garden in contrast to European gardens of the time which consisted of modest herbs planting around monasteries. One of the main feature of Persian garden is lake of 20 × 30 m whereas in other gardens lake was of mile square and planted with fruit plants that gleamed red and yellow as stars of heaven in dusky night. The Persian made following three types of gardens.

1. Palace Gardens. The surroundings of big palaces are beautified in the form of gardens. Big spacious lawns are kept for social, political gatherings, royal ceremonies and royal marriages. The style of Persian gardening dominates. Big shady trees, fountains, water canals are main feature of such gardens *e.g.* The Garden of Golestan palace.

2. Tomb Gardens. The paradise gardens of Iran were recognized after when they were used for setting for tombs as signs of god's bounty to the living. They were modified for the perfect gardens of eternity. The most likely model for muslim tombs were centrally planned, square, round, or Roman Octagonal mausolea. The tomb was a symbol of moderate wealth and decorated accordingly. Some of the imaginary consisted of cypresses which is evergreen, and thus a symbol of eternal life. Peacock and domes are symbol of heaven and paradise. The gardens were made in formal style with dominant planting of cypresses. The important tomb gardens are : Sadi Tomb garden and Aramgah-e-Hafez.

3. Pleasure Gardens. The gardens are quite large in size and need for general public to be used for holding, fairs, social and religious functions. Lawns are big in size so that large number of people can be accommodated.

Styles

There are six primary styles of the Persian gardens which are according to their function and style. Gardens are not limited to a particular style but is mixture of different styles. The styles are as follows : Chahar Bagh, Bagh, Hayat-Public, Hayat Private, Park and Meidan. However, mainly following four types dominate.

Chahar Bagh. Cyrus II the great, is thought to be the pioneer of Chahar Bagh Paradise garden whose basic structure consists of four quadrants divided by water ways or path ways. Chahar Bagh design is a balanced design and central axis plays dominating role and divides garden into two equal parts. To create greenery, planting is done around the periphery of a pool, pathways or waterway. This is a form of garden which means Eden having four rivers and four qanat. The old example of a rectangular canal pattern is at Pasargadae.

A reconstruction of garden of King Cyrus II (550-500 B.C.) who was founder of Persian Kingdom under Achaemenid dynasty in the plains of Meridasht in Southern Iran. Such garden had geometrical plan and stone water courses. Incorporating architectures, planting, water rills and shade giving pavilions, Cyrus's garden offered the back ground to later garden development. It is likely to be planted with pomegranate, cherry and cypress. This is the earliest remnant of what becomes the classical garden of Persia and back ground to later development of such garden. The plan is based on David Stronach's Pasargadae—a report on excavation.

The ancient province of Pars is in central Iran and is known for the being home of sites of immense importance in the history of world. When travelling from Shiraz, the fabled capital of Pars is towards Esfahn, the second largest city of Iran. From here the road goes to the valley of Pasargadae which is the birth place of Persian Empire.

Hayat. The Hayat is a classical Persian layout where man made structures in the garden are made particularly such as pergola, arches and pools. The ground is often covered in gravel and plants are typically very simple, such as line of tree that offer shade. Privately, these gardens have pool in the centre and have structural supports like pergola and arches. The pool serves as a focus and source of humidity for the surrounding atmosphere. In Urban areas, due to availability of limited water very few plants are planted.

Meidan. The meidan gardens have few structure and emphasis is on natural elements. The meidan have trees, shrubs, bedding plants and grasses. There may be an element of pool and gravel pathway which are divided upto lawn. This is a public formal garden.

Park and Bagh. The park and Bagh styles emphasises on the natural and green aspect of the garden. Persian parks are public areas and have strong emphasis on plant life but are usually limited in terms of structural elements. The bagh is private area around a Persians house and consist of lawns, trees ground plants, etc. Waterways and pathways are in this type of garden are largely functional.

Important Gardens and Parks of Iran

Bagh-e-Eram (The garden of paradise). This is of the most celebrated garden of Shiraz and dates back from the reign of Naserod din shah Quasar. It is situated at a distance of less than kilometer to the north- west of Shiraz, on the north side of the dry river bed which makes the natural north boundary of the city. This is a spacious garden which have may cypresses tress and magnificent building and a hall decorated with mirror facets and a fine stone edged tank.

Among the cypresses trees in the garden, there is one very tall and symmetrical cypresses tree (300 year old) which specially attracts attention. It is called Sarv–e-Naz, signifying cypress whose branches stretch out regularly in all directions and there is no tree of its kind to compare with it anywhere else in the country (**Plate 21**).

Shazdeh Garden (Kerman Province). This remarkable garden was made in the 1890 by the governor of Kerman – Mr. Abdol hamid Mirza Naserodol of Qasar dynasty and was left unfinished due to death of Abdol Hamid Mirza in early 1890.

It is a rectangular green oasis surrounded by brown desert. The garden covers the area of 5.5 ha and has wall around it. It consists of an entrance structure and gate at lower end and a two floor residential structure at the upper end. The distance between two is ornamented with water fountains and they work with natural incline of the land. The garden is a good example of Persian garden that takes the advantage of suitable natural climate. Mostly plantation is done with cypresses.

The Garden of Golestan Palace of Tehran. This is the oldest historic monument in Tehran. It belongs to a group of royal buildings that were once enclosed with the mud thatched wall of Tehran's historic Arg (Castle). During Pahlavi Dynasty (1925-1979), Golestan Palace was used for formal royal receptions. The most important ceremony was held in the palace during Pahlavi era where coronation of Reza Khan (1925-1941) in Takht-e-Marmar and the coronation of Mohmmad Reza Pahlavi (1941-1979) in the Museum Hall.

On October 11, 2005 the cultural organization of Iran Submitted the Palace to UNESCO for inclusion into world Heritage list. It is under cultural heritage and being operated by them for 2007 onward.

Niavaran Park, Tehran, Iran. It is a public park and located within Niavaran district and is situated immediately south of Niavaran Park complex. The exquisite scenery and mild weather of Tahran has made Niavaran park as a tourist spot. Tourists and locals choose the park for their week end picnics. It has water tank, fountains and is properly planted with trees and flowers.

Sadi Tomb Gardens (Shiraz, Iran). It was an oasis in desert and it was being celebrated over many centuries. It has a spring and subterranean fish pond supplied by a quanaut. Tomb was built after the poet's death but garden was made in 20th century mostly planted with *Cypresses sempervirens, Borassus flabelliformis* and flowers.

Aramgah-e-Hafez (Shiraz, Iran). Hafez is a revered poet of Iran. Every Friday day many people come at his tomb and pay their respect. There are two memorial structures erected. The open pavilion structures are situated in Musella garden which was built in 1773. The building was built in 1935. Several rectangular pools have been added and well maintained orange trees, paths, streams and flower beds create a pleasant environment. The dome of Hafez's tomb is well lit at night.

List of Trees in Iranian Gardens

Cypresses sempervirens, Juniperous excelsa, Phoenix dactylifera, Alnus glutinosa, Buxus semperviens, Castanea sativa, Betula pendula, Quercus up, Laurus nobilis, Tilea tomentosa, Platanus orientalis, Punica granatum, Prunus, cerasifera, Prunus, dulcis, Prunus persica, Prunus spinosa, Sorbus terminals, Citrus aurantium, Citrus limonium, Acer campesiri, Ulmus glabra, Zelkova carpeivifolia, Populus nigra, Populus alba, Salix alba, Salix ciminalis, Morus alba, Morus nigra.

36

Mogul Gardens

The historical events of development of Mogul gardens have already been discussed in second chapter 'The History of Gardening'. In this chapter various features of Mogul gardens are discussed.

These gardens are of formal style, orderly, symmetrically planned and accommodated in rectangular or square plots. Babar, the founder of Mogul dynasty in India, came from Samarkand. His forefathers *i.e.*, Timur Leng had long association with Persia. Therefore, the gardening style of Mogul was an extended plan of basic form of 'Char Bagh' design of paradise garden of Persia. The primitive plan is simple in which plot is divided into four equal plots by providing water channel in the centre. These channels are raised above ground and used for irrigating the field. These water channels represent the four rivers of life. The plots are planted with trees and flowering plants. The axis is straight and central, cutting at right angles to another axis in the centre. Sometime, the axis is represented by line of trees. These water channels are filled with running water which provides coolness and freshness to the garden. Reflection of blue sky, stars, moon, trees etc. in the water provides a unique attraction to the visitors. With the passage of time, water pool in the centre was created at the same level or at higher level. The central tank remained the main feature of paradise gardens for long. Raised level of central tank gives rise to the concept of *Chadar* or water sheet by overflowing water from the tank. Falling of water in the manner of spring added to the beauty of gardens. The beautification of gardens was further done by providing fountains in the water channel and central tank (Fig. 36.1). With the architectural advancement, water was made to flow down from concrete structures. To break the water into ripples these are made up of different styles. In other architectural modification, the central pool is replaced by main building as it is found in the Sikandra garden *i.e.*, Tomb of Akbar in Agra. In next phase of perfection, the terminus of garden has been made up of strong terminal building (The Taj Mahal—Fig 36.2).

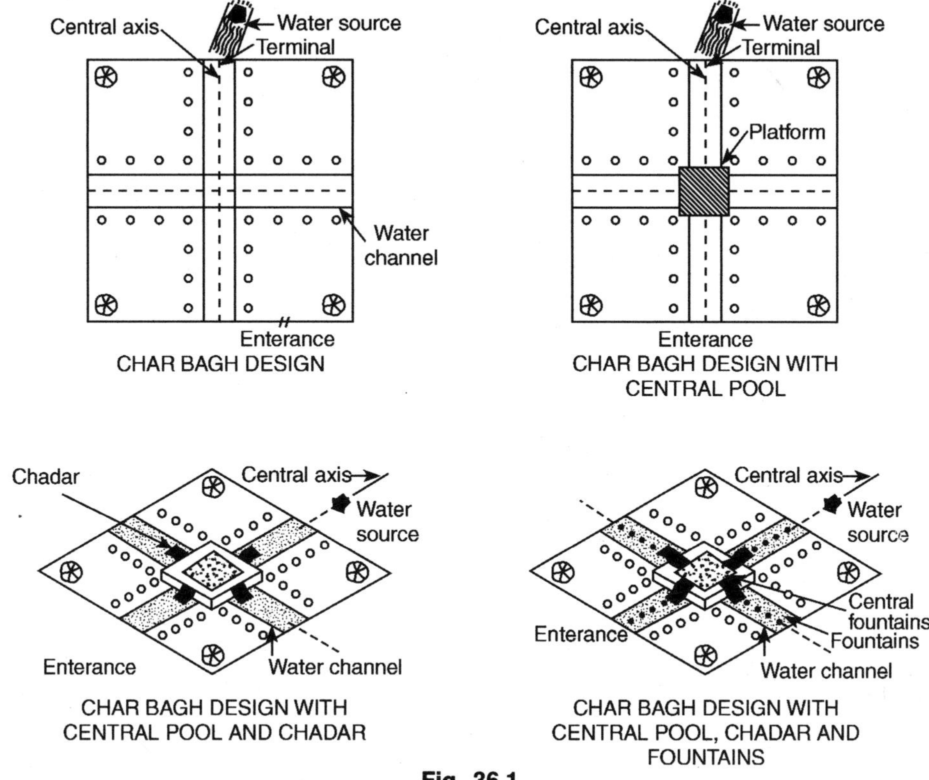

Fig. 36.1

Terraces

To maintain the proportion of land in view of garden, the basic plan is extended into terraces irrespective of topography of the land. The presence of 7, 8 or 12 terraces in the garden symbolises planets or paradise or zodiac, respectively. In many of the Mogul gardens the entrance is at lowest terrace and terminal building at last terrace with the exception at Pinjore garden where entrance is at higher ground and the seventh terrace is at the lowest. On the contrary, terraces are also made on flat ground by dividing into terraces.

Running Water

The presence of running water is the life of Mogul gardens. Broad water canals are made in overall proportion of the area. Central water course is strong and dominant. In the centre there are fountains which makes central axis also. Running water in the form of cascade and *chadar* makes garden splendid. There are different types of *chadars* like silent or ripple chadar, patterned chadar (in water ripples are at level to another level), carved marble chadar or water chute. Another type of cascade is marble cascade made at Agra Fort. In the day gold and silver vases filled with golden are set in niches behind the water. At night these vases are replaced by light forming water lanterns. Such lanterns are in plenty in Taj Mahal.

Fig. 36.2. A view of Taj Mahal
(Showing Terminus Building, Water, Channel, Fountains and Plants)

High Protecting Wall

Mogul gardens are protected by a high strong wall. The top of the wall is adorned by battlements. Such high protecting walls are necessary to provide a protection to the garden against hot, desiccating wind. Trees and other plants are more safer. These high walls also protect against the attack of enemy in those insecure days. Chieftains alongwith wives, and concubines used to take shelter in these gardens in emergency.

Entrance

The entrance of Mogul gardens is an important feature which gives access to the garden. Generally it is a tall and gorgeous. Doors are huge and strong wooden structure studded with heavy iron nails and spikes. Heavy gates provided to protect gardens from enemy attack.

Terminal Building

In Mogul gardens, terminal building is a gorgeous and superb structure. The best example is the Taj Mahal. A great deal of architectural skill has been used in making building. These building have their dynamic force which attracts the mind of viewers. To make the balance, other buildings on the sides have also been made.

Baradari

It is a canopied building with twelve open doors *i.e.* three in each direction. These structures were made to enjoy nature, black clouds, rains and fresh breeze etc. This place was used to hold the parties, singing and dancing programmes. There are also stone platforms or *chabutras* from where emperor could enjoy the breath-taking beauty of the spring blossom or fruit trees. The *chabutra* would probably have been covered with a carpet to give protection from heating sun.

Symbolism in Mogul Gardens

Water is present symbolically and physically as the source of life. The cross, formed in the centre by the intersection of the water channels has been the symbol of meeting of humanity with God as propounded in many religions. Generally gardens are divided into eight parts representing eight division of The Koran. In the traditional planting of alternate cypress and flowering trees along the waterways represented immortality and renewal of life respectively. White *Kachnar* (*Bauhinia alba*) represented the youth and life.

Plant Material Used by Mogul Emperors

The plant material was selected carefully and was planted giving the formal effect. The selection of material depended upon local climatic conditions. They planted fruit trees, flowering plants, shrubs and colourful annuals to give a new life to the gardens. Fragrant flowers provided them the maximum attraction. In Kashmir they planted mostly chinar (*Platanus orientalis*), white poplar (*Poplar alba*), cypress, apple, peach, plum, cherries (sour and sweet), quince, pomegranate and fig plants. In the plains, gardens were planted with mango, grape, fig, gular, aonla, orange, citron, peach, plum, etc. Among spring flowering tress Safed Kachnar (*Bauhinia alba*), palas (*Butea monosperma*), champa (*Michelia champaka*) and Chaltha (*Dillenia indica*), Chandan (*Santalum album*) were planted. Fragrant shrubs like Harshingar (*Nyctanthes arbortristis*), Hari champa (*Artabotris odoratissima*), roses, screw pine, *Jasminum sambac, J. multiflorum, J. grandiflorum,* gurhal (*Hibiscus rosasinensis*) were commonly planted. Flowering plants like Narcissus, Daffodils, Irises, Carnations, Hollyhock, *Tagetes patula*, Larkspur, Stock, Sunflower, Amaranthus, cock's comb, etc. were dominant. The colourful effects were created by massing single colourful annuals.

***

37
Spanish Gardens

The area of Spain is about 200,000 sq. miles. It makes Iberian Peninsula with Portugal and is separated from France by the Pyrenees which is a range of southwest Europe extending along the French Spanish border from the Bay of Biscay to the Mediterranean sea. It makes a gradual slope towards Spain. On all other three sides sea surrounds the peninsula. The alluvial basins of rivers the Guadalquivir, Tagus, Ebro and Duero by group of mountains. In general the climate of Spain is wet in north and west coast and is dry in central plains which has been described as tree less, song less, joyless and without hedges. Due to Spain's geographical situation and orographic conditions, it is divided into three main climatic zones. (1) *The Mediterranean climate*—it is characterized by dry and warm summer, (2) *Semi arid climate*—is found in south eastern Part of the country and dry reason extends beyond the summer and (3) *The oceanic climate*—winter and summer temperature are influenced by ocean. There is also sub types of climate like alpine climate in the Pyrenees and humid sub tropical climate in the canary island. The rain does not fall in the plains but it falls mainly in the north mountain.

Its capital Madrid was founded by Charles V (A.D. 1519-1956) and is described as "Madrid is least influenced by the spirit of beauty and design".

Gardening styles in Spain

Spain has long tradition of making gardens. The development of gardens in different styles have been influenced by various rulers from time to time who ruled Spain. Spain has been ruled by Carthaginian, Roman, Muslims, Turkish, Moors, Nobles, etc. and, hence, the style of gardens have changed from time to time. These are grouped and are briefly described as follows.

1. Gardens of Immigrants from the Carthiginian and Roman Period. Around 1100 B.C, Phoenicians presently Lebanon set up trading colonies along Spanish coast. Greeks also traded along the north southern coast. With the fall of Phoenician regime, the Iiberian Peninsula came under the rule of Carthage (Tunisia) but was then occupied by Rome following the Punic wars. The Roman ruled Iberia for six centuries, laying important foundation of Latin language, Roman laws, the municipality and the Christian religion.

The Palmeral is a landscape groves of date palms and is an oasis. The Palmeral of Elche was laid out at the end of 10th century A.D. where city of Elche was built. Elche is located 23 km of Alcante, and 45 km from Benidorm. It is home of largest date palm plantation in Europe and is unique example of northern African agricultural practice in Europe. Here the date palms are planted alongside the boundaries of irrigated plots of land, helping to maximize the retention of water in arid climate. The same techniques are used around oasis in the North African deserts. Many other flowering plants also have been planted to make it colorful and attractive.

Now-a-days there are 91 orchards in Elche occupying 1.5 km^2 area. This municipal Park is known as the green lung of the area and is most representative of all. The Heurta delcura of Elche is amazing garden with the largest diversity of palm trees.

A famous date palm is the "Imperial palm" (*Palmera imperial*) with 7 stems in the shape of a candelabra and named after Elizabeth, known as Sissy, the Empress Consort of Franc Joseph who visited the plantation in 1894.

2. Nobility, Christians in the Spanish Medieval Period. Spain ceases to be a part of Roman empire and experienced of a series of invasions by the vandals, Suevi, and Alani but was dominated by Visigothic rule. However, Visigoths never got cultural dominion over the region. To a large extent they adopted Hispano- Romans, composing their official documents in Latin and accepting Christianity as official faith.

The Arabs entered Spain from the South at the beginning of the 8th century. They conquered Spain quickly except for a small area of north. The period of Muslim rule is divided into three periods *viz* (*i*) the Emirate (711-756), (*ii*) The caliphate (756-1031) and the (*iii*) Renosade Taifas small independent kingdoms (1031-1492). The year of 1492 heralded the discovery of South America by the Crown of Castile under the command of Christopher Columbus. The canary Islands became a part of Spanish territory in 1495. Cortez, the Spanish adventurer conquest Mexico in 1519 with a small force. Pizzaro defeated the Inca (Peru) and began to rule. They brought back the field crops, like maize, tobacco, cotton, rubber and vegetables like beans, tomatoes, chilli, pumpkin, potato and fruits like papaya, cassava, custard apple, cashew nut and among the flower like marigold, dahlia, zinnia, tuberose, nasturtium, morning glory, heippeastrum, zephyranthes, etc.

On the other hand Spanish travellers and settlers introduced cabbage and peaches and Portuguese travellers introduced lemon, grape, fig, pomegranate, melon, sugarcane, and plantain in Mexico and other South American countries.

Gardens by Islamic rulers, artisans of Al Andalus, the Moorish Iberian Peninsula

Arab civilization when it was at peak but started disintegrating and at last only the kingdom of Granada was survived. Here Moorish castle was built at the hill with a series of gardens and pleasure houses. These gardens give the impression of a number of linked outdoor rooms and corridors. The courtyards strongly reflect the Roman peristyle gardens of Pompei and even the more extended gardens appear to be copy of architecture of a ruined civilization.

Alhambra Palace. There are four principal courts at the Alhambra Palace which illustrates the progressively development of gardens, grow out of the entrails of buildings. The court of the Lions which has lion Fountain in the centre towards which lead four narrow conduits. It does not differ with the spirit of gothic cloister. Modern taste would not call it a garden, but it may not always have been as barren of planting as it is now.

The second court is of Myrtles. This differs from anything of Roman or medieval gardening. The planting is occupied by a long rectangular tank which holds as a mirror of the inverted image of the Arcade and castellated tower at the end.

The third the court of Linderaja, shows how the treatment of these small outdoor room can vary. It is an irregular rectangle, enclosed but not dominated by buildings and is fully planted to provide the shade and provide variety otherwise it would be lacking.

Last of the four courts of the Alhambra is the court of the women, which again takes a step further of development the gardens as a room of out of doors. Here the comparative openness of Linderaja is combined with the reflecting water mirror of the court of Myrtles and a more varied planting. The available space which is not great, split up into a number of rectangular steps showing how the garden retains the essential form and scale of its origin (**Plate 22**).

At some distance of Alhambra there is the garden of the Generalife of a different character. It does not consist of a succession of courtyards contained within building but rather a series of small gardens stepping down in narrow terraces towards the palace.

Architectural ornament dominates in many of these courtyard gardens. Moorish buildings were commonly faced with colored tiles, decorated with arabesques or with poetic inscriptions expressing appropriate sentiments.

Gardens of Post Reconquista Mudejar design and artisans

Mudejar is the name given to individuals Moors or Muslims of Al-Andalus who remained in Christian Territory after the reconquista but were not converted to Christianity. The buildings architect and garden of design of this period is called as Mudejar. The style was strongly influenced by Moorish taste and workmanship. It was highly developed in Aragon especially in Teruel, Zaragoza, Utebo, Tauste, Daroca, Calatayud, etc. Mudejar led to the fusion between Gothic style and the Muslim influences that had been integrated with late Romanesque.

Alcazar. The Alcazar of Sevile is considered one of the greatest example of Mudejar architect and garden style. The Alcazar expresses Gothic and Renaissance styles as well as Mudejar. The gardens of Alcazar are host a few of the treasures that Seville has. Here Arabic ornamentation and Renaissance Baroque dominates. Garden is surrounded by palm trees and abundant vegetation, fountains and jasmine, There are small courtyard with glittery pools, fountains and recessed seats. Decorative grills enhance lush views. There is a tank garden. The planting is of palm, cypress, myrtle, mulberries, magnolia, orange and lemon trees. Glazed and unglazed tiles are characteristics of Arab buildings. In gardens artistic design were made with edge plants which were kept in form by constant trimming (**Plate 23**).

Gardens of Catholic Monarch's during Spanish Renaissance, Spanish Gothic and Spanish Baroque Period

During this period many gardens were made. The important ones are Royal Palace of La Granja and San Ilde Fonso.

Royal palace of La Granja. This palace was built in 18th century in the small town of san Ildefonso in the hills near Segovia, 80 km north of Madrid central Spain. The palace is in restrained baroque style surrounded by extensive gardens in the French manner with sculptured fountains. It is now open to public as a museum.

This is located on the forested northern slope which was the favourite site for hunting by Castlian kings. Henry IV of Castile built the first hunting lodge and along with small shrine dedicated to San Ildefonso. Isabella I of Castile granted both buildings to the monks of Parral monastery in Segovia who built a Granja (farm) and an alm house alongside. The site was purchased from the monk in 1719 by Philip V. who built palace and gardens on the pattern of Versailles. The palace has a **Cour d' honneur** on the approaching side and formal gardens with a main axis centered on the palace. La Granja began as retreat from the court and become a centre of royal government. For the next 200 years La Granja was the courts of main palace where many royal weddings, burials, and political events took place.

Now it is a popular tourist attraction and museum with a collection of paintings and Flemish Tapestries.

The garden spreads over 600 ha, around the palace and is the best example of 18th century European garden. The French designer Rene Carlier used the natural slope from the mountains to the palace grounds both as an aid for visual perspective and a majestic look of 26 sculptural fountains decorated parks. All of the fountains represent the themes from classical mythology including Greek deities, and scenes from myths. The important fountains are: 'Bath of Diana' and 'Fountain of Neptuno'. Rene Fremin was famous sculptor who made several sculptures of fame. The sculptures are cast in lead to prevent corrosion and painted over bronze or lacquered over white oxidized lead to imitate marble.

An artificial lake, El Mar (the sea) lies secluded at the highest point of the park and serve as a reservoir and provide water with pressure for fountains purely on gravity to eject water upto the forty meter high. Now-a-day, only a few fountains are active each day. Twice a year on the feast day of San Fernando and San Louis, all 26 fountains are set to play, providing a memorable show.

The gardens have been planted in formal design with trimmed edges and hedges in different shapes and flowers have been planted in between to make a picturesque effect. On the back, lofty trees have been planted like giant sequoias (*Sequoian dendron giganteum*), sweet chestnut, (*Castanea sativa*).

Most important for the development of the art of gardening was the steady consolidation of the crown of Spain into an absolute monarchy. Philip II continued the policy of his father, Charles V, and consequently great garden could only appear as adjuncts to the royal residences and gardens served more and more as stage for its entertainment. This was chief function of Spanish garden of Buen Retiro and thus theatrical garden was born. It was the creation of the famous Count Duke Olivarez, the powerful minister of Philip IV. It was designed for immediate pleasure rather than future fame. The garden out shone the palace. The principal designer was Cosimo Lotti, a Florentine who had been employed at the Boboli gardens in Florence. There has been a covered walk with an open colonnade (cloister) one side running along the walls of building that face a quadrangle.

San Ildefonso. The last of the great Spanish gardens is San Ildefonso at La Granja. It was built by the grandson of Louis as an echo of Versailles to the design of Boutlet. Its setting 4,000 ft. up among mountains is immeasurably superior to that of Versailles and it enjoys sufficient natural water. However, La Granja lacks the great breath of design and sense of scale. The cascade is fussy, the vistas are dwarfed by proximity of the mountains.

Gardens of Romantic and Modern Period

Gardening hobby is of a rich persons, land owners and business houses as they could spare money and time to passify their passion of gardening.

Park Guell. Very famous garden of this era is Park Guell, Barcelona. It has been designed by Anotoni Gaudi Cormet and was built from 1900 to 1914. This belongs to the court of Guell as an aristocratic city garden with single family residences. At the main gateway there are two administrative pavilions made in stone, catabalan vaults and surface covered with irregular ceramic pieces. The roof of these two pavilions are crowned by original domes reminding mushrooms.

At entering into the park, there is stairway divided by different figures before to arrive, to a colonnade hall with 86 massive columns sustaining the big square. The colored plafonds look attractive. ***The dragon like lizard at the center with trencadis ceramics decorated stair case is the best known symbol of park.***

The big square has a nice balcony for overlooking Barcelona and sea. This square is used as a place for cultural and popular meetings. This square is limited by undulated benches. The park is developed around this square. There is a large network of serpentine paths to walk quietly to enjoy the nature and look at this Gaudi art work. Gaudi was very much interested to have good communication between different parts of parks. For this reason he built a network of path and to connect different levels. He also built some viaducts. These viaducts, paths are the Bugdera path which show a series of different pillars at different levels. All these are built with stone taken from the park land. The portico is supported by a line of inclined columns. The end of portico is formed by a semicircle of columns in two levels. In the opposite extreme of path stands the statue of washer woman as one or more of motifs decorating every column. Each of three viaducts (lower, medium and upper) have been built in a different manner.

Gaudi lived in a house from 1900 to 1925 one year before his death and it was named as Torre Rosa. The house was finally converted into museum, "Casa Museu Gaudi" by 'Association of Gaudi friends'.

The plantation of this park is dominated by date palm, agaves and other garden plants like poplars, pinus, cypresses, etc.

There are other gardens in Barcelona which one should visit. These are Parc di la citronella, parc de Estacio de Nord, parc De Labyrinth, Barcelona.

Gardens of Civic Projects and Exposition

In many countries in recent past many civil projects by central, state, Governments are being taken for landscaping and garden development. Spain is also one of them. One of the such project was Mary Luisa of park. This was built in Seville in 1928 for the Ibero-American Exposition of 1929 which is in a Moorish Revival style in Spanish architecture.

Maria Luisa Park. In 1929 Seville, hosted the Ibero-American Exposition. World's Fair was located in Maria Luisa Park. The entire of southern end of city was redeveloped into an expanse of gardens and grand boulevard. The centre of it is Parque de Mario Luisa, a Moorish paradisiacal style with a half mile of tiled fountains pavilions, walls, ponds, benches and lush plantings of palms, orange trees, Mediterranean pines, and stylized flower beds and with vine hidden bowers.

The Maria Luisa Park with its many tiled landscape elements, continued to influence the landscape design, architecture and urban planning. With tree and shrubs which have reached maturity now makes the perfect landscape design more powerful and dynamic.

The Plaza de E Spano is a huge half circle with building continually running around the edge accessible over moat by numerous beautiful bridges. In the centre there is a large fountain. By the walls of the Plaza there are many tiled alcoves (a recesses rooms) each representing a different province of Spain. Recently many Govt. buildings have been shifted in this Plaza.

Gardens of 21st Century

There are 68 municipal parks, in Barcelona which or grouped into historic parks (12) Thematic park or Botanical Park (5), Urban parks (45) and Forest park (6). The size varies from vest—pocket parks to large recreation parks to large recreation areas. The urban parks alone cover 10% of the city area (549.7 ha). Amongst Barcelona's park (**Montjuick park**) is the largest with 203 ha, located on the mountains of the same area. Other parks like pare de La Ciutadella which houses old military citadel and parliament building, **Guininado Park, Diagonal mar Park, Non Barris Central Park, Can Drago Sports Park, Poblenon park and Labyrinth park** are worth visiting. Apart of these, **Collseroller park** is within city limits.

***

38

Italian Gardens

From the very early time, Italians have been intellectually and expertly cultivating their lands. They displayed their creative nature by combining terraces and walkways with plants, flower, water, fountains and statuary in making the gardens. The use of citrus plants in pots dominated in early Italian gardens. The art of topiary was revived in Renaissance and refined parterre was developed to spread forth from the great plazzos which provides a dramatic view from upper stories. For more than 500 years, the garden of Italy have been irresistibly attractive and have been greatly evolved through ages. Many of greatest artists and architects of Renaissance were employed in their design and ornamentation. Several captivating gardens are spreaded all over in Italy in 11 regions from Lombady and Piedmont in the north to the island of Sicily in south.

Climate

Italy is divided into three main regions according to geographical positions and prevailing climatic conditions. These three divisions are Northern Italy, Central Italy and Southern Italy. The main features of these regions are described briefly.

Northern Italy

As the name indicates it is the northern part of Italy bordered by France, Switzerland, Austria and Slovenia. The vast territory throughout Northern Italy comprises of varied terrain specially the beautiful mountanceous regions of Apennines and Alps which is wonderful place for hikers, climber and skiers. The main citices of Northern Italy are world renouned and recognized in history are : Milan, Venice and Verona. Venice is the country's prominent art hub and is capital of Veneto regions and home of lake Grada.

The Climate of this region is temperate type with chilly winter (0-1-2°C) while summer (30°C) is sweltering hot and humid. The annual precipitation is 100 cm. Snow fall is common during winter and accompanied heavy fog.

The main plants of this region are cork Oak (*Quercus suber*) European olive (*Olea europa*) Cypress (*Cypresses sempeverins*) Chery laurel (*Prunus lauro-cerasus*), Norway spruce (*Picea abies*), Rhododendron, (*Rhododendron arboretum*), Dwarf Juniper (*Juniperous proecumbens*), and wild flowers like Gentian (*Gentiana acaulis*), Rock Jasmine (*Androsace lanuginosa*), Primrose (*Primrose oenothera*) and Saxi fraga.

Central Italy

The weather of Central Italy, which stretches from Bologna to Rome, is more temperate than Northern Italy. It is chilly in winter, and temperature generally remains above freezing point. Summers are much drier, less rain and humidity is significantly lower throughout the summer. Near Apennine mountains there is fog in the mornings throughout the year, but it will be over by afternoon in the summer months.

Central Italy include the area of world famous Tuscany, Marché at Umbria where charming cities run coast to coast, rolling hills offer a pretty back drop. The three most renowned cities in Central Italy are. Rome, Florence and Pisa. Top attractions in Rome include the Trevi Fountain, Revered trastevere neighbourhood, the veteran museum, the seventeenth century Palazzo Doria- Pamphilij and celebrated musei Capitolini, designed by Michelangelo. Marché is a beautiful region bordered by the mountains in Italy and Adriatic sea.

Another important area of Central Italy is Tuscany which exudes a flavour of its own. It is perhaps the most credited for collections of architectural masterpieces, sculptures, paintings and gardens.

Southern Italy

Southern Italy is comprised of Napoli down to the bottom of the country's iconic boot including two islands of Sardinia and sicily. Main cities are Naples, Bari, Toranto, Foggia, Salrerno, etc. Since 1990, the Cilento is second largest national park in Italy and maintained by UNESCO world Cultural Heritage of mankind. It is situated in the province of Campania.

The climate of this region is Mediterranean type. It has most pleasant climate which is characterized with mild winter. Summer months of July-August are characterized by blowing of cold ad dry wind from mountains so the area remains pleasant. The area receives little rainfall during the year and practically none in summers months.

History of Italian Gardens

As the renaissance spiraled out of Florence and interest of the world and the respect for its natural beauty went along with it. Many rich Italians spent enormous money in creating magnificent villas in the country's fertile hillsides. Terraced gardens and giant fountains are found on many of these estates. Luckily a good number of these have been carefully preserved throughout years. Most probably, most scenic garden of Italy is Florence's Boboli garden and the most scenic region of Italy is Tuscany and have many fine Italian gardens. The villa Reale de Marlia and villa Gamberia are important Gardens whose hilly location and accommodating pools are at most superb in Italian garden design. Another major centre of Italians gardens is Lake Como, where the wealthiest aristocrats of the 1800 built their Villas like villa Cicogna Mozzoni and Villa Carlotta. The villa d' Este of Rome in the city of Tivoli is very famous of splitting gargoyles, cascading waterfalls and top heavy cypress trees.

The brief historical facts of Italy reveals that Italians became rich during 1094-1244 after the successful crusades and small group of gifted persons in art, sculpture, poets, architects and gardener got the patronage of rich people, bankers and merchants.

Greek influence was much more prominent before AD. 1453 where Muhammad II, the sultan of the ottoman Turks captured Constantinople, the capital of eastern Roman empire and forced Greek scholars to flee the country and they find refuge in Florence under the patronage of Cosmo-de-Medici (A.D.1389-1464) who was richest bankers and leader of dominant party. Members of platonic academy used to meet at his villa Careggi which had a splendid garden. The garden was planted with box, cypress, myrtle, pomegranate, lavender and other scented herbs, flowers, like carnation, oranges, lemons green in pots grouped around fountains and along the pathways. A vineyard was also planted nearby.

Cosmio's son Peiro continued the patronage of art and learning. Laurentean age began in A.D. 1469 under the patronage of Lorenzo, the eldest son of Piero. This was the era of Leonardo da Vinci (A.D 1452-1519) the greatest Painter and sculptor of renaissance. Other outstanding artists were Michelangelo (A.D 1475-1564), Raphel (1483-1520), Titian (1477-1576), Correggio (1494-1536), Paulo Veronese (1528-1588), Tintoretto and Caravaggio (A.D.1565-1609).

Renaissance Italy produced great art scholars which continued to maintain charisma during machine age. Actually Romans were creators of gardens of Renaissance. The middle age was dominated by the construction of castle, cathedral and monasteries. Its famous style was of Gothic Cathedrals with pointed arches. Its heavenly pointed arches lead the thoughts to the great mysterious power God.

The gardens of Italy can be grouped into six Types *Viz* (1) Roman Gardens, (2) Medieval and early Humanist gardens, (3) Tuscan Gardens, (4) Roman Renaissance Gardens, (5) Gardens of Marche and Veneto and (6) the Gardens of Northern Italy.

The garden style and important gardens are briefly described as follows :

Roman Garden

The Greek culture and gardens spread in Mediterranean and far beyond area after Alexander's conquest was introduced. In Greece, as in Italy, springs and groves of trees had long been dedicated to the gods and temples especially with the deities of nature and fertility. Most famous of these was a park dedicated to Artemis sister of Apollo (sun) at Scilleus. Through their early religious association had now been practically submerged by artistic and social ones, never the less, the component parts of gardens remained the same. The small temple grottoes and nympherems that had originally been shrines dedicated to Muses and Tutelary deities now served as architectural ornaments. One interesting feature appear to have added *i.e.* Topia, portraying its garden architecture in a picturesque setting of rigged mountains and sea side cliffs or on the shore of lakes and rivers. Topia were used to decorate walls of porticos and thus producing the effect of bringing the garden right inside building. It became a distinctive feature of Roman and Renaissance gardens.

The fashion of pleasure was spreading slowly and it provided link between Greece and Rome. Most of early villas and pleasure garden were situated between Rome and Naples near seaside as resorts of Formica and Terracina. These villas resembled those of Pompeian gardens which were of two kinds either a country and garden in the colonnade peristyle of the house where plants were kept in pots or vases or small beds surroundings a pool, or in houses where there is small central courtyard, a portico opens into the garden behind the house. In some cases portico is prolonged and closed from two or three sides. Roman gardens drew the inspiration from such gardens like interpenetration of houses and gardens and the axial planning. Main living room usually opened into courtyard on one side and garden and axial planning on the other side and linking the two. One could see the centre length form the entire of the house.

Apart from paintings and miniature reproduction of landscapes, the Greek word Topia was also used in a Latinized from Topiarus to designate gardener who practiced the art of topiary.

The choice of the site for a villa was given a utmost importance, not only for aesthetic considerations such as view but considerations for health, comfort, sun and wind were given. Thus, one of the main reasons for the choice of a south eastern aspect was that low winter sun would shine into colonnade porticos that formed an important part of Roman villas and garden. These porticos which were called as gymnasia were in Roman gardens only as sheltered walks. Usually they were situated on terraces so that owner could see the view while coming down. Collection of paintings and sculptures formed an integral part of Roman gardens.

The Great Roman Pleasure villa was infact self supporting entity where they grew fruits, vegetables flowers and animals like goat, deer, ducks, Pigeons and thrushes were also bred for eating. Thickets of trees and berry bearing bushehs specially planted to attract wild song birds for nesting, fish ponds and aviaries also existed.

The most celebrated and extensive landscape garden of Rome were those of Nero's Golden House: The garden has an enormous pool, more like a sea which was surround by building made to resemble deities and a landscape garden consisting of ploughed fields, vineyards, pastures and woodland where every variety of domestic and wild animal roamed. It has been described that its wonders was not customary and common luxuries like gold and jewels but were its lawn, lakes, woods, open spaces and view. Nero's architecture outbid nature.

With the passage of time, Italy and Western Europe entered into darkage. The Rennaissance brought human delight in gardens and two fabulous pleasure garden's with statues of heroes and monsters were made with stony gorges of ciminion hills.

Medieval and Early Humanist Gardens

During this period gardening was limited to cultivations of medicinal herbs in pots and flowers for decoration of churches. The plantation of preserved monasteries depict the growing of flowers like roses, gladioli, lilies, amongst scented herbs like rosemary, sage, and vegetable like leeks, lettuce, garlic parsley, were grown.

During this era garden was potrayed in medieval miniature as a grass plot with a fountains in the middle surrounded by a herbs like sage, basil, marjorum, mint and occasional lilies and roses. Turf seats were shaded by vine pergola and sweet scented trees of apples, pears, pomegranate and cypress planted around the well. The second stage had an large orchard and a trellis arbour and was to be fenced around with white sweet briar roses and other prickly shrubs with an outer hedge, of pomegranate, in hotter climate whereas quince and hazel trees in colder parts. The art of clipping of trees to make battlement walls of greenery the garden was followed. Enclosures and bowers of trees planted and trained in such a way that their interlaced branches formed entire walls and roof.

Though most of medieval gardens vanished and found in miniatures medieval manuscripts and in rare frescoes such as 'Triumph and Death' and the 'Garden of Eden' in the Campo Santo at Pisa. But one such pleasure garden survived in 13th century Palazzo Rufdo at Revello.

The 14th century was the period of transitions in Northern Italy. Castles began to evolve into Pleasure Palaces. Galeazzo Visconti II of Milan, who was a keen botanist and collected lot of plants. It was the famous **'Castle Pavia'** that Galeazzo began to build and where Petrarch stayed with him for successive five summers.

The Pavia castle was planned as pleasure garden with revolutionary design with an extensive garden and collection of plants reflected that Galeazzo was a patron of art which were the features of humanist gardens of the next century. Though it was walled and designed like a medieval castle and its main features were such as fish ponds, central lawn, pergolas and fruit trees. Thus 'Pavia castle' is claimed to be first Renaissance villa of Northern Italy.

Cosimo de 'Medici' who used to invite his friends in garden at his villa to discuss the Philosophy and influenced the European civilization for next 200 years.

Cosimo's brother Giovanni de Bicci in 1417, got transformed his old castellated house with changing spirit of time and got planted with box, cypress, myrtle, pomegranate, quince, lavender, scented herbs and flowers.

The intimate classical relation between house and garden was revived under the direction of Albert in 1446. Apart from Loggias linking it to the garden, rooms where also painted with garden subject such as garlands, and fountains. Paths were to be bordered with symmetrically planted flowering trees such as pomegranate and cornelian cherries garlanded with roses, box and scented herbs. There were the other charming habits of ancients like to follow geometrical designs or to write the name of owner. Vine pergolas, supported by marble columns, planted with Juniperous, myrtle, oaks, cypresses garlanded with ivy.

The importance and aspects of gardens in the mind of 15th century humanist has been described by Francesco Colonna in 1467. There was a rose garden planted by a humanist with rose named by Pliny. The Damascene, Pernestine, Companion, Milesean, Lychman and roses of Paestum of classical fame. This was surrounded by priestly garden whose Corinthian columns strike a truly renaissance note in garden design planted by scented shrubs, marjoram, lavender box, juniperous, small myrtle clipped in various shapes grown in antique vases and pots.

In this era, the planting of parterres had become a highly developed art. The use of knots in scented herbs whose contrasting tones of green would have decorated colorful flowers was common. These shrubs had to be clipped constantly and replanted every three years, if they have not become shaggy. It is evident that trained gardeners must have been available at that time.

Interesting aspect of the gardens was the architectural setting of Amphitheatre and peristyle gardens with antique reliefs, statues, herms altars used as garden ornaments. The architectural exploitation of a sloping site was a third dimension in gardens. Thus the early gardens of the quartocento were transitional in characters with flat or sloping enclosures divided into square or rectangular division by covered green galleries and with central fountain and lawn before house. These transitional type of gardens reached at its peak in Florence because they were originated here.

Tuscan Gardens

These gardens have been conceived four hundred years ago by Tuscan Aristocracy. Typical Tuscan Garden reflects the images of formal renaissance gardens, however, in four hundred years a lot of changes had been adapted due to social changes and horticultural advances. The link between man and nature in Tuscany is very strong but social revolutions, political changes and tourism have influenced the style of Tuscan gardens. Due to curious and tense relationship between aristocratic Tuscan land owners and the peasants, have been adopted many symbol that reflected Tuscan culture. For example mulberry tree (*Morus alba*) is almost present in an authentic Tuscan garden. It lives many years and provides leaves for rearing silk worm which served as a cash crop. Similarly wild flower meadows were encouraged for the farm animals. Tuscan's farmer wives can be seen in collecting fresh herbs from the meadows to fatten and flavour the meat of their rabbits and to create natural medicines from these herbs.

During the renaissance period, the style of Italian gardens was formal. Tuscan gardens, water features and clipped hedges influenced the European gardens in following centuries.

There are about 39 gardens in Tuscany Italy and all are worth seeing. The important ones are Villa Medici at Fiesole and Castello, Marlea villa Reale garden, Villa Garzoni, Villa Gamberiza, Giardino de Boboli, Gardino Bardini, Plazzo Piccolomini garden, Villa Poggio Torselli, Peyron Villa garden, Villa Torrigiani garden, Poggio a cajano, Villa La Force, etc. For better understanding the style of few Tuscan gardens are described hereby.

Villa Medici at Fiesole. This villa at Fiesole was built during the mid 15th century when Elder employed Michellozo di Bartolommeo to design the villa for his second son Giovanni dei Medici. It was intended to be a setting for a intellectual life rather than a working villa. The quadrangular building is a typical 15th century edifice, with square pietra serena window and broad loggia looking out over the surroundings. The villa remained the property of the Medici until 1671. After changing many hands it was purchased by lady Sybil Cutting.

The Geographical position of villa shows that it is situated on gently sloping land suggested the layout of garden on three terraces. The first of these at the end of an avenue lined with cypress trees that runs underneath a holm-oak wood has rectangular lawns with potted lemon trees. The second terrace is over looked by the near of the building and is reached by indoor staircase. This is the least altered part of the garden, has flower beds lined with box hedges with a large fountain in the centre and is laid out in the shade of large magnolia trees. The third terrace was created between 1911-1923, by Ceil Pinsent and Geoffrey Scott, is aligned longitudinally with the first but is 11-12 m lower down. It is laid out in Italian style with a fine pergola positioned mid way between two levels.

Boboli garden. It is another magnificent garden of Tuscan. From the painting of Uttens Lunette of Boboli which is in the Museo Topografico. It is vivid that Tribolo's layout of this original part of the gardens which was continued by Ammanati was different from what it is today. The contour of the small valley behind the palace was modified to form a definite horse shoe shape but more than half was filled by regular plantations leading a medium sized square lawn with a fountain in the middle as a central motif of the whole layout. From this long alley, forming the main axis is led through plantations to a regular tank on the hill above, where the fountain of Neptune stand, and symmetrical arranged walks delineated the outline of the horse shoe shaped valley. By the first half of the seventeenth century the Boboli garden had already under gone many changes under the direction by Baccio. The great amphitheatre has been cleared of the plantations and wall with tiers of seats reflect an ancient Roman hippodrome. This innovation provided a special setting for the lavish pageants and entertainments and grand ducal marriages which were very common during 17th century.

Garzoni Gardens. Garzoni castle dates back from early1600 and was built by Garzoni family form Pescia who were sent into exile. They decided to built a villa in Collodi near ancient border of the Republic Lucca. By 1652, the garden laid out in present form.

Garzoni is a beautiful Baroque garden. Such gardens were made for show purposes. High society used to gather and admire. The characteristic of baroque gardens were a centrally positioned building, elaborate parterres, fountains, basins, canals, etc.

Garzoni garden has pools, waterfalls, statues and a maze. **Maze** is an intricate usually confusing network of interconnecting path ways with the help of clipped edge plants. From the entrance the garden unfold before your eyes creating a magnificent scene. The Garzoni garden is considered to be one of the best garden not only in Tuscany but in Italy too. Whimsical real statues such as monkey adorn Garzoni gardens. The Villa Garzoni sits above the garden and can be seen from several vintage points of the garden (**Plate 25**).

Roman Renaissance Gardens

During renaissance, the development of art, sculptor, music and gardening got influenced by scholars of Italy, Greeks or muslim rulers time to time. The other factors which stimulated the apt for gardening and style of gardening was the constructions of big villas or huge mansions and planting of gardens was due to comparative peace and maintenance of law and order in Italian cities. The architecture of fear gave place to architecture of pleasure and happiness. Though that villas had no heating arrangement, they were well lighted and decorated with paintings and linked with formal gardens. The gardens of renaissance era were characterized by terraces linked by magnificent staircases, often decorated with statues and green walls or rooms of clipped evergreens. The design and planting of separators become a highly developed art. Scented herbs, planting of clipped myrtle and cypress along with covered walks were very impressive.

During this period various botanical gardens were established in Italy. The first of such botanical garden was established in 1543 A.D. at Pisa followed by at Padua (A.D. 1545). These botanical gardens become a place of introduction, multiplication and distributing centre of improved plant species. In Pisa botanical garden, plants introduced were horse chestnut (*Juglans nigra*), *Cinnammum camphora*, *Flowering Quince*, (*Chanomeles japonica*) *Magnolia grandiflora* and tulip tree (*Thespesia populnea*).

At Padua, there is a green house which shelters a palm specimen of Zingaro palm or Mediterranean palm (*Chaemerops humilis*). The other plants which were first time introduced in Italy and Europe were *Begonia radicans, Cederus deodara, Cyclamen persicum Jasminum nudiflorum, Robinia pseudoeasia* and number of Bamboo species.

After Florence, Rome became a central place for art, literature and gardening. Cathedral of St. Peters was built by Pope Julius II under the patronage of Michelangelo.

He engaged Bramante a famous architect for the design of Vatican garden. Bramante set the trend of shaping of whole hill side into garden by making of series of terraces and open air theatre as a part of ever green garden. The spectacular staircases made to link the parts of the garden. Thus the Vatican gardens combined the beauty of architecture with that of plants. Graceful sculptors are placed in niches whose wall covered with figures of birds and plants provided a pleasant frame work. Entrances and staircases were decorated with clipped hedges and topiary.

The important gardens of this era are Villa d Este Tivoli, Medici Villa lante at Bagnaia, Villa Madama, Villa Caprasola, Villa pamphilji, Villa Ruspali at Vignanello, Villa at Frascati, Villa Aldobrandini, Villa Borghes, etc. Few in brief are described hereby :

Villa d' Este at Tivoli. This is considered as the most Roman as well as most typically renaissance garden of all Italy. The Roman influence is seen as hillside converted into garden. Renaissance effects are visible as every inch of space within boundaries was subject to manmade order and symmetry. Terraces of hill side are crossed at right angle. The crossing and subsidiary paths were marked by open spaces and all of them were planted in a relation to each other and accentuated by fountains and stone work of same kind. Each pause in the ascent of hill side garden is marked by a new and varied use of water, by fish pond at the lower level, glittering water staircase as climb begins there by great monumental fountains enclosed in garden room among the trees. The superb promenade of the hundred fountains provide a breathing space. The terminal and junction points of ramps are marked by grottos and fountains. Finally last silver jet sprays the water upward to the sky as the great open terrace before the villa is reached. And at last the climber turns to gaze at vast panorama of the Roman Campagna. There is a great fountains of the ovata in the villa is one of the most beautiful representation of an Italian garden scene (**Plate 24**).

The fountains of Villa d' Este have for centuries been one of the sight of Italy visited by every foreign traveller. Villa d' Este was planted with number of trees like cypresses, bays (*Laurus nobilis*), *Lentisus arbutus*, myrtles and Holm-oaks and the ground underneath was carpeted with soft green moss.

The Villa d' Este was commissioned and built by cardinal Ippolito, II d' Este the son of Alfonse I d' Este and Lucrazia Borgia and grandson of pope Alexander VI Pirro Ligorio worked 17 years in designing the garden.

Villa Medici. The Villa Medici is founded by Ferdinando I de Medici grand Duke of Tuscany. It is an architectural complex centered on the villa whose gardens are contiguous with the larger broghese garden. In ancient times, the site of the villa Medici was part of gardens of Lucullus. The villa Medici on the Pincian hill is the only great Roman Villa of 16th century whose garden layout has survived almost intact. The design of garden villa is essentially that of town villa. The most striking feature is the clear space immediately adjacent to the house behind which lies a well drilled garden of floral parterres, while the shady walk and cypress planted terraces stand at a respectable distance from it. The reason for placing the parterres behind the house was that they can be seen from above and full effect of their design would be appreciated.

The Great persons associated with it were Micheloangelo, Vasari, Ammanati and Vignola. The villa originally was superb but it has been shorn of its garden over a large area in all directions. The planting was of 36000 trees of elm, chestnut, cypress, bays, Myrtle and pomegranate and it was filled with fountains, fish pond and grottoes, statues, aviaries and pavilions.

The central focus of the whole layout is the magnificent horse shoe shaped court which has been divided by a porticoes screen.

Napolean Bonaparte in 1803 moved the French Academy to Villa Medici from Rome to protect it from French revolution.

Villa lante at Bagnaia. The main features of villa lante are cascades, fountains and dripping grottoes. The quarato is a perfectly square parterre. The twin houses stand on one side and on the remaining three sides the garden is enclosed by high hedge of box. In the centre, low box is sculpted and formed into decorative pattern around small fountains and sculpture. The feature of this parterre is the complex fountain at its a centre, formed four basins and separated by parapeted walks. At the heart of the complex, a centre basin contains the *'Fontana dei Mori'*, four life sized moors stand square around two lions. They hold high the heraldic mountain surmounted by star shaped fountain jet, the *Montello Coat of Arms*. This is the focal point.

In the first of ascending terrace lodged between two stone staircases is the fountain of lamps. On the next terrace, there is large and long stone table with a central channel with water flowing. At the back of this terrace are sculpted river gods flaming a fountain. The rill of small basins allow the water to ripple down to arrive at the fountain between sculpted crayfish claws.

On the next upper terrace are further fountains and grottos and two small houses of the Muses. The sides frame of the large fountain of the Deluge that terminates the axis of garden. The main façades of these small houses feature Serliana loggias articulated by ionic columns. One house gives access to a small secret garden, a garden of hedges and topiary with a line of columns.

A perspective plan of 1609 shows a wooded area of walks and vistas to obelisks (obelisks is a tall narrow four sided topiary monument which ends in a pyramid shape) and maze.

Gardens of Marche and Veneto

Region of Marche and Veneto lies in the centre on the coast of Adriatic Sea. The area consists of rolling hills offer a pretty back drop. Thus, the gardens are terraced gardens.

Gardens of Marche. There are three famous gardens namely Giardino Buonaccorsi, Caprila villa and Villa Imperiale. These are described briefly.

1. *Giardino Buonaccorsi.* This garden was first made in 15th century and has been maintained in its original form in 18th century by Contessa Giuseppina Buonaccorsi. It is considered an extraordinary and stunning garden full of fascinations. Its superb distant vistas are attractions for visitors and garden designers. Its main attractions are mosaic stone work, statues and sea views. The house stands on the crest of small hill with a delightful landscape of garden hills and valley extending between it and the sea. The hill behind the house and its southern slope have been laid out as a series of walled and sheltered terraces, open to the sun and with a lovely view. There are five of them in all. The highest which lies between the houses and the chapel has open forecourt and a slightly raised Giardino Segreto

2. *Caprile Villa.* It is a walled terraced garden with restored water games. Its surprising jet's of water is more popular with the visitors.

3. *Villa Imperiale.* It is renaissance villa on a steep slope of mountainous terrain. It was damaged during Second World War and has been restored now.

Gardens of Veneto. In the northeastern corner of Italy, Veneto is beautifully located between the mountains and the biggest lake in Italy, Lake Garda. The landscape of Veneto is varied. To the north it is dominated by majestic mountains and flat landscape in south. The climate is pleasant all year which favours good growth of plants.

Veneto has about 13 gardens and important ones are : Giusti Giardino, Villa Pisani, Valsanzibio gardens, Villa Emo Rivella Garden, Botanical garden, Padua, etc. These Gardens are briefly described hereby.

Giusti Giardino. It is one of the Italy's famous gardens and dates back from 16th century which was modified in 18th century and restored in 20th century. The garden is on a hill behind the villa, has terraces, parterres and a grotto. Planted with cypress, maze edge plants, statues, fountains and lush green lawn. Decorated with flowering plants in pots at appropriate places.

Villa Pisani. A 18th century palace with a baroque garden was made in 1730-40. The central canal which now dominates the layout was made in early 20th century. Planted with maze of hedge and edge plants, have statues, and arbors.

Valsanzibio Gardens. It was laid out is 1669 and it is liked due to its site and perfect planning. This is considered as an amphitheatre of hills and house stands sideways in an amphitheatre. The end is linked by a great avenue flung across the valley. The garden has sculptors, pools, fountains, lemon, trees and box, hedges. The baroque garden have seventy statues, cascade, fountains and water features. The trees cypresses and parterres with hedge is worth seeing.

Villa Emo Rivella Garden. A Palladian style villa of 16th century attributed to Vinceza'O Seamozzi, with an extensive flower garden made after 1966. Though a recent creation is still classically Italian in conception uniting typically Italian renaissance element such as pergola, orchard and fish ponds and includes an exceptional variety of flowers.

Botanical Garden, Padua. This is one of the earliest botanical gardens in Europe for collection of plants, maintenance and teaching to the students. In Europe potato was grown first in 1590. The German poet, Goethe visited this garden in 1786 and planted a palm tree and anticipated the theory of evolution. The garden is protected by a circular wall, rebuilt during the 18th century, with statues and bust. The circular wall encloses four square gardens, two of which have circles within. The use of circles and square symbolize the perfection.

Gardens of North Italy

In North Italy, gardens were made by great Genoese admiral and statesman Andrea Doria in the beginning of 16th Century and existed for three hundered years. In 1561, Jan Massys artist painted **"Cytherian Venus"** in which he portrayed the Garden with a wonderful setting, terraced gardens with marble fountains and pergola's sweep down to the water's edge where porticoes palace to look distant views. At distance he painted cliffs and sunlit sea. Such gardens became popular in Genoa. The beautiful site near sea shore of Genoa and luscious vegetation was called paradiso. So many gardens came up with similar design.

In North Italy the important gardens are : Doria garden, Palazzo Podesta, Villa Cicogna Mozzoni at Bisuschio near lake Lugano; Castello Balduino at Monalto Pavia; Plazzina de Caccia of stupinigi, Garden near Turin, Villa Carlotta and Villa Crivelli Sormani. Out of these few are described hereby.

Villa Cicogna Mozzoni. It was established before 1440 as a hunting lodge of the Mozzoni family. It was completed between 1530 and 1559 with frescoes and wall paintings by the Campi brothers. The most pleasant characters is merging of lodge with garden. One enjoys two flight stair way which leads directly to the halls of villa and also enjoy from above the sight of beautiful panorma. The wall in garden is typical example of formal Italian garden with box hedges, water basins, fountains and a grotto with remarkable water works. The garden and villa were built simultaneously to have continuity between outside and inside. The first garden (near to the partico) with two large fish ponds and many statues is like an open air room with the bay hedge on other side. The two other terraced gardens looking north are in perfect continuation of the house. Finally the English park with green houses complete the garden. There are many trees like *Fagus sylvecta atropurpurea* (*beach*), *Taxus baccata*, *Cedrus libani*, *Quercus suber*, *Thuja occidentals*, *Ilex aquifolium*, *Quercus ilex* and *Camellia joponica*. The view from the top towards lake Lugano and mountains is beautiful.

Now-a-days villa is used for social gathering cultural events, shooting, exhibitions, marriages etc.

The Plazzina di caccia of stupinigi. It is one of the residence of Royal House of Savoy in Turin Northern Itally. The new palace was designed by architect Filippo Juvarra for Vittorio Amedo II. It is a hunting lodge and symbolized by the bronze stag perched at the apex of the stepped roof of its central dome. The building has a saltire plan: (it is a heraldic symbol in the form of a diagonal cross like letter X). Which has four angled wings project from the oval shaped main hall. The extension of this resulted in separate pavilions by long angled galleries and long octagonal forecourt enclosed by wings, extended in two further entrance courts. It was the preferred building to be used for celebrations and dynastic weddings by the members of House of Savoy.

Castillo Balduino at Monalto Pavia. It is located at Montallo Pavia in 1909, the Balduino family bought the garden and castle and invited Giovanni Chevalley to restore the garden. The garden after restoration includes fountains and garden sculpture along a narrow walk. The shapes are geometric mostly of box and yews terraces that follow the bastions of the original 13th century castle. The garden has majestic entrance and potted lemon trees and topiary.

Villa Carlotta. It was completed in 1745. It was presented as a wedding gift by Princess Marianna of Prussia to her daughter Charlotte, after whom the villa is named. It is on lake Como in Northern Italy. It covers an area of 7 ha. in Tremezzo facing the Bellagio Peninsula. It is a typical Italian garden with steps, fountains and sculptures. The statues of mars and venues are very attractive.

By the beginning of 18th century, France rather than Italy had become the acknowledged centre of garden design. On Northern Italy, French gardens had become the fashion and many gardens were built on French style. French gardens influence was also reached to other European countries like Spain, Sweden, England, Russia.

Though the Italian gardens are still centre of attractions in its original form, but after 19th century informal style of English garden started influencing the Italian gardens too with meandering path ways, asymmetrical water bodies, irregular flower beds, planting of flowering trees, etc.

***

39

English Gardens

Changes are nature's rule and are inevitable. The change in weather, season, climate, human likings are bound to occur. The change in garden style of English gardens with the passage of time and human likings is the best example of changing phenomenon. English gardens are most beautiful gardens amongst all European gardens. There are three main factors responsible for making gardens so beautiful. They are : (1) England enjoys the typical grassland climate with showers of rains at frequent intervals. This is highly favourable for the growth of herbaceous vegetation. (2) Gardening is the hobby of rich people. Englishmen in due course of time became very rich and were ruling over many countries. With the help of money, collection of valuable plant materials from various countries was done. In making of gardens and maintenance of gardens, good amount of money is spent. Money is also used for various garden ornaments, artificial waterfalls, fountains, statues etc. (3) Englishmen are very fond of flowers.

In England's history of gardening, there are three distinct phases after passing through which gardens have reached at its peak stage.

I. Search of Sustenance (1066-1500 A.D.)

During this period, gardens were attached with monasteries and were planted with fruit crops, vegetables and other shrubs with the sole purpose to provide the food to these monasteries. These gardens were mainly utilitarian type. In these gardens, roses, carnations, violets, lilies, holyhocks, and other flowers were planted. Amongst fruit plants, these monasteries were planted commonly with vineyard, apple, pear and peaches whereas fruit trees like mulberries, cherries were restricted to the gardens of nobel persons.

The English gardens during 13th to 15th century were orderly and very formal. Though it looked unnatural. During this period, the population of England was sparce and mostly big houses were planned in the center of garden, orchard or farmyards.

II. Elizabethan Period (1500-1700 A.D.)

During this period very famous Hampton Court was built by C. Wolsey. It has a geometrical plan and walks were planted with fruit plants. It had resting places. Henry VIII (1509-1597 A.D.) converted Hampton Court into one of the royal palaces and garden was enlarged with fresh plantings. He built a mount of geometrically artificial hill. It had a pavilion, statues and effigies of hearaldic beasts. Flower gardening got a special stimulus and reached the age of symmetry at the time of Henry VIII. There were symmetrical walks of clipped evergreens, statuary and masonry. Apart from these, metal stags, statues of Greek Goddess and nymphs were scattered on architectural features. Overall, there was only single design followed in making gardens with local variations.

Introductions of various exotic plants took place in the era of Queen Elizabeth (1558-1601 A.D.) when new trading companies were formed. Alongwith raw goods, plants were also collected in big way from various countries like India, China, Asia Minor, Africa, Mexico, etc.

There was outstanding change in garden style and utilitarian gardens were converted into flower gardens. These flower gardens were divided into rectangles or square by broad walks. Plant material comprised of fragrant herbs like carnation, rosemary, lavendar, thyme, lilies etc. It also had singing birds in cages hanging on the trees. Gardens were decorated with garden ornaments like fountains, sun dials, statues, wooden figures of animals. Topiary was very popular and different images were created. Plants used for making topiary were juniper, yew, rosemary or privet.

Le Notre's influence. English gardens reached at their peak of symmetry and formality under the influence of Le Notre—a French garden architect. According to Le Notre's method, trees and shrubs were planted very closely and strictly according to geometrical designs. Ultimately these gardens after growth of trees or shrubs looked like rooms or vast halls. Therefore, they looked dull and monotonous. These gardens were afforded by big land lords. For keeping the plants in formal shapes they were clipped and mostly were not allowed to grow in their natural way and produce flowers. Plants were mutilated and tortured. Ultimately English gardens looked artificial. This forced to generate the new thinking for allowing the plants to grow in natural state and develop gardens in informal style. For this, the landscape painting of Claude Lorrain (1600-1682 A.D.) gave inspiration to many landscape gardeners to develop the gardens in informal style.

III. Return to Nature (1720 A.D. to Today)

Henry Hoare was the first who started changing the landscape gardens in 1720 A.D. of Stourhead estate according to the taste and skill which became the basis of present day gardens. His son also called Henry worked on the layout and planting of the great landscape gardens from 1741 to 1771 A.D. During this period, the landscape paintings of Claude, Salvator Rosa and Zuccarelli were getting popular. Painters, poets and naturalists were spreading the message of love and world of nature to their countrymen. All this had the impact on the art of gardening too.

During middle of 18th century, the English gardens were further redefined under the leadership of William Kent (1604-1748 A.D.), Lancelot Brown (1755-1783 A.D.) and Humphrey Repton (1752-1818 A.D.). William Kent was a painter and sculpturist who worked on stone, wood or plaster. He modified number of gardens before making a great Rousham garden. Rousham is equidistant from Oxford and Banbury. His object was to create a dream like parkland scene, utilizing existing plantation wherever possible and planting a new wherever necessary linking walks and vistas and eye catchers in the form of statues, garden seat and temple of Mill. Like formal gardens, most of Kent's garden creations *i.e.* Store clarement, Gunners bury, Carlton House, etc. have either disappeared altogether or altered beyond recognition. He thought garden as a part of whole landscape view and introduced the concepts of harmony, balance and proportion. Formal lines were broken into gently curvaceous lines as prevailed in nature. Formal water bodies were converted into lakes of irregular shapes. Straight paths were replaced with winding paths. Instead of clipped hedges, informal groups of shrubs were planted. The basins of trees were planted with bulbous plants. The lawn was introduced. Major elements of gardens were trees, grass, water and open spaces. Another important garden architect, Lancelot Brown, had a major influence in modifying more than hundred gardens or parks. His ideas were based on using water to make lakes, walks and paths like serpentine. According to his planning, lawn used to bring upto the house so that the house stood in a grass. This concept was not observed earlier. Brown planted famous garden of Luton Hoo covering about 640 ha. The main attractions of the garden are the water features in rivulets and waterfalls were so arranged that they could be enjoyed in long sweeping vistas from the window of the country houses. Garden was planted with flowering shrubs and trees like double cherry, Rhododendron, azaleas etc. Bulbous plants were planted in basins of trees. There were forest garden, rock garden, water garden and rose garden.

Humphrey Repton (1752-1818 A.D.) in the contrast of earlier garden architects, modified garden designs from picturesque to the natural. This was done by planting trees in group as they would grow if left in the nature. These groups consisted of combination of trees of different age, size, character, etc.

Herbaceous Border. William Robinson who is credited with the invention of the herbaceous border (though he did not use the word herbaceous) mentioned in his book, "The English Flower Garden" published in 1893. In this book, he described the arrangement of flowers in borders. By the time of Gertrude Jekyell (1843-1932) the word herbaceous border was accepted and became essential feature of an English garden.

Royal Botanical Garden at Kew. Royal Pleasure Gardens at Kew was established in 1757. With the establishment of Royal Horticultural Society in 1804, the royal Pleasure Gardens at Kew assumed the role of botanic garden. Vast number of plants from abroad were brought and maintained. Hortus Kewensis—a catalogue of plants cultivated at Kew was published in 1789 A.D. in three volumes and 5500 species were listed. Later on, Hortus Kewensis was published from 1810-1813 A.D. in five volumes containing the description of 11,013 species for the use of practical gardeners. New palm House was constructed. The garden grew in its size tremendously from 15 acres to 250 acres in five years from 1841 to 1844 A.D. In 1848 A.D. Museum of Economic Botany was opened and in 1853 A.D. Herbarium and Library were founded. By 1861 A.D. lake garden was completed. Rock garden and water garden were also made. Plant materials like Rhododendrons and orchids from India and Sikkim; Chrysanthemum, Lilies, Cherries from Japan brought revolution in English gardens.

Chelsea Physic Garden. Every year there is a great spring show for four days in May, organised by Royal Horticultural Society. This show has been held at the Royal Hospital Chelseas grounds since 1913 which covers the area of 11 ha. The RHS renders free guidance to gardeners of every degree. There is a display of vast varieties of flowers like roses, orchids, rhododendrons, daffodils, larkspur, lilies, flower arrangements, apples, tomatoes, asparagus, etc.

Cottage Gardens. There are small gardens which are made around cottages of lower and middle class people. Depending upon their economic conditions, planting material differed. Fruit trees and vegetables including potatoes were grown in abundance. These gardens were developed by Gertrude Jekyell by planting shrubs, herbaceous plants, bulbs, rare and common plants together with the main idea of beauty. Hedge of yew and topiary are common features alongwith fragrant annuals like sweet william, pinks, lavender, tulip, etc. arranged in herbaceous border and roses of different classes.

Gardening Now-a-days

With the change in political systems there are mainly two types of gardens :

1. Public gardens and parks made by Government agency at city, state and national level. These gardens are maintained by trained personnels.
2. Gardens in surroundings of houses, estates and greater residential blocks. Planning and selection of plant material is mainly dependent upon individual choice.

Under the national garden scheme, there are several hundred gardens worth seeing and are open on one or more days in every year, when these gardens are at their best. Following is the list of gardens and months when to be visited.

April	—	Three great gardens Kew, Seville, Wisley
May	—	Bodnant, Ness, Wood bridge
June	—	Chatsworth, Harlow car, York gate
September	—	Stour head, Wilton, Furry, Compton Acres
October	—	Westonbirt, Westbury, Hideote, Kifts' gate

In every spring, several thousand persons come to see the parks and gardens to admire nature's paradise of flowers and foliage with sincere joy.

***

40 Landscaping of Homes

East or West, home is the best. This common saying reflects the sense of security, shelter, rest and enjoyable living which one gets by residing in one's own home. Man being a civilised animal enjoys his living in privacy. Therefore, the homes are constructed for private living. Labour and pains taken during construction, later on give him immense joy and satisfaction on occupation. There is lot of sentimental and psychological attachment with the homes. These homes are constructed according to the individual's liking and needs. Similarly the landscaping of these homes differ according to individual's choice especially in the adoption of style and selection of plant material.

Landscaping of homes involves the consideration of inside and outside in totality. Two main principles are kept in mind while deciding the landscaping plan.

***(i)* Utility and beauty.** It is individual choice to give preference for utility or beauty. In general utility refers to convenience, comfort, workable or productivity whereas beauty means pleasant, inspiring or relaxing. In limited space of home gardens, it is a ticklish point to select either growing of vegetables or cut flowers.

***(ii)* Relations.** Home and garden should be closely linked and both should be complementary to each other. Different parts of the home and garden should be merged in such a way that it creates harmony.

The planting scheme of the house will also depend upon the surroundings (neighbours), orientation of the house to get the sun during winter and shade during summer and location of the house *i.e.* whether it is on the national highway, other main road or sub-road.

The planting scheme is influenced by the planting plan adopted by the neighbourers. Some unusual or different trees should be planted which differ from the plantation of the surroundings. House should be oriented in such a way that during winter there is a plenty of sunshine and shade during winter and summer, respectively. To facilitate this, winter deciduous trees should be planted in eastern and southern sides so that it check minimum sunlight; on the contrary, shady trees should be planted in western side. The house located on national or other highway needs screening with tall trees to check dust or sound pollution. On sub-roads, flowering trees should be selected.

TYPES OF HOUSES

Urban houses can be broadly divided into two main categories *i.e.* (*i*) old houses (*ii*) new houses.

Old houses

These houses are characterized by the construction of houses very closely, practically without any open area. Area in densely populated lanes are narrow. The beautification of such houses is done only by growing plants in pots or in big containers.

New Colonies

For better living, new colonies are being planned by Improvement Trusts/Development authorities/boards. The main objectives are to provide better environment, sufficient open spaces and wider roads for free air circulation. Such settlements are mainly of two types : (*i*) multistorey flats and (*ii*) individual houses.

1. Flats of multistorey building. There is no private land in multistorey buildings for personal use where planting could be done. Therefore, the beautification inside the flat is only possible with pot plants and cut flowers. The selection of pot plants mainly depends upon individual liking and availability of plant material. Window or sill gardens are of immense importance in flats which can be decorated. The windows which open towards east, south or west where sunlight is abundantly available at least for 4-5 hours in a day are suitable for flowering plants. For other type of windows shade loving plants can be selected.

Inside the house, plants can be selected for various spots according to the intensity of light available *i.e.* sufficient, medium or sparse. There is a wide choice of selecting plant material to suit one's need. Ideal plants for interior decoration are : *Araucaria* sp., *Ficus elastica decora, F. elastica Black Prince, F. elastica, F. benjamina, Pothos* sp., *Monstera* sp., *Philodendron* sp., *Aglaonema* sp., *Diflenbachia* sp., *Maranta* sp., *Dracaena* sp., *Crotons, Coleus, Portulacaria afra, Asaparagus* sp., other succulents and cacti. Flowering plants like Geranium and Begonia spp. can be kept near windows where sufficient light is available. Seasonal flowers like cineraria and salvia in the pots plants can be kept where medium intensity of light is available. These can be used as specimen or can be kept in groupings. For hanging baskets, suitable plants are *Asparagus* sp., *Tradescantia* sp., *Zenbrina pendula, Pothos* sp., *Portulacaria afra,* etc.

2. Individual houses. Such houses have limited space in front and backyard. These can be suitably planted with, depending upon their sizes.

Planning of Landscape plan

There should be a complete planning of building and garden. Both should be considered as a unit of each other. Plan should be conceived in totality and thoughts should not be given in pieces. Three dimensional approach with all facets, factors and considerations must be given.

First of all paper plan according to scale should be made. Requirements such as patio for outdoor living, space for drying clothings, beds for cut flowers and place of pet, etc. should be accommodated on paper plan. Total area can be divided into three functional units *i.e.*

(*i*) Approach or public area. It comprises of front yard, drive way and porch. if area permits, this can accommodate grassy lawn, flowering trees, shrubs and climbers. To break the monotony of straight wall, curves should be given while planting trees, shrubs and flower bed. The planting of bottle brush or weeping willow with its hanging branches near the gate will act as symbol of welcoming to the visitors.

(*ii*) Service area. This area is comprised of kitchen garden, garage, badminton court, animal, poultry shed, etc. Fruit trees like papaya, guava, baramasi lime etc., vegetable and cut flowers are accommodated in this area. The area should be separated from private area by planting shrubs in the row or hedges. A path studded with stones to and through kitchen garden is required.

(*iii*) Private garden area. This area is known as outdoor living area. It should be designed according to the taste and need of individual family. The area should be screened off from undesirable sites by planting shrubs or hedge to have complete privacy. Grassy lawn adds beauty to this area. The design of garden, *i.e.* formal or informal should be decided so that mason work in garden can also be done. The concept of gardening in houses depends upon the plot size and area available for making gardens.

Execution of the Plan

It should be done carefully and systematically. If the building is newly built, collect all the garbage and clean the area. If some filling is needed, fill it with good quality soil and level it. Now mark the position of trees, shrubs and flower bed. Wooden pegs can be used for such marking. If any alteration is required, it can be easily done.

Following are standard sizes of plots and area to be kept as open spaces as advocated by Improvement Trusts/Development Authorities.

Small plots

These plots are measuring 100, 125 and 150 sq. yds. (Fig. 40.1). Making of gardens in such small plots is not possible. Owners prefer *pucca* ground in front and back sides which can be used for various purposes. Beautification is only possible with the help of pot plants. The selection of pot plants needs to be done judiciously.

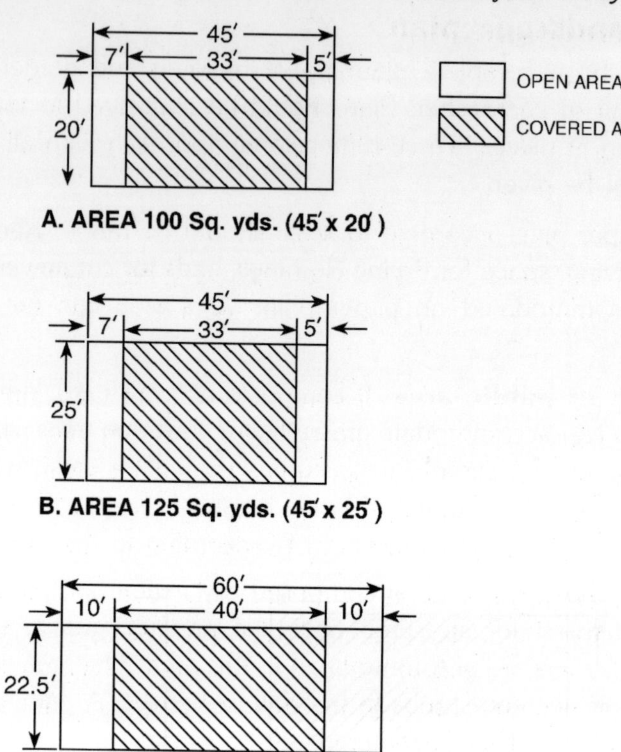

Fig. 40.1. Area of Small Plots.

Medium plots

The plots measuring 200, 250 and 300 sq yds are grouped into medium category (Fig. 40.2). Generally, owners of plots of these dimensions also afford to leave only a small area for garden and prefer to make *pucca* flooring so that it can be used for parking vehicles or for sleeping in open. There are, however, good number of owners who prefer to make gardens, whatever it is possible in the small area available. The making of lawn enhances the beauty of house. Planting of one or two specimen trees, shrubs and climbers in public access area break the monotony of concrete structures, In the backyard, the planting of fruit trees like baramasi lime or guava or papaya or a grapevine along with few vegetables like radish, bhindi or bitter gourd, bottle gourd, etc. can be done. Cut flowers like gladiolus and chrysanthemum can add to aesthetic beauty and also serve the utility, Following is a detail of plots and area allowed for construction and planting.

The planting plan of 250 sq yds is shown in Fig. 40.3. In addition to the plan proposed, the pot plants can be kept in porch, drawing room and lobby to suit the individuals likings.

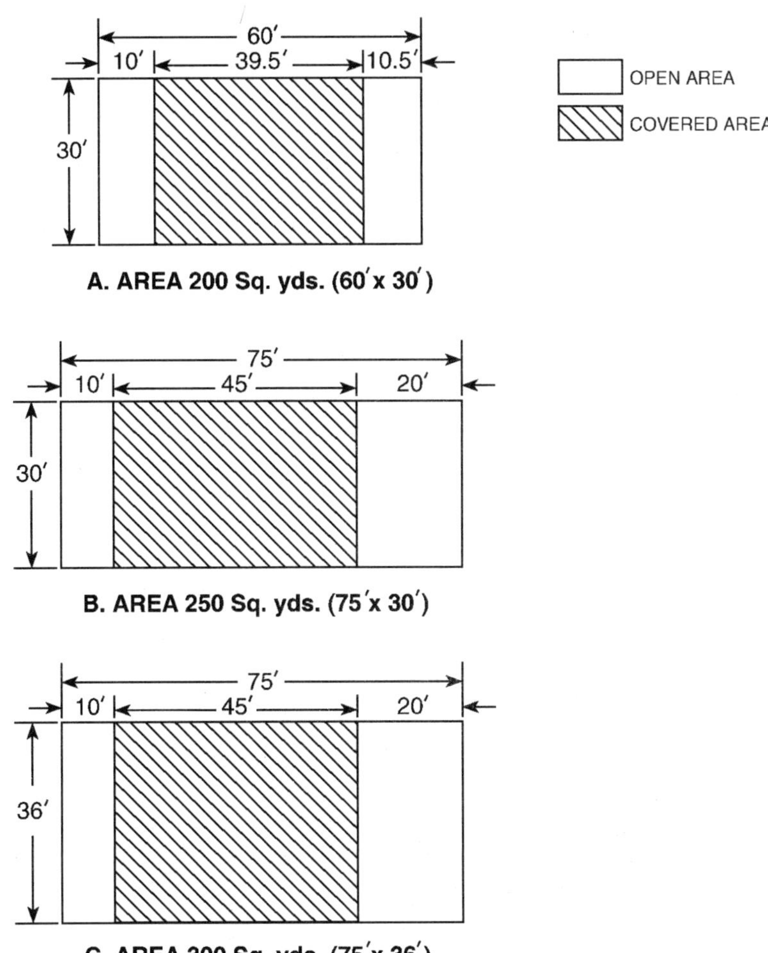

Fig. 40.2. Area of Medium Plots.

Large plots

In this category, plots measure 400 and 500 sq yds (Fig. 40.4). From gardening viewpoint, these plots are also not very large to accommodate massive landscape plans, nevertheless, such plots can be beautified tastefully and the urge of gardening can be satisfied. The standard dimensions showing covered and open area are as follows :

The basic requirements of these categories are also similar to that of medium plot owners. However, more number of plants can be accommodated. A fairly large number of pots can be used to decorate interior and exterior. There may be several variations in selecting the plant material for various situations. The details of landscape is given in Fig. 40.5.

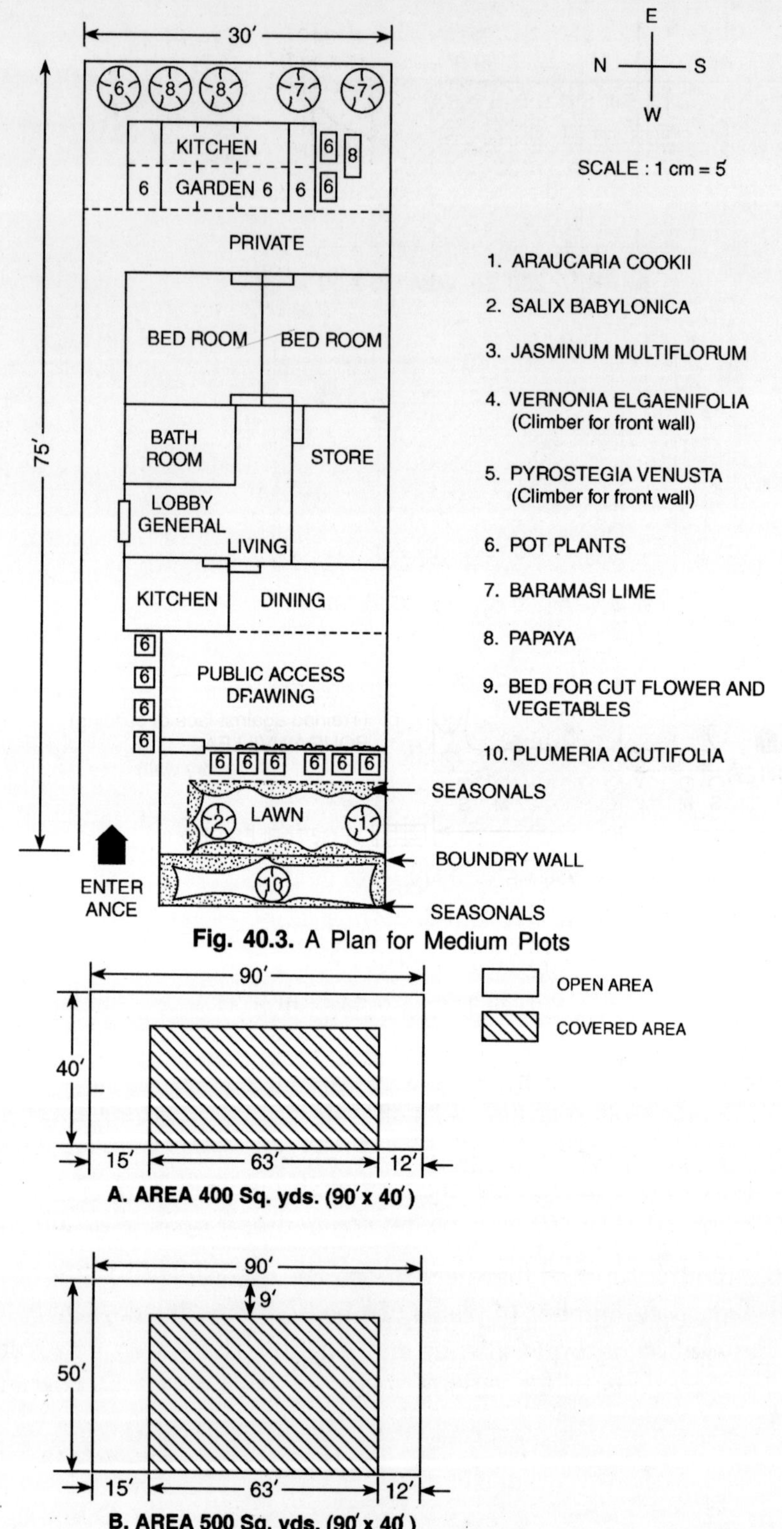

Fig. 40.3. A Plan for Medium Plots

Fig. 40.4. Area of Large Plots

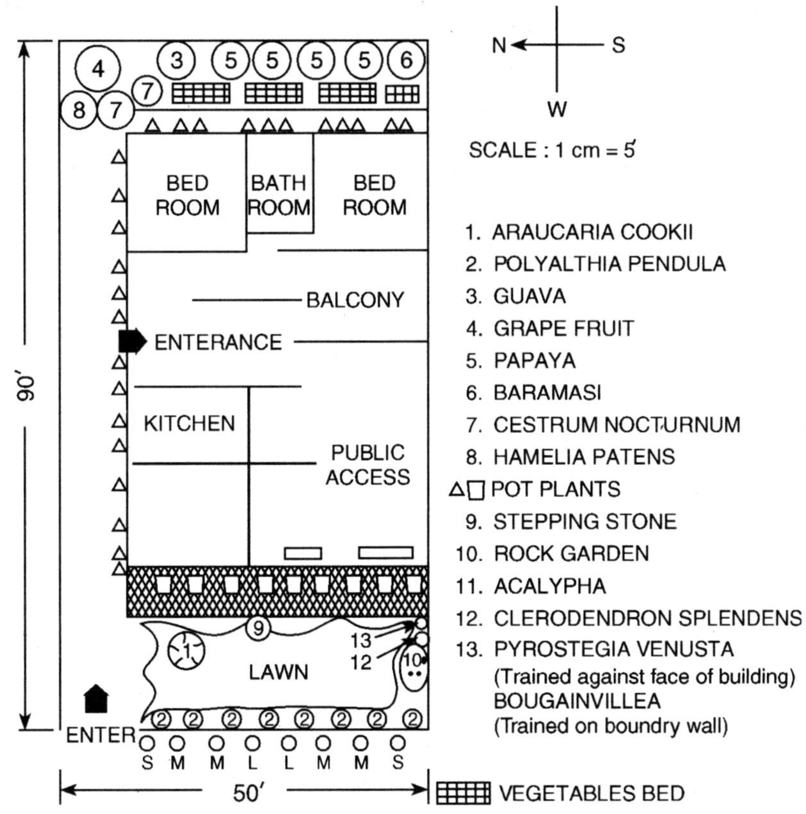

Fig. 40.5. A Plan for Large Plots

***

41

The Terrace (Roof) Garden

Due to limited space available for gardening on the ground level in small or medium sized houses or flats, roof is the only place which can be utilized to satisfy the urge of gardening. It can be converted into an interesting garden. For this one has to plan several innovations in the design suiting to the site. Thus, planning should be done meticulously and much before the start of construction of the house.

Planning

The parapet is mostly kept about 75-90 cm in height which is comparatively low. To soften the harshness of strong wall wooden slats each about 5 cm wide and 150-180 cm long can be set vertically or angles can be fixed at appropriate distance and stretching of iron wire can be done in two-three tiers so that suitable climbers at appropriate places can be planted to add the greenery and colourful flowers. Very recently green plastic net is available which can also be used in place of wooden slats or wire. Then the making of cemented beds, pots or containers of different sizes are done keeping the choice of plants in mind.

Before constructing the structures for beds or pots, one should be satisfied with the strength of roof and load bearing capacity with the help of structural engineer and architect. Similarly, the arrangement of automatic drip irrigation of individual pots or sprinklers should be included in the planning. This will save the lot of labour required for watering the pots, beds or container separately. The arrangement of draining water from the roof into gullies or outlet should also be planned.

Various ideas can be conceived to create surprises. If, it is possible a water feature can be created. A central bed is also created with a large sized fibre glass boulder with light inside. Now-a-days small to medium sized water falls fitted with light and pump for circulating the water are available and suitable type can be selected and can be kept at appropriate place.

Making of pots, beds and containers

To avoid the seepage of the water and moisture in the roof, efforts should be taken to collect all the extra water into out let from where it passes easily from the roof. For making lawn, plastic pots of brick size can be used or plastic sheet of 100 micron having 5-7.5 cm thick layer of garden soil and grounded well rotten FYM or light weight compost (3 : 1) is used for planting grass. Though it is some what difficult to maintain good lawn thus an astro turf can be rolled out and trimmed like carpet to get the desired effect. (**Plates 17 & 18**)

The cement and concrete pots, beds and containers can be architecturally designed according to the planning and made. Care should be taken that there are sufficient drain holes in them to remove the extra water from these structures. Basal layer of gravel of sufficient thickness should be provided and then garden soil or light weight compost (3 : 1) should be filled.

Selection of plant material

This is very ticklish point because out of wide range of plants, suitable types are to be selected according to the choice and design. Ideal plants for roof garden are which do not shed the leaves, architecturally superb or can be trimmed into different shapes. Plant should be hardy so are able to withstand harsh climate of open conditions like too much sun-shine, temperature variations, windy, dry conditions or excess of watering during rainy season, etc. Such plants are limited in member. Keeping the requirements of such climatic adversaries, plants like *Juniperous chinensis, Juniperous prostrata, Juniperous horizontalis, Thuja orientalis* var. *Compacta, Cupress sempervirense, Ficus deltoidea, Ficus panda, Ficus benjamina, Beaucarnea recurvata, 'Duranta goldiana, Acalyplea wilkesiana, Artabotyris, odoratissima, Murraya panniculata, Lantana sellowiana, Lantana depressa, Thunbergia erecta, Pentas carnea* etc. can be selected. Recently dwarf type of shrubs like *Hamelia patens, Ervatamia divaricata, Hibiscus rosasinesis, Nerium indica, Lagestroemia indica, Dwarf bougainvillea, Cuphea hyssopifolia*, etc. are available which are most ideal plants for roof garden. Climbers like *Jasminum humile, Clitoria tarneata, Clematis panniculata, Vernonia elganaefolia, Clerodendron thomsane, Ipomea chearteruse, Ipomea tricolour*, etc. can be used for parapet or trellis or for fall.

Roses can find place in a sunny corner to create colourful effect. Miniature roses should be preferred. Amongst bush roses less vigorous varieties of flori bunda can be selected.

Some of succulents and cactus can also be selected. These are very good for sunny situation. Plants like *Fercarea, Pedilanthus, Sensevieria* species, *Agave americana goldiana, Euphorbia splendense, Aptenia cordifolia, Sedum, Yucca*, etc. can be selected.

Similarly for shady situation various kind of foliage plants *like Asparagus meyri, Nephrolepsis exaltata, Pilea cadieri, Tradescantia bicolour, Exocaria bicolour*, Coleus, Croton, etc can be selected.

Maintenance of plants

Plants need regular watch for their various operations which are to be done time to time. Plants should be kept in certain limit shape as desired by regular trimming. Regular feeding and watering will keep the plants healthy and of shinning foliage. Timely spraying of insecticides and fungicides should be followed to keep the plants free from attack of insect, pests and diseases.

Sitting arrangement and Swing

To enjoy the beauty of terrace garden under open conditions a suitable arrangement for sitting should be planned in groups. The furniture may be of stone, steel, or wrought iron which can be left in open conditions whereas the wooden furniture or of cane or plastic has to be kept in the shade in the day and in the evening or night it can be brought back as planned. Steel swing can also be fixed at appropriate place, and thus landscape of terrace garden, sun, moon, clouds, stars can be enjoyed in motion.

42

Water Garden

The evolution of life began in the water where simple form to complex from of life of flora and fauna developed. Water allows to perform various physiological and biochemical actions necessary for the growth and development of life. Therefore, the settlement of human life started near the water *i.e.* near rivers, lakes or seas. In water ecosystem, various forms of plant developed according to the depth of water, pH, presence of nutrients and salts, etc. In different situations like salty water of sea, sweet water of rivers and lakes, marshy lands, various kind of vegetation developed. Amongst various kind of water plants, few plants produced most exquisite flowers which caught the attention of plant and garden lovers. The most common with beautiful flowers are lotus and water lilies. Lotus is the national flower of India and is associated with Lord Vishnu who appeared in sitting posture on lotus during the process of churning of sea or holding of lotus flower in his hand has been depicted in most of the paintings.

The presence of water in the garden in the various styles and forms add the beauty of garden tremendously. It symbolizes and signifies the importance of physical and spiritual life in various ways. It performs following functions in the garden.

1. It serves as source for watering, irrigation and is used for running fountains or creating water fall.
2. It provides freshness and coolness in the garden.
3. Standing water in the form of pool or vast water body provides quiteness and calmness ideal for meditation or for watching-reflections of big building, clouds, blue sky, sun, moon, or stars, lofty trees, etc.
4. Running water in the form of fountains, water fall, chadar, canals, signifies the movement of the life, happiness and joyousness. Ripples of water creates ethereal sound which gives unique heavenly feelings.

Different forms of water in the garden

Basically water is present in the garden in the following forms :

1. Standing water. It is very common feature of Japanese gardens and is present in the form of small or big lakes, or informal water body. This gives impression of vastness and mysticism. It provide ideal environment for meditation and worship. The reflection of trees, blue sky, clouds, buildings, moon, sun, stars, etc. add the beauty of the garden.

2. Running water. It is present in the garden in the form of canal, fountains, water falls, etc. Since water is always in movement which signifies movement of life, vibrancy, happiness and joyousness. There is a great amount of variation of fountains are available and suitable types are selected according to the design and requirement.

There are various situation in the garden where water pond of different sizes can be made. In certain situations, there may not be sufficient space to make a good size pond however still it can be made in the form of small, sunken or trough to hold some amount of water sufficient to plant a dwarf size water lily and it adds tremendous charm to a garden.

Shape, Size and Depth

These are three main parameters which are very important and to be decided with great care. The shape depends upon garden style. For formal garden the shape can be of any geometric shape like square, rectangular, circular, star shaped, etc. whereas for informal gardens the irregular shape suiting to surrounding are made. The size of water body depends upon the size of garden and it should be in proportion to the size of garden so that it harmonizes well with surroundings. Out of proportion will create visual tension and irritation.

The depth of pool depend mostly on the effects desired. On an average depth of 60-70 cm is considered to be an ideal but some time it is upto 1.20-1.50 metre. Shallow ponds are easy to clean and maintain in comparison to deeper ponds.

Very recently the concept of rock garden and water pond has been in practice. Large variations in size and designs are being constructed.

In deeper pool, two levels are created. At the bank shallow planters 30-60 cm wide, are made all along the perimeter and filled with garden soil + FYM and planting is done suitably.

Two types of water bodies are generally constructed keeping the level of ground in view.

***(i)* At ground level.** This is most common and widely used. In this, level of water pool and ground is the same so that construction becomes very easy, less expensive and easy to maintain. This type of water easy to maintain. This type of water pond merges very well comfartably with the design of garden and soundings. However, during cleaning operations pumping out the water from the pond may be some what troublesome.

***(ii)* Above the ground level.** The construction of such type is done above the ground therefore the construction becomes little more expensive. Though these type are not very common but are constructed in limited garden despite of its high cost of construction. In this case cleaning is very easily done as water is drained out due to gravity and this water can be used for irrigating garden beds. It is also constructed to break the plain level of garden and extra water is made to fall from one side or all four sides in the form of chadar or small fall. The beauty is enhanced tremendously.

Fountains

Fountains are the main attractions in the garden. It creates coolness in summer by throwing the tiny droplets of water in the air ultimately through evaporative cooling system it lowers down the temperature of surroundings. The fountains of various designs are available to create enchanting effects in the garden. Water proof lights are also fitted in the pool or under fountains so mystique effect is created when water and light intermingled with each other. Earlier fountains with single spout or rose head were very common which used to work on natural gravity system of water. So in these type, the height of water head used to depend upon the height of water tank. Now with the invention of pumps of various horse power, water force can be regulated as per requirement. Hence, now various kind of fountains are available. Different effects are created depending upon the nozzle type used and their arrangement. Very popular types are *Geyser, Bell, Finger Jet, Bubble* and *Floating* type fountains. These names have been given on the ultimate effects of fountain created by falling of water in particular fashion. Other type of fountains are *Double Crown, 2 stage jet, 3 stage jet, Flower jet, Vertical jet, Foam jet, Ball* with Crown, Umbrella and Lily fountain. To create special effect, Curtain fountain is also used.

Recently musical fountains are getting very popular. In this type various kind of fountains are used and fixed at different positions in certain sequence. These fountains are then played in accordance with the rhythm of music played. The height of water, speed and interval of water fall is regulated in synchronization with melodies of song. Simultaneously, lights are put on or put off in sequence so ultimate effect is very mystique and combination of fountain, light and music enthralls the spectators.

Making of Water Body

Shape, size and depth. After making decision regarding the type, location, design, size and depth of pool, digging of ground is to be done. The position of fountains is to be earmarked and galvanized pipe of 2.25 to 2.50 cm diameter alongwith 'T' junction and pipe of 1.25 cm diameter of suitable length is to be fixed at appropriate distances

DIFFERENT KINDS OF FOUNTAINS

Bell Fountain

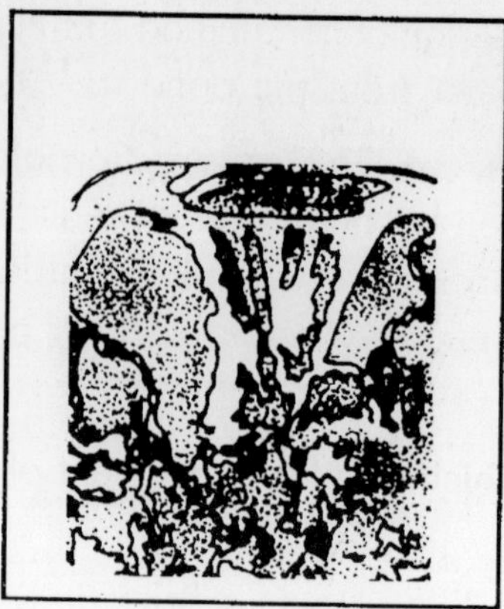
Lilly Fountain

3 Stage Jet Fountain

2 Stage Jet Fountain

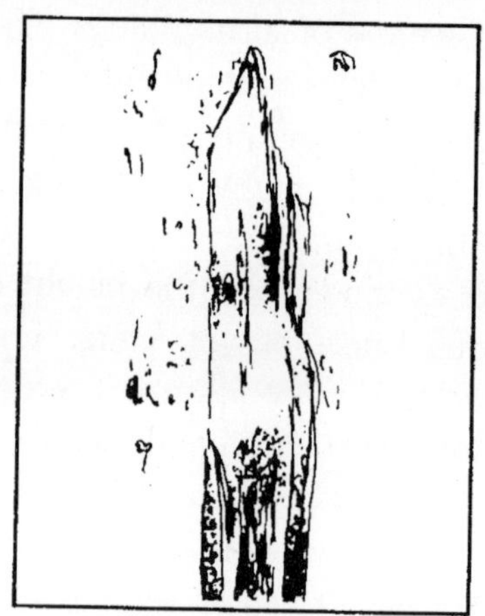
Vertical Jet Fountain

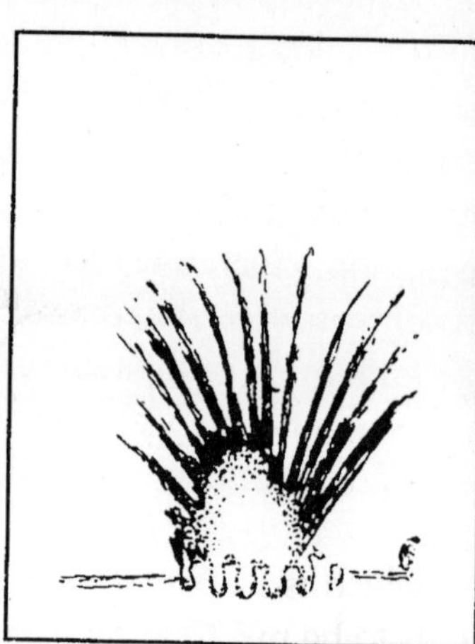
Finger Jet Fountain

Flower Jet Fountain

Foam Jet Fountain

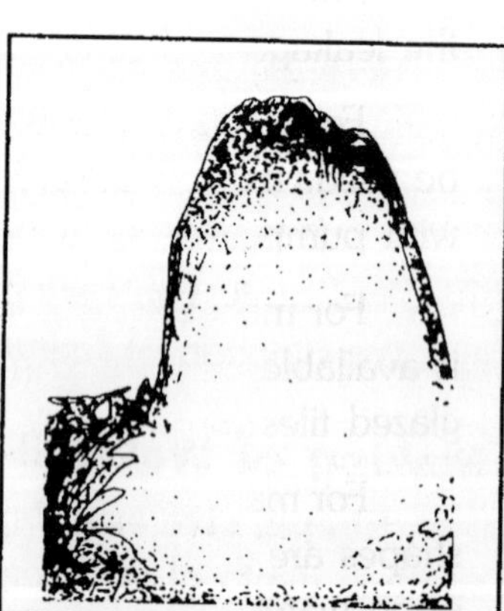
Geyser Jet Fountain

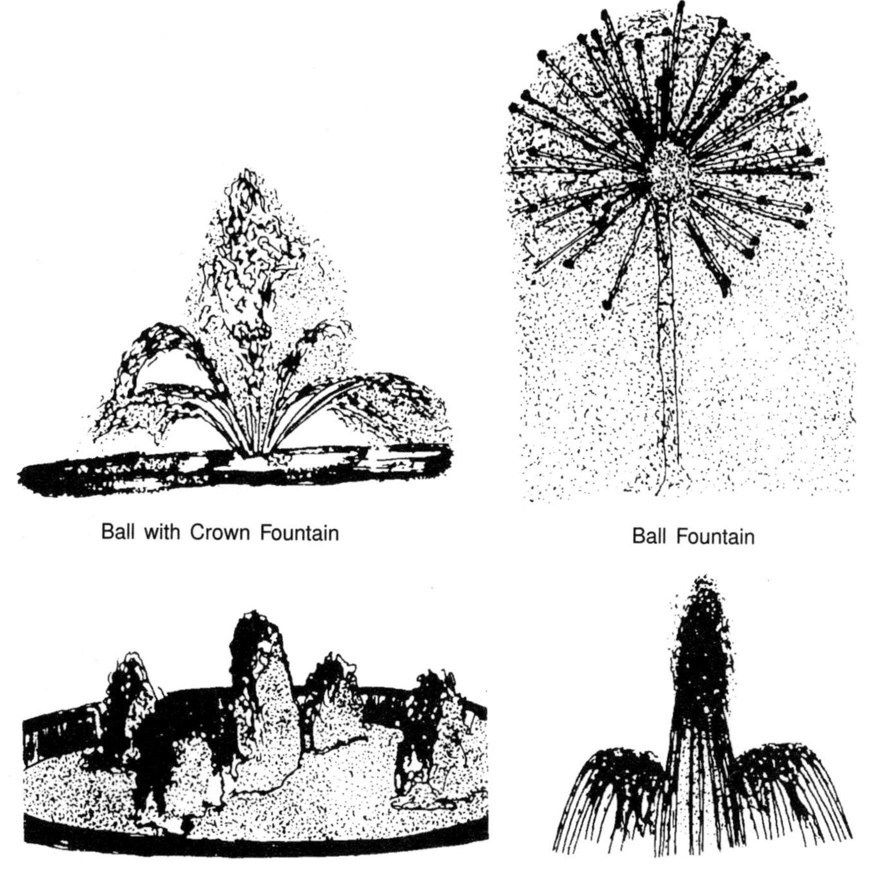

Ball with Crown Fountain

Ball Fountain

Bubble Fountain

Double Crown Fountain

which is for final fixing of fountains. This pipe is to be buried in the ground and floor is to be made pacca with crusher, sand and cement in the ratio of 3:1:1, respectively and should be compressed well. Retaining wall of 25 cm thick is sufficient. Care should be taken during the plastering with cement, corner should be well rounded off to prevent the leakage from these area.

For filling the tank and draining the tank, the pump of suitable horse power should be used. If the same water is to be re-circulated then inlet and outlet should be joined with pump.

For making pond attractive, different type of material for flooring and retaining wall is available. Keeping the liking and cost, it can be selected. However now-a-days mostly glazed tiles of many colour combinations are available.

For making water fall, natural boulders of different sizes and colour are used. Different shapes are created suiting to site and general character of garden. In such type mostly same water is re-circulated. Therefore the pipe of inlet and outlet are connected with pump of suitable horsepower, keeping the amount of water to be re-circulated.

Planting

For planting water plants old tank earth is highly suitable or compost of equal parts of FYM and clay soil is good. It can be filled in basket and plant the bulb in it and sub merged into shallow water. Care should be taken that compost does not get washed by placing coarse gravel or clinker as first layer near the periphery of the basket. Sink the basket a few centimeter deep so that bulb dips in the water and as soon as new leaves appear lower the basket to full depth.

Pests and Diseases. Water plants practically are free from diseases. Caterpillars and snails some time eat away the foliage of water lilies. Some time under water mossy algae chokes the growth of plants.

Plants for Water Pond

There is a wide of range of plant material which can be planted in the water pond and around pond where marshy or bog type conditions exist. Most common types are described here.

***(i)* Lotus.** *Nelumbo nucifera* (Nelumbonaceae). It is commonly found in sub-tropical and tropical part of the world. It is national flower of India and highly sacred to Indians due to its association with Lord Vishnu who has been painted holding pink colour lotus flower in thy hand. It is affected by frost and hence, it flowers during rainy season. It produces very showy flower on long stalks. Fruit seeds and roots are edible and are consumed. It prefers heavy and highly fertilized soil. It is easily propagated through seeds after scarification and by bulbs. *N. lutea* produces rose like yellow flowers.

Important varieties are Alba grandiflora (white) Alba striata (white with red margins), Charles Thomas (Lavendar Pink), Momo Botan (Dark rose pink petals yellow towards the base).

***(ii)* Water Lily.** *Nymphaea* spp. (Nymphaecae). It is commonly called as water lily and very popular as a water garden plant. It is hardy and plant produces different colour of flowers like white, blue, yellow, maroon, etc. These are propagated through separation of off shoot in spring and planting in muddy soil rich in humus in spring. Broadly, there are two types *i.e. N. pubscens* which bloom in cold months whereas *N. stellata* which is deciduous and bulbs remain dormant in dry ponds for months together without any harm. They come to life during rains and blooms onward. *Nymphaea odorata* produces white flowers.

Important varieties are Marliacea Albida—pure white with distinctive yellow stamens, scented; Escarbouncle—red colour with golden stamens; Gonnera-white; Gaziella—cup shaped flower of maroon colour; Vensusta—warm pink with golden stamens. Other hardy varieties are : Charles de Meurville, Chrometallo, Gold Medal, Indian, Lucida, Hollandia and Morning glory. *Nymphaeae caerulea* produces beautiful blue flowers.

***(iii)* Victoria regia (Nymphaceae).** It is very attractive and magnificent water lily. It has very large leaves measuring approximately 1.5 m in diameter so the size of water pond should be sufficiently large and deep (1.20 m). It produces large cream coloured flowers with the base of red colour which are highly scented. Seeds are hard and in large number which are to be chipped or rubbed to facilitate their germination.

***(iv)* Pontederia cordata (Pontederiaceae).** It is very pretty herbaceous, hardy, vigorous and perennial plant. It has glossy green and heart shaped leaves pointing upward. It produces spikes of purple blue flowers during summer and rainy season. It has creeping rhizome which can be cut with a knife and divided.

***(v)* Caltha palustris (Ranunculacae).** It is commonly called marsh marigold and perennial in nature. Leaves are round in shape and dark green in colour. Flowers are cup shaped, yellow in colour and 5 cm wide. Easily propagated by division of roots and through seeds.

Caltha palustris 'Flore Pleno' is a double form of the marsh marigold forming splendid yellow colour clusters.

***(vi)* Eichchornia grassipes (Pontederiaceae).** It is native of south America. It is perennial floating plant with thick floating stem. It produces long purplish green roots hang down 30 cm in water. It is produces light violet flower having yellow markings on upper petals.

(vii)* Cyperus alternifolius (Cyperaceae) :** It is a perennial rhizomatous plant with densely tufted branches. It has green stems each bear a terminal whorl of 10-30 deep green leaf like bracts. It is easily propagated through seed as well as division of rhizomes. ***Cyperus papyrus: It is commonly called as Egyptian paper rush. It is a perennial clump bearing, with tall, 3 angle pithy, leafless stems. Umbels at the end of stems are globe like with 100-200 thread like rays. It needs high humidity. It is also easily propagated through division of rhizomes.

Bog Plants

Such plants tolerate high degree of moisture and standing water for limited period.

***(i)* Typha angustifolia (Typhaceae).** It is commonly called as Indian Bul rush. It is perennial plant with strap shaped leaves. Leaves have sheath at the base. It produces from brown flower spike during summer.

***(ii)* Astilbe simplicifolia (Saxifragaceae).** It is a rhizomatous, perennial plant from moist areas of mountains, wood lands of South East Asia and North America. It produces plum like panicles of tiny, red, white, pink and purple flowers mainly during summer. Propagation is done by division during winter or spring months. *Astilbe japonica* is another good species. Important varities are; Aphrodite, Bridal Bell, Peach Blossom, Venus, etc.

(*iii*) **Grasses.** These are a group of grasses which is very commonly grown and if used judiciously, creates good variation in water garden. Grasses like *Miscanthus sinensis, Zebrinus* and *Carex elata Aurea* have variegated foliage and makes good contrast with other plants.

(*iv*) **Iris.** This is very good small plant which have dark green sword like foliage and produces very colourful flowers during spring. The common species which are used near the water tank or big lake are : *Iris Kaempferi*-deep purple; *Iris sibrica*—soft blue and *Iris missourenourensis*-lavender colour.

(*v*) **Ferns.** Mostly ferns do their best under cool and humid climatic conditions and considered moisture loving plants. Ferns like *Osmunda regalis*, *Adiantum pedatum* and *Matteccia struthiopterus* (ostrich fern) are ideal types near water pond.

***

43

Vertical Garden

It is a recent concept of gardening in urban area where northern walls of house, are covered with suitable small plants, climbers, twinners or runners. Such planting helps in improving the aesthetic look as well as improvement in general environmental conditions. With the increase in high rise buildings in urban areas, skyscrapers, multistory buildings, big hotels, etc., the environment is being degraded and air pollution is increasing tremendously in big and mid towns. Use of high numbers of air conditioners in offices and residential areas is also responsible for poor environmental conditions. Therefore, to reduce the ill effects of pollution, vertical gardens play a vital role. There are many other situations at outdoor as well as indoor where green patches in the form of vertical walls, or horizontal green patch can be created. Such effects can also be created very easily in public buildings, apartments, street scapes, pillars of bridges, fly over, airports, hospitals, universities, etc.

Advantages of vertical gardens

There are several benefits of vertical gardens which are summarized below :

- It improves aesthetic look.
- It acts as natural insulation for hot and cold air and save energy.
- There is reduction in carbon dioxide level and increases oxygen level and thus quality of air get improved.
- It conserves water and watering takes less effort.
- It improves thermal insulation and energy efficiency.
- It helps in sound absorption and thus reduces noise pollution.
- It holds rainwater, provides food and shelter to wildlife.

TYPES OF VERTICAL GARDENING

Green facade. The principal front of a building that faces toward street or open space, is beautified with the help of plants. Green facades are a type of green wall system in which climbers or cascading ground covers are trained to cover specially designed supporting structures. In this system, plants are growing in ground or in elevated containers where they are watered and fertilized.

Living green walls. It is composed of pre vegetated panels. Vertical modules or planted blankets are fixed vertically to a structure, wall or frame. Such panels are made of plastic, expanded polystyrene synthetic fiber which provides support to various types of plant species *e.g.* Ferns, Syngonuim, Asparagus, *Zebrina pendula*, *Rheo discolour*, Pedilanthus, etc.

Modular green walls. Vertical green modules are made up of regular polythene material. It has an attractive look, is highly durable in nature and can be easily installed. It provides a quick solution for making vertical garden. Now-a-days, empty plastic bottles of different sizes are being used as containers for growing plants, which are fixed on a panel.

Before making green wall, there are certain considerations which should be kept in mind. Selection of plant material will depend upon availability of sunlight and local climatic conditions. The ideal plants for vertical gardens are dense, compact, and low growing. If there is a sunny spot, select hardy, drought tolerant plants. Whereas under shady situations, shade loving plants like Ferns, Asparagus, Taradesentia, etc. can be selected. In the vertical garden, top, middle and bottom areas have different moisture level. Micro-climate of the area should be considered specially for exposed area, for high speed winds which can cause some injury.

Plants Suitable for vertical gardens. Out of several plants, the most suitable plants for green walls, are listed below :

1. **Indoor green walls/for shaded areas.** *Pepromia* sp., *Syngonium podophyllum, Chlorophytum comosum, Pilea mucosa, Rheo discolour, Fittonia verschaffeltii, Spathyphyllum wallisi, Schefflera arboricola, Nephrolepis exaltata, Zebrina pendula,* etc.
2. **For outdoor green walls.** *Asparagus densiflorous, Pilea macrophylla, Alternanthera, Cuphea hyssopifolia, Ophiophogon, Dianella tasmanica, Callisia repens,* etc.

STRUCTURE AND COMPONENTS OF GREEN WALL SYSTEM

A vertical garden can be installed in locations like railway stations, subway stations where horizontal space is difficult to spare. Plant material can be selected according to availability of sun light, wind and direction.

There are different parts of green wall such as front panel, bottom drainage tray, stabilizers, side panels if required and hanging hooks. A structure should be strong enough to tolerate the weight of plants, containers, media, watering system etc.

Design

The design of vertical garden will depend upon where it will be built. Keeping in mind the local and micro climatic conditions of the area, sunshine hours, direction, wind velocity, etc. the size and design will be decided. Similarly, water source, water quality in terms of pH and TDS and its drainage system has to be considered.

Growing media

Media for growing plants in green wall need to be light weight, having high water holding capacity, rich in nutrient, porous and preferably neutral in pH. Keeping these characteristics in view, combinations of cocopeat, vermicompost etc. are used. There are three types of growing media which can be used *i.e.* loose media (mostly preferred), mat media and structured media.

1. **Loose media.** This can be of two types *i.e.* soil on a shelf or soil in a bag. Latter type is more common and easily applicable. Now-a-days plastic bottles are commonly used as a bag and connected with micro irrigation tubes for watering. However, fresh media is needed to be added after one or two years.
2. **Mat media.** For making mats, coir fibre or felt mat of very thin layer is used. They have life of 10-15 years. According to type of plants, water and pH can be customized. According to requirement, long pieces of mat are cut and used. Similarly, layers of mats are decided.
3. **Structured media.** These are solid blocks of different sizes, shapes and thickness. They have long life and do not break or disintegrate upto 10-15 years. According to plants requirements of pH and water content, bricks can be made accordingly. They are easy to handle and carry. In light wind area, these are the best. Initial cost of making may be somewhat more but in long term maintenance cost is much less. These structural media are better than others in terms of longevity and performance in various situations.

Plant nutrition and irrigation system

The lush green effect of plants depend upon nutrients availability, water and its quality. Loss of water and nutrients is a regular phenomenon in plants and hence, these should be compensated regularly. Manual watering will be a cumbersome task for green walls whereas drip irrigation system or other automated irrigation system can be used efficiently. In such systems, mixture of nutrients is also added in irrigation water. So, it makes available water as well as nutrients to the plants. Though these systems require technical know-how.

The water application system is designed to minimize water consumption. It consists of an automation unit with equipment for control of nutrient injection and irrigation cycles. The irrigation system is divided into segments so that it can be programmed separately for each part. Water consumption varies with heat and sun exposure, however, normal water consumption varies between 0.5-20 liters per day/sq.m.

In soil-based system, mixture of slow release fertilizers is added to the growing media instead of using fertigation system at the time of planting. For regulation of water supply, water meter is required to monitor the amount of irrigation water. If stored water is used, pump is necessary to apply water.

Light

It is very important for the growth of plants. On an average in offices 300-500 lux of light is available. Therefore, provisions should be made for artificial light to meet the light requirement of plants. It can be measured with lux meter. Placement of green wall should be such that it gets light according to the requirement of plants.

Insects, pests and diseases

Regular inspection should be done of green walls for the attack of insects, pests and disease. Systemic insecticides or fungicides should be used to avoid foul smell of pesticides.

44

Xeriscape Garden

The word Xeriscape was coined by the Denver Water in 1981. This is combination of Greek words *i.e.* ***Xeri*** means dry and ***Scape*** means view or picture of a scene usually used in combination like landscape.

Now a days, keeping water constraint in view and desire of living in a healthy and green environment, dry landscape strategies are in a great demand. For this there are five fundamentals which should be followed for xeriscape gardening.

Selection of plant material. Care should be taken while selecting the plant material for xeriscape garden. Planting of trees, shrubs and other plants like succulents can be done which require less water or are drought resistant so that water demands will be less and plants will grow happily. Easy to grow perennials are available for a diverse climate conditions in a variety of colours and forms. Thus, selection can be done according to choice for various locations keeping climate conditions in view.

Mostly native plants should be selected because native plants survive better.

Few evergreen trees and shrubs for xeriscape garden are :

Acacia auriculifromis, Aganies flexuosa (Pepper Mint), *Callistemon viminalis* (Weeping bottle brush), *Calocedrus decurrens, Cupressus* sp., *Juniperous* sp., *Sechinus molle, Lagerstroemia indica* (crape myrtle) *Prosopis juiflora, Punica granatum, Vitex agnuscastus, Leucophyllum fruitescense* (purple sage), *Tecoma stans, Tecomella undulata, Tecomaria capensis, Bougainvillia* sp. *Hibiscus syriacus*, (Rose of Sheron), *Lavandula* sp. , *Plumbago auriculatum* etc.

There are many other plants which can be selected for making Xeriscape's look attractive. These are *Beaucarnea recurvata*, *Cycas revoluta*, *Cycas circinalis,* Yucca, *Furcraea*, etc. Other plants which are useful for wildlife can be also be planted like *Peroskia atripliciefolia* (Russia Sage), *Lavandula angustifolia*, *Festuca glauca*, Iris, *Poinsettia pulcherimma*, etc.

In such garden, other material like stones, steppingstone, boulders can also be used.

Efficient water application. The different methods for application of water in gardens of dry area can be used. The different ways like drip, sprinklers, water guns, furrow irrigation system, etc. can be employed and proper selection can be made by efficiency of system and by keeping cost effectiveness in mind.

Soil. Soil conditions are very important with reference to the preference of the plants grown in Xeriscape Garden. In such conditions, plants should have deep root system. For this, soil should be light, deep and friable which helps in root penetration to great depth which will result in better survival. On the contrary, if soil is heavy clay type, root growth and penetration to great depth will be checked. To make soil fit for growing plants, soil amendments can be added according to the need.

Watering of xeriscape gardens. In world, wherever there is arid climate or areas adjoining deserts, gardens are being made on xeriscape principles mainly keeping in view the availability of water and prevailing drought conditions. Plants whose requirement for water is less for growth and flowering are selected. In such a landscape garden, the lawn area is reduced and use of native plant is done as they are best adapted. Such green beautiful areas give a high sense of relief which is essential for healthy living.

It is not only for arid areas, but it should be applicable to all areas as water table is falling alarmingly low. Care should be taken that conservation of water should not take back seat to aesthetic and joy of gardening. Now-a-days keeping water constraint in view, planning and execution should be done carefully.

Sound landscape planning and design. In such designs, lawn area is kept considerably lowest so that water requirement is less. Use of drought resistant trees, shrubs and succulents is preferred which results in low cost of maintenance, in terms of growing and management of plant material as insecticidal and fungicidal sprays also get reduced.

Use of stones, boulders and pebbles is recommended which can enhance the look of the garden, check erosion of soil and guide the direction of movement.

Mulching. Mulch may be an organic or inorganic layer spread on the ground to conserve the soil moisture and prevent erosion of soil. It improves soil fertility and health by reducing weed growth. It improves the beauty of landscape area. Amongst organic mulch, tree bark mulch is very commonly used all over the world. Availability of tree bark depends upon location and climatological conditions which ultimately results in dominance of species. Trees of temperate climate like Pine especially maritime pine bark are used and mulch layer of 5-7.5 cm is spread. The leaves of different trees can also be used. Other organic mulch like compost, straw, wood or landscape fabric are used but seldomly.

After the discovery of polythene (plastic), it is extensively used. It is available in different colours like white, black, green, etc. but black plastic is commonly used. It conserves soil moisture as well as helps in sterilization of soil. Due to high temp of the ground in summer, it kills all harmful foreign bacteria and insects. In the market, plastic mulch unrollers are available through which larger area is covered with the help of tractor mounted plastic. The conventional polythene is preferred over paper and other biogradable materials. Plastic sheet of 150 to 200 gauze is used.

Suggested Readings

Conder, J. 1964. *Landscape Gardening in Japan*. Dover Publication Co. 180., Varick Street, New York.

Crowe, S., S. Haywood, S. Jellicoe and C. Patterson. 1973. *The Gardens of Moghul India*. Vikas Publishing House Pvt. Ltd., New Delhi.

Eckbo, G. 1956. *The Art of Home Landscaping*. McGraw Hill Book Co., New York.

Gopalaswamiengar, K.S. 1953. *The Complete Gardening in India*. The Hosali Press, Bangalore.

Hyams, E. 1966. *The English Garden*. Thames and Hudson, London.

Nelson, P.V. 1991. Green House Operation and Management. Prentice Hall, Englewood Cluffs, New Jersey, U.S.A.

Randhawa, M.S. 1950. *Beautifying India*. Tajkamal Publications Ltd., Delhi and Bombay.

Randhawa, M.S. 1971. *Beautiful Gardens*. ICAR, New Delhi.

Randhawa, M.S. 1976. *Gardening through the Ages*. The Macmillan Co. of India Ltd., Delhi.

Saito, K. and S. Wada. 1964. *Magic of Trees and Stones*. Dai Nippon Printing, Tokyo.

Stevens, D. and U. Buchan. 1994. Garden Book. Conran octopus Limited, 37 Shelton Street London WC 2 H 9 HN.

flowers have six long, narrow segments united at the base by a thin membranous cup or corona. It is commercially grown around Mumbai where flowers are sold in the market. Individual flowers are used in making garlands.

Iris sp. (Iridaceae)

It is a native of South East Asia and Southern Europe and is known as Flag flower. It is suitable for group planting near lily pools, rockey and in shady situations. Leaves of light green colour are pointed at the apex. Different species of Iris are grouped into two main classes *i.e.*, tuberous and bulbous types.

The species is native to Himalaya like *I. kashmeriana*, *I. nepalensis* and *I. kamaonensis* are rhizomatous which are comparatively easy growing. The rhizomes are planted in October and plants flower in March. Every year fresh planting is not necessary and plants can be left in the ground. The colour of *I. nepalensis* is pale lilac whereas that of *I. kamaonensis* and *I. kashmeriana* is white and purple mottled. Other important species are *I. cristata*, *I. susiana*, *I. siberica*; *I. kaempferi* (Japanese iris) etc. The commercial varieties of German Iris (rhizomatous) are King Midas, Petrea, Black Panther, Sahara etc.

The bulbous Iris is native of Holland, Spain and England. These are highly suitable for group planting, borders and cut flowers. It is available in wide range of colours. The bulbs are planted in September-October about 10 cm deep. Flowering occurs during February-March. The important species of this group are : *I. tuberosa*, *I. hollandica*, *I. danfordiae*, *I. reticulata*, etc. The important varieties of Dutch Iris are; Blue Champion, Harmony, Lemon Queen, King Mauve, White Excelsior, Bronze Beauty whereas Cajanus, Enchamress, King of Blues, Le Mogol, Summer Time are common varieties of Spanish Iris.

Narcissus and *daffodils*—*Narcissus* sp. (Amaryllidaceae)

It is commonly called as Nargis and is native of Asia, North Africa and Europe. Botanically both belong to same genus but differ in the size of cup or corona of the flower. In daffodils the cup or corona is large whereas in Narcissus it is small. They are highly suitable for growing in beds, pots, under the deciduous trees, window gardens, etc. They are very good cut flower and are sold in the market. There are several species but common species are *N. triandrus*, *N. bulbocodium*, *N. poeticus*, *N. aurantiacus*, *N. albus*, *N. tazella*, *N. tazetta papyraceus*, *N. odorus*, *N. jonquilla*, *N. cyclamineus*, *N. obvallaris*.

In the plains of north India, the bulbs are planted in October. Soil should be well fertile, sandy loam. Planting is done 20 cm apart and 7-10 cm deep. Flowering occurs during February-March. After flowering water is withheld. Leaves dry and bulbs can be left in soil for 2-3 years as such without disturbing them. After three years, bulbs are lifted from the ground and separated during April-May and are stored at cool place. Again fresh planting is done.